D1266425

Quality Assurance Management

Quality Assurance Management

Second edition

Michael J. Fox

Independent consultant, and Associate of Durham University Business School, Quality Consulting Services Ltd., University of Paisley Quality Centre. IRCA-registered Lead Assessor of Quality Management Systems.

CHAPMAN & HALL

London · Glasgow · Weinheim · New York · Tokyo · Melbourne · Madras

**Published by Chapman & Hall, 2–6 Boundary Row,
London SE1 8HN, UK**

Chapman & Hall, 2–6 Boundary Row, London SE1 8HN, UK

Blackie Academic & Professional, Wester Cleddens Road, Bishopbriggs, Glasgow G64 2NZ, UK

Chapman & Hall GmbH, Pappelallee 3, 69469 Weinheim, Germany

Chapman & Hall USA, 115 Fifth Avenue, New York, NY 10003, USA

Chapman & Hall Japan, ITP-Japan, Kyowa Building, 3F, 2-2-1 Hirakawacho, Chiyoda-ku, Tokyo 102, Japan

Chapman & Hall Australia, 102 Dodds Street, South Melbourne, Victoria 3205, Australia

Chapman & Hall India, R. Seshadri, 32 Second Main Road, CIT East, Madras 600 035, India

---

First edition 1993
Reprinted with corrections 1994
Second edition 1995

Typeset in 9.5/10.5 Meridien by Pure Tech India Ltd, Pondicherry, India
Printed in England by Clays Ltd, St. Ives plc.

ISBN 0 412 63660 3

A catalogue record for this book is available from the British Library

Library of Congress Catalog Card Number: 95–67901

∞ Printed on permanent acid-free text paper, manufactured in accordance with ANSI/NISO Z39.48–1992 and ANSI/NISO Z39.48–1984 (Permanence of Paper).

Contents

Foreword

Quality is an essential ingredient of business success. But, until comparatively recently, the perception of quality and the means of achieving it were subjectively defined.

Since the early part of the 20th century great efforts have been made to define quality in a workable way and to achieve both a commonality of understanding and of practice. In the development of a global trading environment this commonality is essential. It enables trade through customer/specifier confidence and through compatible systems.

At the core of this very practical book is an exploration of the quality standards in the BS EN ISO 9000 series which evolved from a British Standard, BS 5750, published in 1979. It was a synthesis of ideas from many defence and company quality standards, all used for supplier assessment. This plethora of standards caused a great deal of unnecessary expense with many large suppliers maintaining large teams to manage the constant throughput of inspectors. BS 5750 enabled a simple but thorough third-party assessment to be carried out and certified for all potential purchasers.

So successful was this standard in improving the UK's practice and reputation for quality that it was adopted by the international community in 1987. Its requirements are a proven formula for helping companies of all sizes to develop their management system.

Today around 100 000 companies world-wide are registered to the standard and, through the lingua franca that it has established, there is an ease in international trading and a high degree of confidence that what is specified is what is actually delivered.

This is a subject which repays study by anyone in business. It is also a subject which bears out the old adage – the devil is in the detail. I hope that this book will be used by serious students and by newcomers alike and will help play its part in cementing a genuine commitment to quality throughout the business community.

Sir Neville Purvis KCB
Chief Executive
The British Standards Institution

Preface

Aims of the book

The book is aimed towards students of quality assurance as well as active practitioners. Moreover, since the advantages of formal quality assurance, and the benefits of practising it are becoming more widely appreciated in service companies, it should be of value to non-specialist executives in smaller companies. It will help you manage your business responsibilities to ensure that you satisfy the expectations of your customers as economically as possible, and in a way which gives you visibility of the status and progress of your 'quality'. It will direct your attention to the writings of the most influential interpreters of quality concepts, especially in the USA and Japan, and how their lessons have been adopted and adapted by successful businesses throughout the world.

In order to maximize the usefulness of the book to students studying for professional qualifications in Quality Assurance, the requirements of the following examinations were taken into account when planning the scope of the book:

The Institute of Quality Assurance's Associate Membership examination syllabus (as revised April 1990) for papers:

A1 Introduction to Quality Assurance
A2 Specifications, Standards and Quality Audit
A3 Quality Assurance Management

The City and Guilds of London Institute's syllabus for course:

743 Quality Assurance

Origins of the book

This book is adapted and expanded from the course material which I created for the Quality Management module of the Durham University Business School's distance learning MBA degree. This had itself evolved from a distance learning course prepared earlier for use by the Rapid Results College.

As well as the cited authorities, the book draws on my interpretation of my own experience in quality assurance management within the microelectronics industry, and latterly as a consultant, a trainer and an assessor of quality systems.

Structure of the book

The book is intended as an introduction and convenient reference to its subject, and will cite other works which explain certain topics in more detail.

Books on Quality Management tend to fall into one of two groups:

1. There are a limited number of reference books which aim to be comprehensive. These assume no initial knowledge, and yet try to cover each topic in depth. The two best-known of these are Juran's *Quality Control Handbook*, and Feigenbaum's *Total Quality Control*. Both these books originated from material written in the early 1950s, but because of their popularity, they have been regularly updated and reissued. Of the two Feigenbaum is the shorter, available as a paperback of 851 pages, still less than half the length of Juran's tome.
2. Other books either concentrate on a particular element of Quality Management (e.g. Quality Circles or Quality System Audit) or else they focus on a particular way of looking at Quality Management, addressing their comments to people whom they see as already practising quality management.

The present book aims to be more concise than Feigenbaum or Juran, yet broader in the topics covered than most other works, so as to be of value to beginners as well as practitioners with Quality Assurance experience. It is divided into four sections as follows:

Part One: The meaning of quality, and how it must be managed if it is to be achieved.
Part Two: Specialist 'Quality Control' and 'Quality Assurance' techniques.
Part Three: Getting a whole workforce to contribute to the maintenance and improvement of quality.
Part Four: The monetary dimension of quality; quality and competitiveness in world markets.

Acknowledgements

My thanks go first to Durham University for their permission and encouragement to develop course material, of which they hold the copyright, into this book.

Secondly to Mr Barry Reavill, FIEE FIProdE FIQA MBIM of Portsmouth Management Centre, who provided the material for the DUBS course which forms the basis of Chapters 2 to 6 of this book.

Thirdly to the British Standards Institution and to the Department of Trade and Industry for permission to quote extensively from standards and booklets which are BSI or Crown copyright.

Fourthly but not least to my DUBS students in Europe and Asia, through whose tutor-assessed assignments and dissertations, I too have learned more about quality, as well as about the topics where students can find difficulties.

Part One

What Managing Quality Means

1

What is quality, and how can it be managed?

Introduction

In this first chapter we shall define the word **quality** in its industrial context, together with other terms which include the word 'quality'.

We shall indicate how this understanding of the word 'quality' differs from some of the ways quality is regarded informally; we shall show that the way in which quality is defined in industry invalidates many everyday assumptions concerning the nature of quality.

In particular, the level of quality is measurable; and through being measurable it can be managed and controlled.

Where quality is today

About 20 years ago, a senior executive of a major and successful American electronics corporation was regularly visiting a Japanese company active in the same field, negotiating a technology cross-licensing agreement. In fact, so frequent were his visits that he was given a pass which enabled him to go in and out via the employees' entrance. As they entered each day, there was a wall facing them across the lobby which bore a slogan in enormous Japanese characters. After a few visits he asked what the sign said, and was told:

Quality is the battleground of tomorrow!

What was then tomorrow is now today, and this book will survey the battlefield, and arm you to play your part in the struggle for prosperity; through the supply of goods and services which best meet the needs of your customers.

Even one of the more cynical British exponents of quality improvement has said that when he started work, 'quality control' was a despised discipline, practised by despised people. Now by contrast, he claims, the techniques are honoured (though the practitioners are still despised).

The object of the book is to enable you to truly understand what quality is, and aid you in promoting and achieving improved quality during your career. It is written under the assumption that the majority of readers are practising managers at some level. It also assumes that you want to supplement your work experience with an understanding of the principles of quality management and how these ideas are currently evolving, so that you will have the breadth of vision to be an effective manager honoured not despised, on account of your concern for quality.

Quality is not an optional 'bolt-on accessory' to make your business go faster. It is integral to all management planning and decision-making.

Whereas I hope this book may attract readers who expect to focus their interests on 'quality assurance', every manager should be exposed to the concepts it expounds. Those who need it most are the ones who don't realize it. Thus it is that all quality specialists have themselves to be educators, and so another reason for the value of this book is that it will help you to preach what you practise!

The meaning of quality

The ISO definition of quality

It is time for us to nail our colours to the mast and say what quality is. The International Standards Organization definition is given in its document ISO 8402 *Quality Vocabulary*. This international vocabulary also forms Part 1 of BS 4778, of which Part 2 is 'national' definitions not defined in Part 1. We shall turn to this standard again whenever we want definitive meanings of other quality-related terms.

ISO 8402 says quality is:

> The totality of features and characteristics of a product or service that bear upon its ability to satisfy stated or implied needs.

Other definitions of quality

Other definitions have been coined earlier by various writers, and also in some National Standards which pre-date the ISO standard. Here is a selection:

> Fitness for purpose or use. (Juran)

> The total composite product and service characteristics of marketing, engineering, manufacture and maintenance through which the product and service in use will meet the expectation by the customer. (Feigenbaum)

> Conformance to requirements. (Crosby)

Links between the definitions

These four definitions are worth considering together; they are complementary, since each emphasizes a particular point which is only implicit in the others.

Juran's is a brief definition. The ISO description elaborates on Juran's 'fitness for use'. It emphasizes the totality of quality considerations which together satisfy all needs, whether these are expressed or taken for granted.

Feigenbaum's definition reveals whose criteria of need, fitness and expectation are being addressed in the definitions; they are the customer's. Feigenbaum also names the key divisions of the business enterprise, each of which has a critical role to play in the achievement of quality. We shall devote a chapter to examining the quality role with each of these divisions; marketing, engineering, manufacture and maintenance.

Both ISO and Feigenbaum stress that the definition applies equally to a manufactured product and to a delivered service.

Finally, Crosby's definition implies that the customer's requirements can be documented, conformance to the requirements investigated independently, and hence measured.

The customer as arbiter

You will see from these definitions that in matters of quality, the customer is indeed always right. What the manufacturer feels the customers should want, or what the supplier believes they are getting, is irrelevant. There is no place for the alleged riposte of Ettore Bugatti, when teased about the poor cold starting of the exclusive cars his company manufactured, that every **gentleman** had a heated garage; nor of Lord Nuffield who, when asked why Morris cars did not have heaters as standard equipment, is claimed to have answered that personally he always wore an overcoat when driving in cold weather.

Common misconceptions

The definitions we have been examining run counter to many everyday perceptions of quality. Widespread beliefs include:

1. Quality is difficult to define, but you can recognize it when you see it;
2. Quality is expensive;
3. Quality is craftsmanship;
4. Quality is luxury;
5. Quality is exclusive;
6. Quality is in short supply.

Ask the average person what product epitomizes quality and the answer is very often the Rolls-Royce motor car. Certainly the Rolls fits the above imagined criteria. But is that the whole story?

I doubt if any one of you is a 'Roller' owner; (please forgive me if that was an unwarranted assumption) but if you could exchange your present car for a Rolls to use in its place, would you take up the option? Probably not, because your present car is more practicable transport for you and your family. There is nothing paradoxical about this; your present car is better adapted to your personal needs; it is therefore the better quality car for you.

Quality 'grade' and quality 'level'

Part of the confusion that arises when people compare 'everyday' and industrial perceptions of quality derives from failure to recognize and distinguish between the concepts of quality grade and quality level.

Quality Grade is defined in ISO 8402 as follows:

> An indicator of category or rank related to features or characteristics that cover different sets of needs for products or services intended for the same functional use.

There are footnotes to this definition, one of which (note 2) is also quoted here:

> A high grade article can be of inadequate quality as far as satisfying needs and vice-versa. E.g. a luxurious hotel with poor service or a small guest-house with excellent service.

Quality Level is not defined in ISO 8402. The following definition, which is the sense in which we shall use 'level' throughout this book, comes from British Standard BS 4778 part 2.

> A general indication of the extent of departure from the ideal; usually a numerical value indicating either the degree of conformity or the degree of non-conformity, especially in sampling inspection.

Thus we would say that the luxury hotel cited in the footnote to the ISO definition of 'grade' had a high quality grade but a low quality level.

Our quality expectations are higher, the higher the grade aspired to by the supplier. A Rolls-Royce considered by a customer to be unacceptably noisy would in all probability be noticably quieter than a 'quiet' Ford. However Roll-Royces seldom disappoint. Their distinguished reputation results from delivery of a high quality grade combined with the simultaneous achievement of a high quality level.

Industrial quality: specification and contract

We are now developing a clearer picture of what a customer may be looking for. We have so far taken our examples from consumer products; such examples as motor cars. Were we to switch our attention to industrial items such an aircraft radar system, an oil tanker, or a nuclear power station, we would find that:

1. The quality requirements were specified in great detail;
2. Adherence to the quality requirements formed part of the contract between contractor and customer.

Explicit needs

You will appreciate that if an enterprise is supplying a consumer product or service, a quality specification will not have been provided by the customer, who may have only a vague notion of what he or she is looking for. In these situations it is a task of the marketing department to establish, for the company's internal use, a target or marketing quality specification on behalf of the customer. This activity will be examined more closely in Chapter 2 dealing with quality in marketing.

Implicit needs

Prompted by the ISO definition of quality we have recognized that customers' expectations may involve implied needs as well as explicit needs. Safety is a need which is implied but not usually expressed in customers' requirements, thought it may be expressed in national consumer safety legislation.

Latent needs

There are also customer needs even harder to identify than the implied needs. These may be referred to as latent needs, which are not perceived as available features of a product or service but a market would exist if they could be made available.

Thus there was not a market demand for home computers until manufacturers had shown that it was possible to offer a computer at a price a family could afford; even then it was necessary to market 'computer games' to create the demand; but once home computers were available and potential purchasers were aware of the facilities they made available, sales developed rapidly.

Equally, there may be a latent desire for a performance enhancement beyond the present 'state of the art'. Thus not many years ago 30 mpg would have been considered a normal fuel consumption for a medium-sized family car. Nowadays that would be considered unsatisfactory, and a target figure of 40 mpg would seem to many purchasers a reasonable expectation. We could say that when 30 mpg was the norm, this figure was accepted as representing a serviceable quality specification, but there was a latent demand for improvement if that were possible.

Features of an effective specification

So, the grade of quality offered or demanded can be documented in a specification, with these provisos:

1. The specification must take into account the purchaser's implied needs, and also statutory requirements (for safety etc.);

2. If the customers are not in a position to formulate their needs accurately, the manufacturer must elicit these needs (e.g. by market research) in order to provide guidance to its own organization;
3. The specification must be clearly and unambiguously expressed;
4. User-expectations evolve with time and a specification may need to be periodically reviewed and upgraded if it is not to become obsolete and misleading in terms of its market acceptability.

Quality as conformance to specification

Despite these limitations on the utility of a quality specification, you will see that such a document is a necessary and powerful tool for managing quality. With it, Crosby's definition of quality as 'conformance to requirements' becomes 'conformance to specification'.

So the specification defines the quality grade being offered. The target quality level defines how closely, or how frequently the grade is to be met. Again quoting from Crosby, the goal for quality level should be 100% conformance to the requirements of the quality grade, every time.

As quality grade can be defined, so it follows that quality level can be measured.

Measuring quality

There is a shortfall of quality level every time a discrepancy is found between any item of product or service and the grade set in the specification. Such discrepancies can be recorded and counted either as:

1. Number of defects recognized;
2. Number of items found defective;
3. Measured deviation from the standard.

The responsibility for quality

From the time of the industrial revolution to the present day, there has been an evolution in the way in which we have controlled quality level. Initially craftsmen-workers applied their own appraisal standards to their work; they learned how to avoid errors during their apprenticeship, and if any were made it was a matter of pride to recognize and correct the defect.

The mediaeval guilds formed a regulatory body for craftsmen working in competition with each other in the larger towns; the craftsman had to produce a 'masterpiece' for inspection and approval before he became a full member of the guild and, as a Master, was allowed to operate his own business, employing his own workmen and training apprentices of his own.

Assembly line manufacture separated the tasks of production, inspection and rectification, and during the 1930s sampling plans were developed which enabled a statistical level of defects to be predicted, together with the concept of an 'acceptable quality level', i.e. a level of defects which manufacturer and customer alike accepted was low enough that it was not economical to improve it further.

With increasing system complexity the wheel has in effect come full circle; acceptable quality levels are too low (in the region of parts per million) for sampling to be effective. The emphasis is on discovering and eradicating the causes of defects, and once again the roles of builder, inspector and problem-solver merge.

Employees are now encouraged to recognize that they are linked, one to another, in internal supplier-customer relationships.

Quality in service industries

Techniques to control quality were developed in a manufacturing environment, and only gradually was it realized that the concepts and disciplines formulated for that environment could be applied with equal advantage to the indirect, non-productive parts of the company's organization; and indeed to organizations supplying a service, not a product at all.

In truth the differences between a product manufacturing organization and a service providing organization are harder to pin down than you might at first think. Within the manufacturing company the majority of employees are providing services: marketeers, technical writers, designers, clerks, secretaries, store-people, expediters, accountants, personnel officers.

Figures 1.1 and 1.2 are extracts from BS 4778 intended to illustrate the determinants and measures of quality in (respectively) a manufacturing and a service organization.

Crosby (*Quality Without Tears*) suggests that an automobile manufacturer (fictitious name given) has 345 000 employees of whom only a third are directly involved in the manufacturing operations. He cites a typical bank as having similarly two-thirds of its employees providing analogous service functions to the one-third who actually deal with customers over the counter.

Because we are conditioned to think of quality in terms of manufacturing we expect to measure quality non-conformance in terms of 'scrap', 'defects', 'malfunction', 'out-of-spec', etc. But we can recognize poor quality in a service organization more easily if we look for 'errors', 'wrong entries', 'missing information', 'communication failures', 'time-wasters', etc.

The special features of organizations engaged in providing services will be examined more closely in Part Three, Chapter 22 of this book.

The management of quality

Having expressed our understanding of the meaning of 'quality' in business, we are going to examine how quality can be managed. That is to say, what responsibilities does the management of a company have for ensuring customer satisfaction with the product or service? (and hence ensuring the continued survival and prosperity of the company). Also, how can the management discharge these responsibilities?

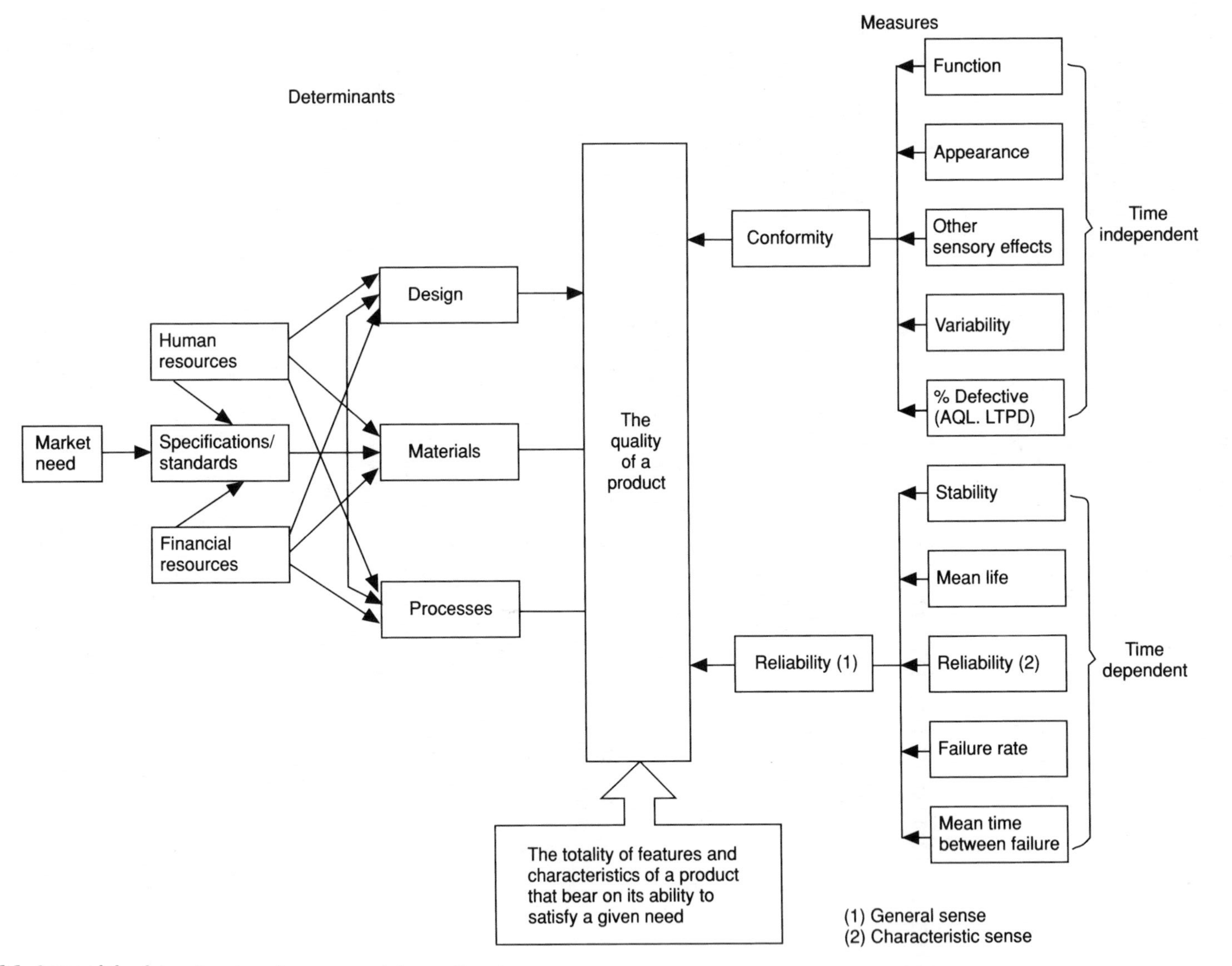

Fig. 1.1 Some of the determinants and measures of the quality of a product.

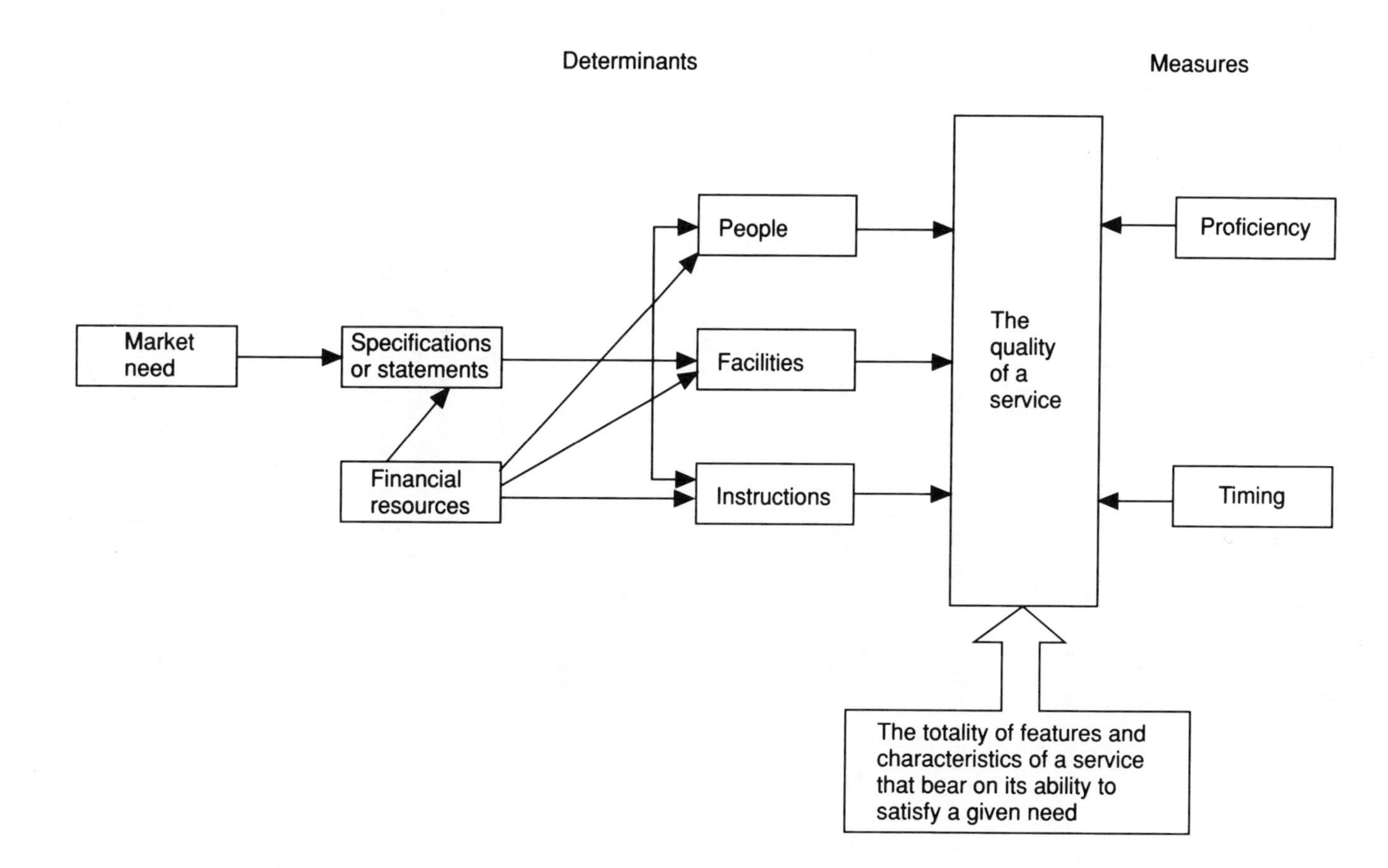

Fig. 1.2 Some of the determinants and measures of the quality of service.

Management of resources

The general principles of management are outside the scope of this book. Here we will simply quote Drucker's definition (in his book *The Practice of Management*) of the manager's job as 'the systematic organization of economic resources', these resources being:

- Money;
- Materials;
- Manpower;
- Time.

The end use of these resources is the profitable provision of the intended product or service.

The manager's job

Thus management is not a job which exists in isolation, but in relation to the resources available, and the nature of the desired end product. It has been described as 'The art of getting things done through other people'.

The elements of the manager's job can be classified as:

- Planning;
- Organizing;
- Motivating;
- Controlling.

Managing for quality

Quality is vital to the prosperity of an enterprise, and all areas within the organization have some influence on quality. The management functions – planning, organizing, motivating and controlling – must be exercised in pursuit of quality in every department and activity of a company. In order to achieve this, it is necessary for a company to have a conscious and coherent 'quality policy'.

The quality policy

ISO 8402 defines Quality Policy as:

> The overall quality intentions and direction of an operation as regards quality, as formally expressed by top management.

It is much more illuminating to read the quality policies formulated by actual companies. In *Quality Without Tears* Crosby says that 'Policies should not be hidden in books for just a few people to read, they should be painted on the water tower'. (3M United Kingdom PLC prints its quality policy on the back of all its business cards.)

In Chapter 11 of the same book Crosby gives examples of quality policy statements from various companies. He recommends that any policy should say the same thing, but in the company's own style, namely 'We will deliver defect-free products and services to our clients, on time.' Other authors such as Feigenbaum and Oakland also give examples of actual company policy statements.

Definition of quality management

ISO 8402 defines Quality Management as:

> That aspect of the overall management function that determines and implements the quality policy.

The 'quality loop'

Figure 1.3 illustrates the concept of the 'quality loop', and is taken from ISO 9004–1. Starting with the customer's perceived needs, the activities essential to the quality of the product or service impact the following stages:

1. Marketing and market research;
2. Design/specification engineering and product development;
3. Procurement;
4. Process planning and development;
5. Production;
6. Inspection, testing and examination;
7. Packaging and storage;
8. Sales and distribution;
9. Installation and operation;
10. Technical assistance and maintenance;
11. Disposal after use.

The 'quality spiral'

A related diagram, termed the 'quality spiral' has been used by certain authors. This version serves to emphasize that the quality expectations of the market-place are never static; it is always appropriate to take stock of what has been learned about quality in one product cycle, and use it to launch another, enhanced, product.

The quality management system

The provision made by management to ensure that quality is protected and promoted throughout all an organization's activities is referred to as the 'quality system' or 'quality management system'.

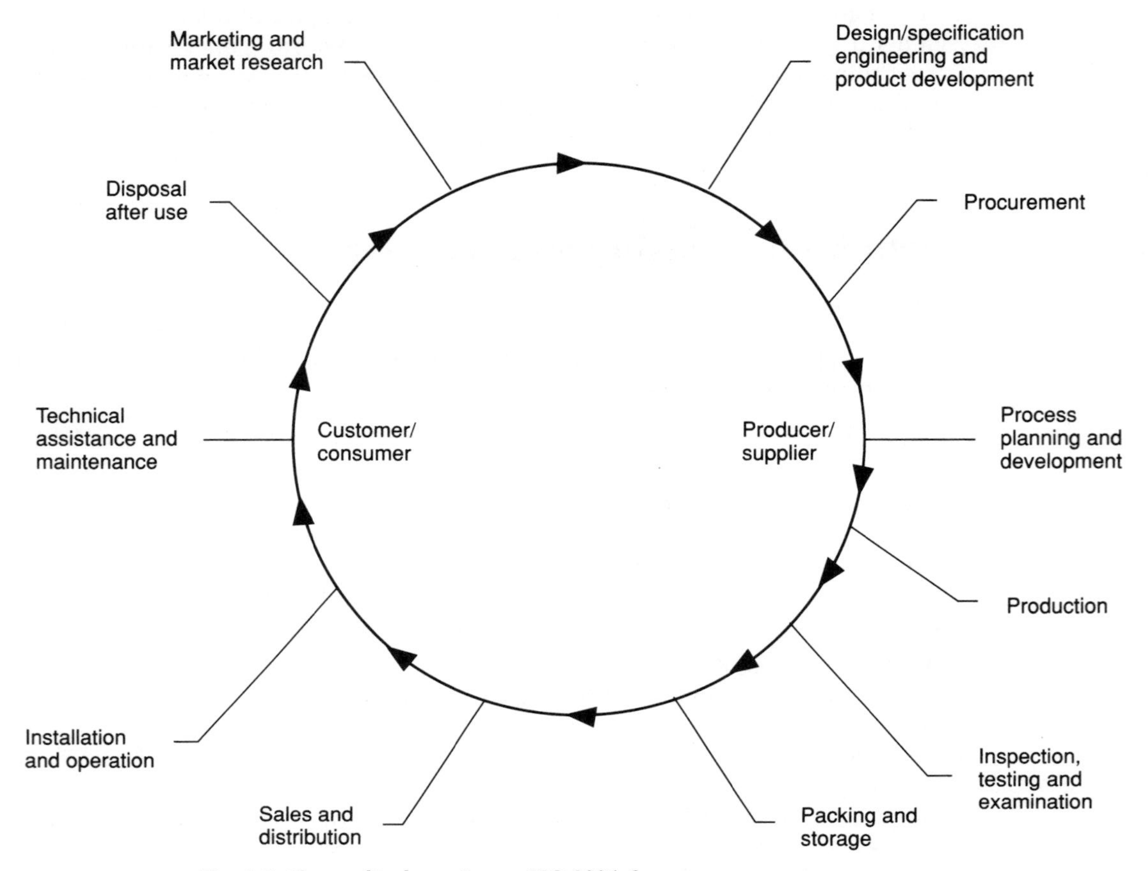

Fig. 1.3 The quality loop. *Source*: ISO 9004–1.

Definition of 'quality system'

Quality system is defined in ISO 8402 as:

> The organizational structure, responsibilities, procedures, processes and resources for implementing quality management.

Inspection, quality control and quality assurance

The development of quality management has been presented as an evolution from inspection, to quality control, to quality assurance. You will meet these terms constantly, so before we leave this preparatory chapter, we can start by giving the ISO 8402 definitions:

Inspection is:

> Activities such as measuring, examining, testing, gauging one or more characteristics of a product or service and comparing these with specified requirements to determine conformity.

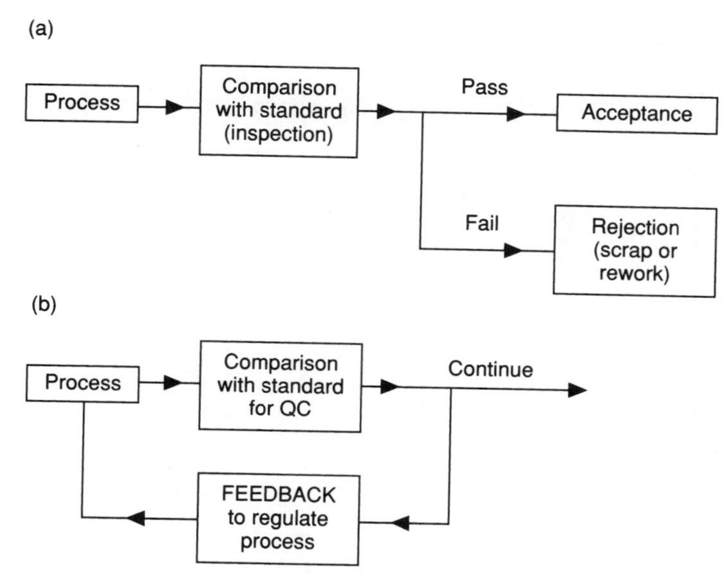

Fig. 1.4 (a) inspection, (b) quality control.

Quality control is:

> The operational techniques and activities that are used to fulfil requirements for quality.

Whereas quality assurance, again according to ISO 8402 is:

> All those planned and systematic actions necessary to provide adequate confidence that a product or service will satisfy given requirements for quality.

These definitions do not sharply delineate between these activities, so let us distinguish between the purpose of inspection and quality control by comparing diagrams of the inspection and QC activities. This is demonstrated in Figures 1.4a and 1.4b.

Inspection is concerned with **sentencing** the product as good or bad, by comparison with the standard. On the other hand, quality control is concerned with **feedback** of the comparative information in order to **regulate the process**. Sentencing **may** also be involved, but ideally the limits are set so that the process can be adjusted before product from the process reaches the limit where it has to be rejected. This is illustrated in Part Two, Chapter 10, which introduces statistical quality (or process) control.

Whereas both inspection and quality control concentrate their attention on the processes of creating the product or service, quality assurance addresses the whole of the quality loop. It attempts to anticipate and prevent errors at all points in the organization which could have an impact on the quality of the complete, delivered package associated with the product or service.

Usage of the terms QA and QC

Unfortunately, some commentators have used the terms quality **assurance** and quality **control** in a way which makes QA a subset of QC; others in a way which

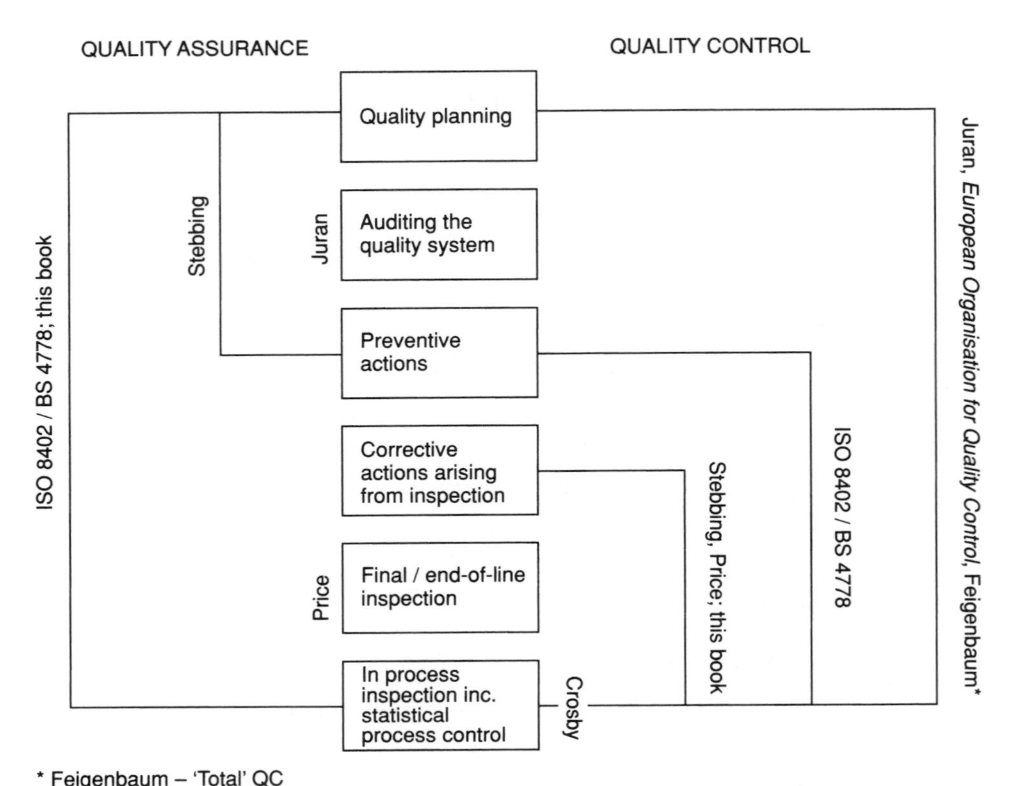

Fig. 1.5 Quality assurance and quality control.

makes QC a subset of QA. I have tried to illustrate the differences for a number of well-known writers in Figure 1.5.

If you refer to Feigenbaum's book, you will soon realize that in talking of **total** QC he is describing essentially the same thing as what we call QA. To minimize confusion, we will use the term 'quality assurance', and so far as possible avoid the use of the term 'quality control' without qualifying it in some way.

We shall follow the ISO 8402 definitions in which QA includes planning, and the quality system, and all the actions arising from them, i.e. the whole scope of this book. For us, QC on the other hand is the specific techniques and activities employed.

Summary of conclusions

At the end of each chapter I shall summarize the major conclusions that have been reached. In the case of Chapter 1 these are:

- Quality has a clearly defined meaning which is bound up with the concept of satisfying the customer's needs.
- Ideally, these needs are expressed as a specification, mutually agreed by customer and supplier.

- The idea that customers and supplier are mutually bound to define and provide these needs can be applied within an organization just as much as between an organization and its partners.
- Once quality can be defined, with grade and quality level specified, then quality can be measured. If measurable it can be managed, controlled and improved.
- In this book quality control represents the actions taken to prevent faults and errors affecting the customer; quality assurance focuses on the actions taken to prevent them happening.

While quality means meeting customers' expressed needs, excellence lies in exceeding them. It often makes better business sense to exceed customers' expectations. Why this is so, and how it can be achieved, is revealed in the rest of the book.

2

Quality in marketing

Introduction

The remaining chapters in Part One of our book will examine the demands that the pursuit of quality places on the marketing, engineering, manufacturing and maintenance functions of a company, i.e. those that figure in Feigenbaum's definition of quality, together with the company's relations with its suppliers and sub-contractors in the quest for quality.

Defining quality

In Chapter 1 you were given a number of definitions of quality. The ISO 8402 definition, 'the totality of features and characteristics of a product or service that bear on its ability to satisfy a given need' is a mouthful – but nonetheless correct.

A number of very short definitions have been proposed over the years by a number of quality professionals. These include:

1. Conformance to requirements;
2. Value for money;
3. Fitness for use;
4. Meeting customer requirements.

This latter definition is interesting in that it mentions 'the customer'. In other chapters you will find 'the customer' has many connotations. Everyone, in every walk of life has 'a customer' – usually numerous 'customers'.

In the simplest terms, in the commercial field the customer is the person/organization who pays for the product or service offered. The customer is the one who must be satisfied if he or she is to return on subsequent occasions to make further purchases or orders. There are many clichés in quality, one which comes to mind is:

> Price is negotiable, quality is not.

If you decide on the make of car you want, taking into account your requirements, you can visit many showrooms and you will be quoted many prices. The price will be negotiable, but you will still get the car you negotiated for!

So we must get the quality right from the beginning. Quality cannot be built in or on later. It has to be planned for from the earliest possible stage.

Another principle which must be mentioned at this stage is that it is cheaper to put problems right at the earliest stage than it is later.

'Ten times' rule

As a rough guideline when considering the cost of putting right problems at appropriate stages the 'Ten Times' rule is worth remembering. As a rough approximation, it costs ten times as much to resolve a problem in the production phase as it does in the design phase, and it costs ten times as much again to wait to resolve that problem once the product is in the hands of the user then to put it right in production. Therefore the moral is – resolve problems at the earliest stage. Also ensure quality is thought about and planned from the earliest stage. So what is the earliest stage?

Basic manufacturing process

Figure 2.1 shows the basic structure of an organization. It is clearly a manufacturing organization, but this is convenient because the disciplines and techniques of quality management were developed in such an environment. As you progress through the book and come to understand the principles better you will appreciate how they can be applied to service organizations also. We shall then focus on the special features of service organizations in Chapter 22, the final chapter of Part Three of the book. Figure 2.1 will be referred to again in Chapters 3–6.

Quality aspects must be considered in the areas above, but from our initial argument we must start planning quality aspects from the beginning when it has the biggest impact. Where is the earliest stage? It is in the marketing organization.

Vital role of marketing

It is the manufacturer's marketing department which is charged with finding out what the customer wants, by means of market research, and in effect, writing the customer's quality specification for him/her. It will then be the design department's task to incorporate the requirements in the detailed design, and the manufacturing department's task to produce it with minimum scrap and rework from the drawings, specifications, etc. issued by the design department.

Naturally, some of the objectives of the initial customer's requirements may be contradictory, and the departments concerned will have to arrive at a specification

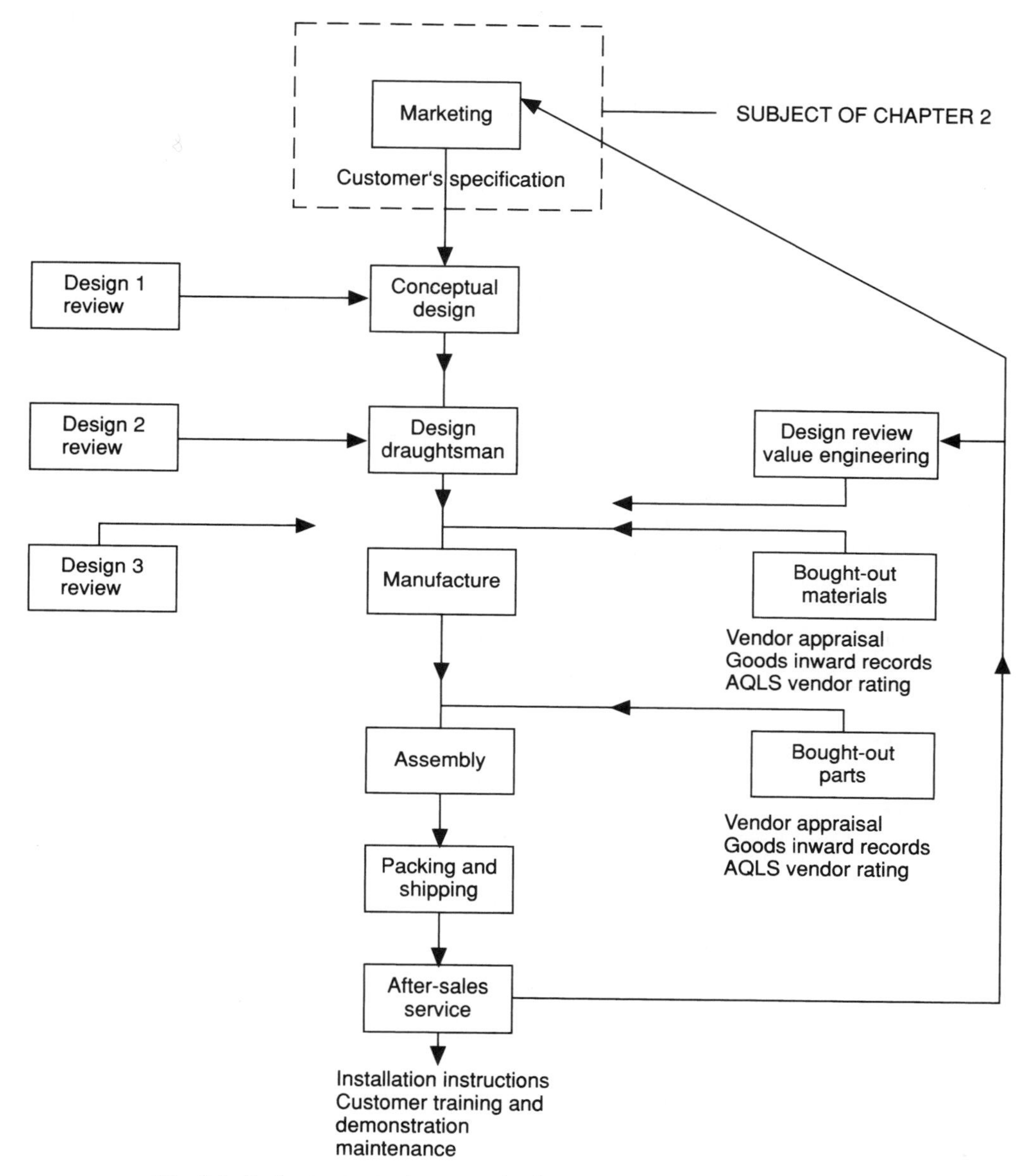

Fig. 2.1 Basic structure of an organization.

which is functionally serviceable, manufacturable within the target price and still attractive to the customer.

The risk in this situation is that the consumer's quality priorities may not have been correctly gauged. This can only be tested by maintaining the sales level and evaluating any customer complaints. If the initial customer specification is not an accurate representation of the customer's needs and perception, then no matter what efforts are made by the rest of the organization, quality problems will be encountered.

Planning is essential

It is essential that the marketing operation is planned meticulously and this may be seen through a marketing plan, the provision of which is necessary to ensure the quality of the marketing operations is planned, implemented and maintained.

This may cover the following areas:

1. The financial quality and other targets for several years ahead for 'order' taking and delivery schedules;
2. The resources required by marketing to achieve the plan (space, manpower, equipment, capital investment, training, etc.);
3. The commitment required from other areas of the organization;
4. The plan would contain sufficient detail of the main parameters – specification, delivery, quality, external events (currency, political, etc.) so that risk analysis can be performed: and so assess the effect of variations in these parameters.

The firm should demonstrate that it has organized itself to meet the requirements of the marketing plan. Particularly in the numbers and skills of its personnel, which includes those quality personnel necessary to implement and maintain the quality system in marketing.

The quality system itself must be reviewed at regular intervals to ensure that it can assess the validity of, and the constant updating of, all the marketing documents and procedures affecting quality.

Planning

The firm should demonstrate that there are written plans in place to produce on an ongoing basis:

Market research and analysis

Note: If sub-contractors are used for market research, then they should be subject to:

1. Vendor appraisal;
2. An assessment of the efficacy of their market research against subsequent events and present knowledge;
3. Any corrective action on their services put into place.

Specification of market requirements (the market specification for a particular area of the market)

This should include (where appropriate) the nature of the market(s) – in time, demographic factors, geographical location, features required (including reliability/quality), assessment of competition, etc.

In particular the change of market requirements with time must be assessed – particularly if there is a 'time window' in which the product(s) or services must be offered/delivered.

This must include also the maturity and life cycle of the products necessary, together with any future enhancements. (Note: The market life cycle is distinct from the product life cycle.)

The market specification should be subject to change control. Any change should result in a reappraisal of the products offered, and under development to meet that market.

It is important to distinguish between 'a market requirement' and a supposed product solution to that requirement.

The supplier must maintain a register of standards and legal requirements pertinent to his/her markets.

The product specification (or requirement specification)

This is the specific definition of a product to meet a particular market area and should include all the developments required (not just the hardware/software/system), e.g. it should also include:

1. A literature specification (manuals, etc.);
2. A marketing specification, including brochures, training, test marketing, promotions, etc.;
3. A 'launch' specification, including publicity, brochures;
4. A specification of how the product is expected to work, what functions, properties does the customer expect (ease of use, repair, installation, environmental factors – space requirements, transport requirements, legal/safety requirements, geographical and language requirements, etc.) if it is a complex product. Just technical details of the parts are not sufficient;
5. The planned life-cycle of the product.

The product implementation specification (i.e. what it is agreed to develop manufacture, purchase)

This includes similar entries to the above section (1–5) and also includes test procedures and product verification procedures. This document may change as the development proceeds and it is important that marketing know of, assess, and approve changes (e.g. in timescale, performance, price).

Any change in the market requirements may of course result in changes to this document including the test and verification. It is not unusual for this and the above section to be combined in a customer specification.

Discovering customer requirements

As stressed previously it is very important that the marketing department discover the customers' expectations which may be expressed precisely in a specification, implied or latent.

If the producers are in an environment where their customers furnish them with a detailed specification, e.g. defence industries where the explicit requirements are given them, in many ways they are lucky, but in the commercial field the producer has to determine his customers' requirements via the marketing department through market research, taking into account the producer's own organization's strengths, weaknesses and capability.

Let us look at this latter situation in a little more detail. Customer information can be gleaned from a number of sources. These may include:

Feedback of information from customers

One of the main objectives of quality management systems is to produce information and then use that information for improvement. Customers can produce vital information, some of it from customer complaints, warranty and service calls, spares usage and even in some cases via telephone calls and surveys initiated by the producer. The marketing function must establish an information monitoring system to use this information produced. All information pertinent to the quality of the product or service must be analysed, collated, interpreted and finally communicated by well defined procedures.

Use of experienced marketing/sales staff

Marketing and sales staff can be used in a variety of ways. They will be in constant contact with possible customers, and they can produce ideas via the following:

1. Brainstorming sessions;
2. Interrogation of trade associations;
3. Carrying out surveys.

In fact the number of methods and techniques for researching market demand is almost unlimited and offers much scope for individuals with flair and initiative. The important point to stress is that the producer keeps very close to the customer, and in the end a specification is produced which reflects the customer's requirements. Only in this way is there any chance that a product or service will be offered which meets the perceptions of the customer.

Good communications are important

It must be obvious from what has already been stated that good communications with potential and past customers is essential if the marketing department is to be able to define future needs, but what about internal communications within the producer's organization? These too must be well defined and used, and two-way. The marketing section must be able to communicate customer requirements clearly, unambiguously and accurately within the company, but they themselves must be the recipients of information from within the company which will aid them in producing a realistic requirement.

The marketing department must have good communications to receive information on the following:

1. Internal estimating and costing procedures;
2. Cost of bought-out items;
3. In-house production capability;
4. In-house test capability;
5. Cost of necessary bought-out services.

With this information they will be able to gauge if the market requirement can be met considering the two main parameters – 'cost' and 'time'.

It is extremely important to offer the product or service to the potential customer in the correct time frame and at a realistic cost. These aspects are as important as offering the correct (as specified) product or service.

Customer's specification

I have implied from the beginning of this chapter that getting the specification right during the marketing phase is extremely important.

The amount of detail offered in the customer specification will be dependent on the qualifications and experience of the marketing staff. If they are well qualified the marketing staff will be able to provide a very detailed and realistic specification in which they will have every confidence that the design/manufacturing departments will be able to produce. If they are not so well qualified and the information available from their own organization is not as good as it should be, then they should aim at a minimum standard for their specification which should include:

1. Performance characteristics (e.g. environmental and usage conditions and reliability);
2. Sensory characteristics (e.g. style, colour, taste, smell);
3. Installation configuration or fit;
4. Applicable standards and statutory regulations (implicit requirements);
5. Packaging;
6. Quality requirements, including information on 'quality grade' (Chapter 1).

The characteristics/requirements shown above are the minimum requirement. To help market staff produce adequate specifications for designers to work on, certain areas of industry issue their marketeers with 'check lists', which are a list of the most important characteristics necessary for designers to work on. This system has many advantages, but of necessity it has to be limited to very specialized areas if the check lists are to be meaningful.

The above will give the basic necessary information which can be used in the design department. There is still, of course, room for misinterpretation no matter how carefully the specification is written and at the earliest opportunity a design review should be scheduled (Chapter 3) so that marketeers and designers can get together to ensure that the design is meeting the requirements and there are no misconceptions. Trade-offs may also be discussed.

The loop starts all over again

An early warning system should be established by the producer for reporting instances of product failures or shortcomings, if these unfortunately surface for the newly introduced product. This system should ensure rapid corrective action. A feedback system regarding performance in use should exist to monitor the quality characteristics of the product throughout its life cycle. This system will analyse, as a continuing operation the degree to which the product or service satisfies customer expectations on quality, including safety and reliability.

Information on complaints, the occurrence and modes of failure, customer needs and expectations or any problem encountered in use should be collected, collated, interpreted, analysed and made available for review and the information used for subsequent corrective action.

Market readiness

One aspect of the analysis of the market demand which extends back into the organization is the review of market readiness of a new product or service. This determines whether field support (Chapter 6) is adequate and provided for, for the new or redesigned product. Depending on the product, the review may cover the following points:

1. Availability of appropriate user, maintenance and installation manuals;
2. Existence of an adequate distribution and customer service organization;
3. Training of appropriate installation/maintenance personnel;
4. Availability of spare parts;
5. Availability of appropriate test equipment.

An example

When the Nimrod aircraft project for the UK Ministry of Defence was cancelled, the cry from the designers/manufacturers was that the customer's requirements were always changing, so what chance had they of meeting these requirements within the initial time scale and cost? It is easy to see their problem. Plans had to be constantly updated and much work had to be scrapped or added to. In the fiercely competitive commercial market this type of indecision can be disastrous. You can miss the timescale 'window' and allow a competitor to scoop the market, or you deliver a product or service which is underdeveloped with the fear that your customers will not be satisfied and will not return to make further orders.

Under these circumstances it is easy to see that getting the customer specification right first time is important and plans should be put in motion to achieve that. If the customer specification is wrong, everything which follows will be a waste of time. If you have a good marketing organization and they can produce a sound specification which reflects the customer's needs there is every possibility (see subsequent chapters) that a 'quality' product or service will be produced.

The key phrase is 'get the goalposts firmly fixed in concrete! It's very difficult to score if the goal posts are forever moving!'

Summary of conclusions

Marketing and quality are both concerned with 'giving the customer what he wants'. There should thus be very close communication between the marketing and quality assurance functions of a company. If the links are not there, or if either function is not performing correctly, the results in terms of interpreting customer needs, and hence in providing customer satisfaction will be disastrous.

One symptom of the kind of problem which can arise is failure to agree on the specification which is being used, or the product or service's ability to meet the specification, and how this is to be tested. This often also involves failure to meet projected costs or time-scales. Many major projects will doubtless spring to mind, especially in the defence, nuclear power, aerospace, computer software and construction fields.

3

Quality in specification and design

Introduction

In Chapter 2 we stressed the importance of looking at quality aspects from the earliest stage in the product cycle. In this chapter we shall show that planning the design programme is essential and some of the techniques used in the design process will be highlighted.

Figure 3.1 shows the basic structure of an organization which we looked at in Chapter 2; we saw that the marketing department must define what the customer wants. If they cannot clearly state what the market requires then the design department cannot be expected to complete a satisfactory design.

In this chapter we will be looking at the area highlighted in Figure 3.1 in more detail.

Design

From what we have said in the previous chapters, it should not come as much of a surprise to find, as a definition of design:

> The translation of customers' requirements into instructions for manufacture of the product.

It takes account of:

1. Reliability;
2. Ease of maintenance;
3. Interchangeability;
4. Standardization;

5. Ease of production;
6. Environmental requirements.

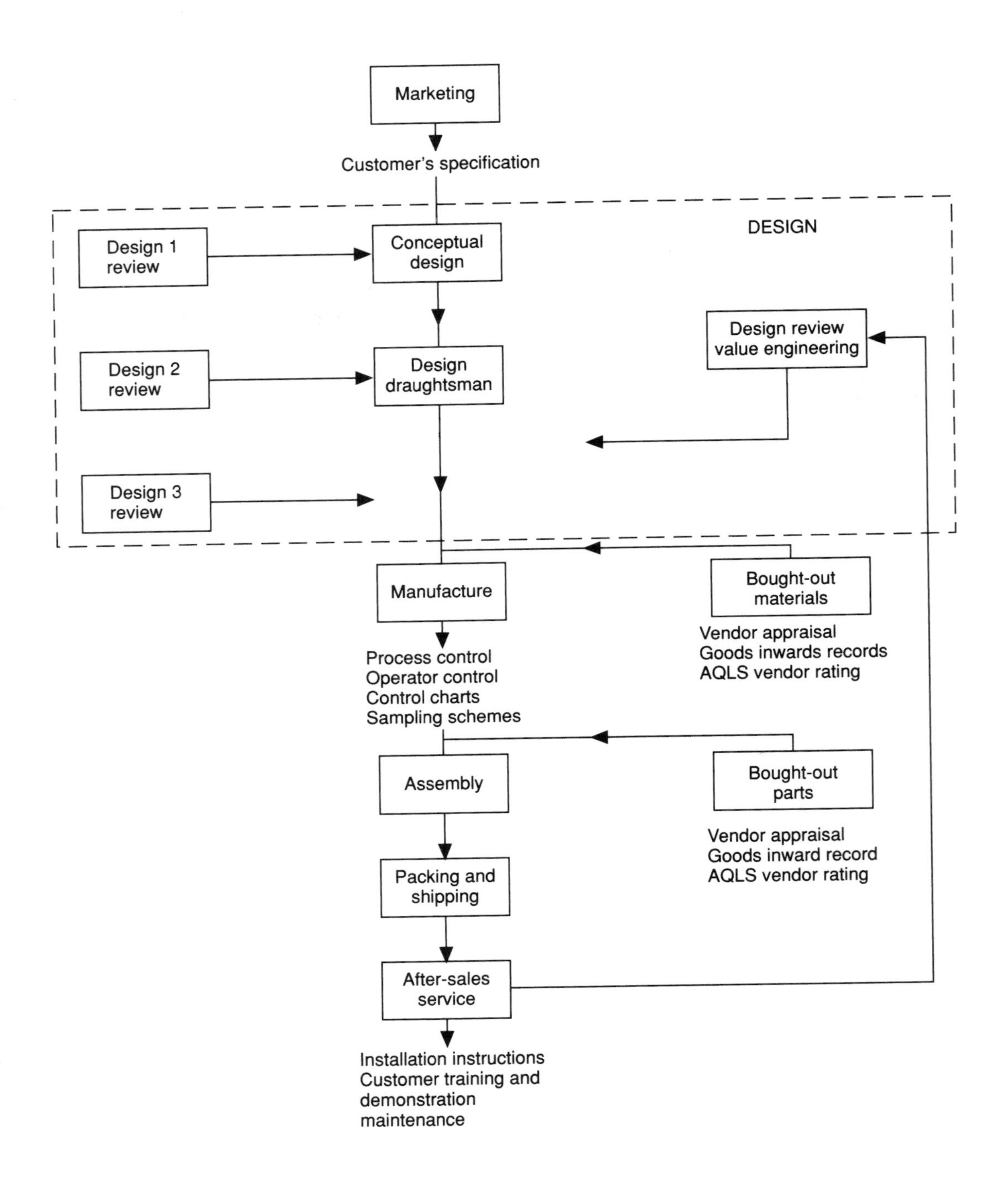

Fig. 3.1 Basic structure of an organization.

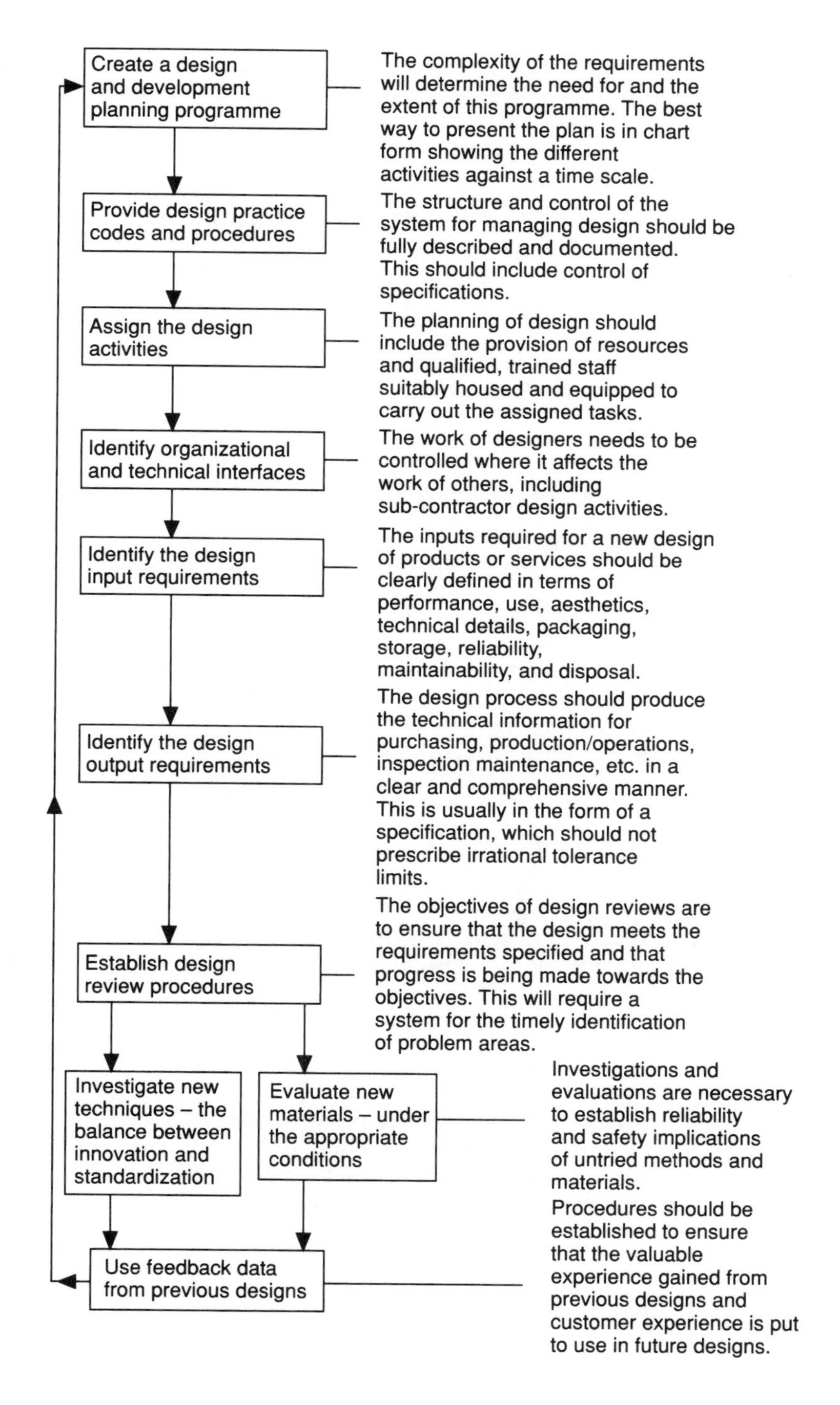

Fig. 3.2 The design control process.

The phases of the activity shown in Figure 3.2 are:

The design and development planning programme

The complexity of the requirements will determine the need for and the extent of this programme. The best way to present the plan is in chart form showing the different activities against a time scale.

Creating the design practice codes and procedures

The structure and control of the system for managing design should be fully described and documented. This should include control of specifications.

Assigning the design activities

The planning of design should include the provision of resources and qualified, trained staff, suitably housed and equipped to carry out the assigned tasks.

Identifying organizational and technical interfaces

The work of designers needs to be controlled where it affects the work of others, including sub-contractor design activities.

Identifying the design input requirements

The inputs required for a new design of products or services should be clearly defined in terms of performance, use, aesthetics, technical details, packaging, storage, reliability, maintainability, and disposal.

Identifying the design output requirements

The design process should produce the technical information for purchasing, production/operations, inspection and test, maintenance, etc. in a clear and comprehensive manner. This is usually in the form of a specification, which should not prescribe irrational tolerance limits.

Establishing design review procedures

The objectives of design reviews are to ensure that the design meets the requirements specified and that progress is being made towards the objectives. This will require a system for the timely identification of problem areas.

Investigating new techniques – the balance between innovation and standardization

Evaluate new materials – under the appropriate conditions.

Investigations and evaluations are necessary to establish reliability and safety implications of untried methods and materials.

Feeding back information from previous designs

Procedures should be established to ensure that the valuable experience gained from previous designs and customer experience is put to use in future designs.

The concept of 'design' is complex and not easily defined. For our purposes it is understood to include the following activities.

1. One or more customers formulate a set of requirements (possibly with the assistance of sales or marketing organizations);
2. After negotiations, calculations, and sometimes experiments, a specification is agreed which describes a self-consistent set of requirements acceptable to the customers;
3. There are then a number of creative activities (inventions and bright ideas) followed by calculations, simulations and tests of models or prototypes. ('Development' is that part of the design process during which models and prototypes are built and tested);
4. The end product of the design activity is a set of drawings, specifications and other descriptions (collectively referred to as the 'design package') which together will describe a product which will meet the customers' requirements. In general the product may include equipment, computer software and procedures. The design package is sufficiently detailed to ensure that a product conforming to it will invariably meet the requirements.

Good design requires good management, which implies recruiting staff of the right ability, creating an organization which exploits their strength, and supporting them with the right facilities. Control also implies a feedback system which is capable of detecting and eliminating errors and their causes. The quality staff in both supplier's and customer's organizations have important functions in the design area, especially in making sure that requirements are properly stated and in maintaining communications.

The design control process

This is shown in more detail in Figure 3.2 and each of the categories there will be discussed in more detail in the text later, but at this stage let us consider software. This element of design deliverable software is becoming more important as the years pass. In fact there is little difference between the process involved in hardware design and those involved in software design. Once these two parts have been designed and tested they are integrated, tested and finally emerge as the system envisaged.

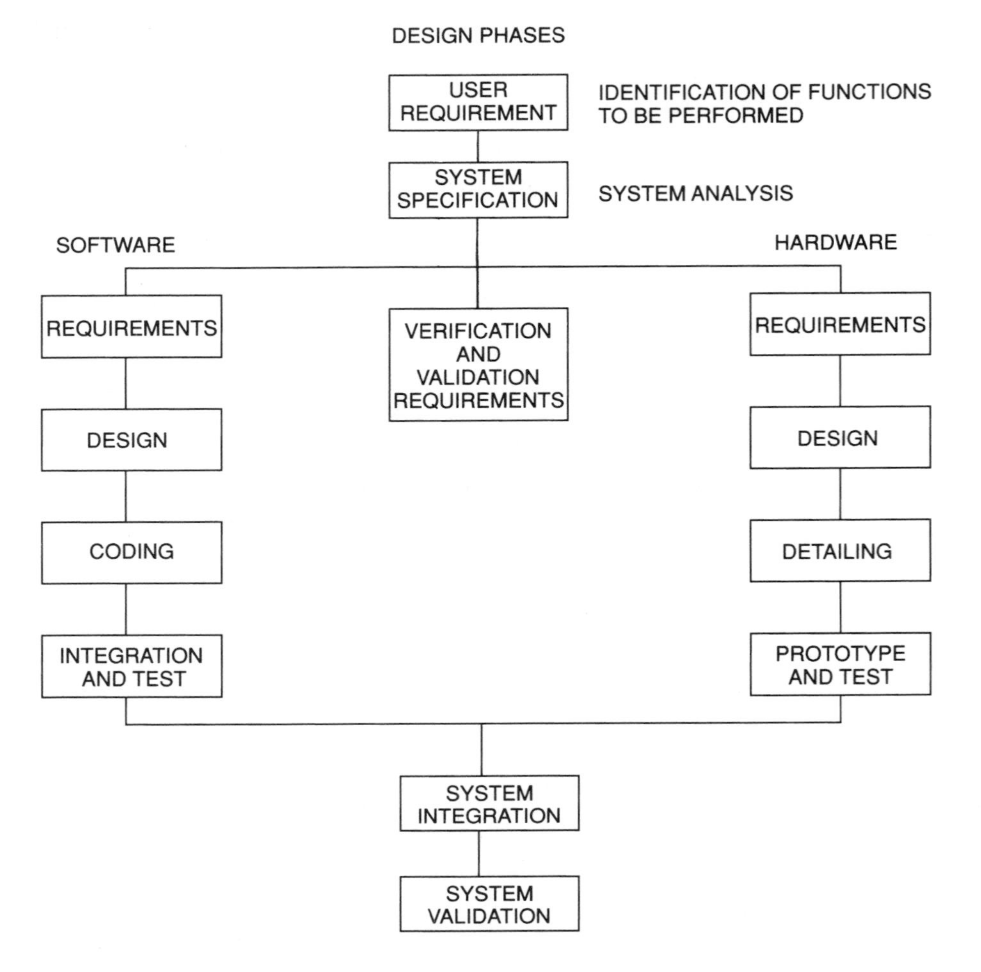

Fig. 3.3 Design phases.

The two processes are shown side by side in Figure 3.3 and in more detail in Figures 3.4 and 3.5.

Figure 3.4 shows parallel development paths for 'hardware' and 'software'. You may be inclined to assume that software refers specifically to computer software; indeed computer software design and development is discussed in Chapter 13 of Part Two. However, do not ignore other forms of product-related 'software' such as installation instructions, user guidance, maintenance and repair manuals, and spares lists which will be discussed in Chapter 6 of Part One.

Quality requirements of specifications

As mentioned earlier, quality can only be 'designed in' if the specification being worked to is complete, unambiguous and reflects the customer's requirements.

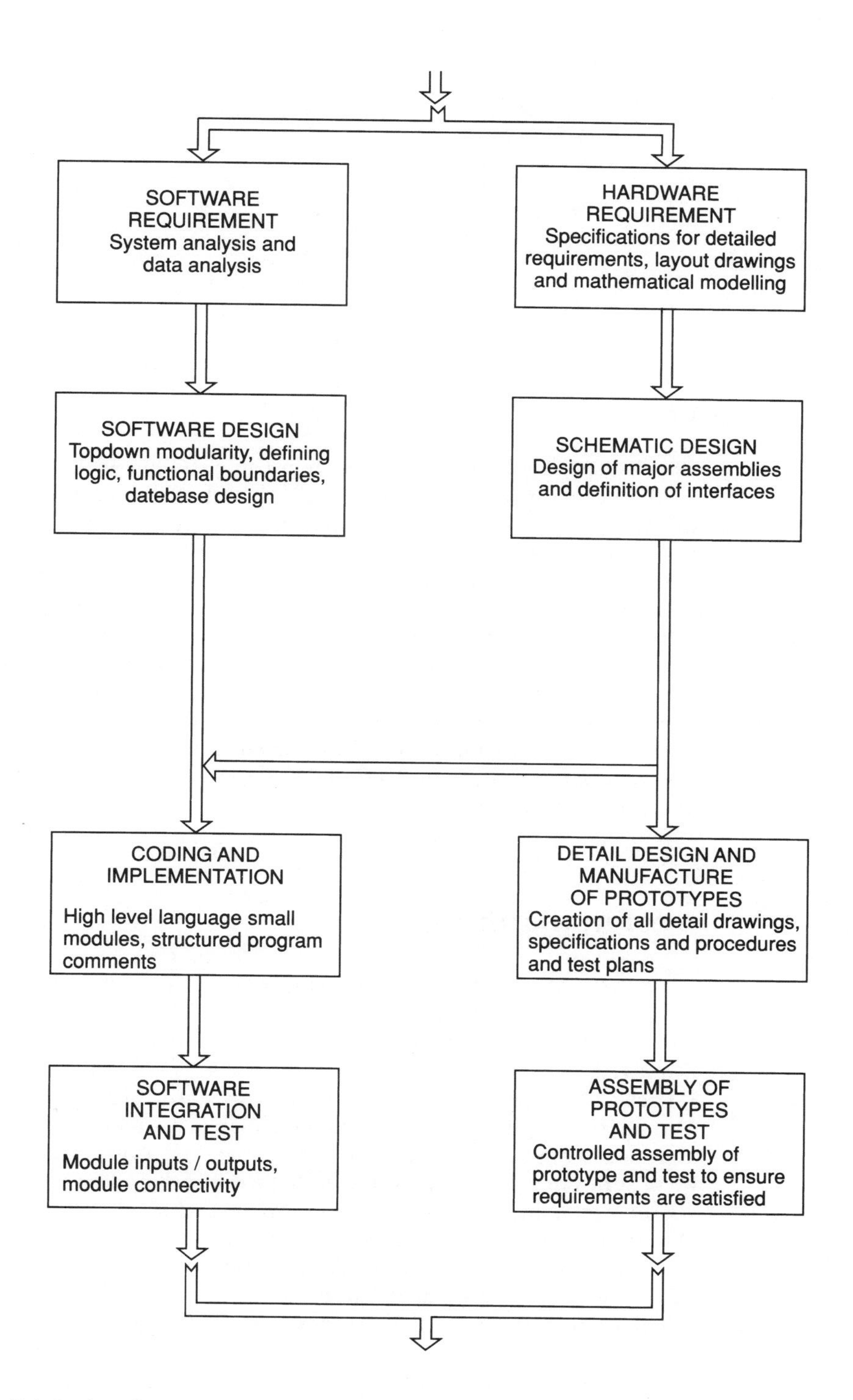

Fig. 3.4 Design phases.

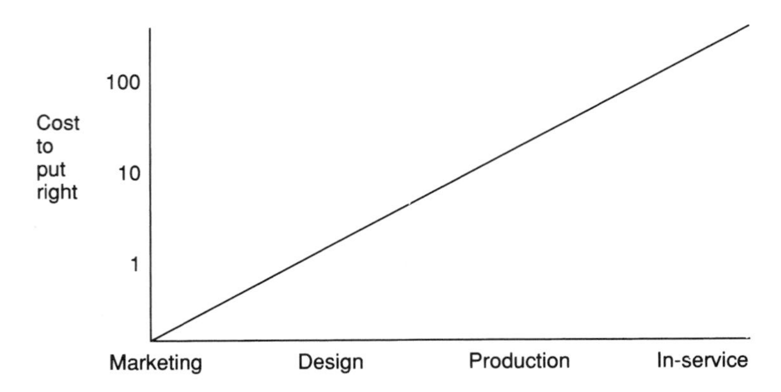

Fig. 3.5 Cost benefit of early correction.

Check-lists may be generated to help the customer or the marketing department in drawing up an initial specification, but it is still too easy for designers to misinterpret the requirements. To overcome this it is necessary to incorporate in the design phase a control to prevent much work being expended on a design which has no way of meeting the customer's expectations. This is the initial design review. The result of that review may vary from a complete re-design to permission to carry on to the next stage.

Design is frequently an iterative process. The amount of effort a firm devotes to this activity will be entirely dependent on the nature and complexity of the product and where it anticipates the market potential to lie; it is the designer's own business how the appraisal is conducted, but an enquiring attitude towards the market is an indication of a rational approach, which is likely to lead to a successful and profitable design.

Design planning

To control any activity, it is necessary to plan. In design, we will need to estimate the human resources, equipment, and training required to complete the work. Senior management will be keen to impose constraints, namely time and money.

These items must be considered at this early planning stage and the project Manager of the design must produce a plan. This will normally take the form of a chart showing activities (usually costed) against a time scale. Activities will either be in series or parallel, i.e. some activities cannot be started unless another one (or more) are completed beforehand. Such a chart is often called a PERT chart (Programme Evaluation and Review Technique).

In many firms senior management pay excessive attention to the time and cost constraints, and under these circumstances designs can often be offered to the market without sufficient tests and trials, particularly with respect to reliability.

To help overcome this problem it is often desirable (and sometimes contractually obligatory) to produce a special 'quality plan'.

Quality plan

The quality plan is (ISO 8402): 'a document derived from the quality programme setting out the specific quality practices, resources and activities relevant to a particular contract or project'.

Note that the plan is contract or project specific.

At this stage we must be clear what are the differences between the quality plan and the quality manual.

The quality manual is (BS 4778): 'a document setting out the general quality policies, procedures and practices of an organization'.

In other words the procedures used on a day-to-day basis. The quality plan takes precedent over the quality manual on the project on which it is used.

The third set of documents which we can see in Figure 3.2 are the Codes of Practice. Codes of Practice are the designer's standard procedures and you would expect to see volumes of these in every design office. Again they cover the day-to-day standards to be used in that office. Possible topics which may be covered in a hardware design office are listed below but note that this list is not exhaustive. The quality plan is a very important document, which is often contractual in practice.

The quality plan and quality planning documents should identify the product or service required under the contract and invoke the systems and procedures of the quality manual in so far as they apply to that contract. They should identify all quality requirements and/or procedures that are mandatory under the contract and any necessary amplification or variation of the procedures of the quality manual.

The quality plan and quality planning documents should include the definition of planned quality tasks; identification of the responsibility for all tasks; the time scales to be met; the method of task implementation; the provision for monitoring implementation; the need for additional or special capabilities and training; and the allocation of resources.

On larger contracts where a quality plan is required there may be a need for specialist activities to be defined in separate but associated plans to the quality plan, these may include reliability, maintainability, software, configuration management, installation and trials.

The quality plan and quality planning documentation should include the arrangements for providing assurance of satisfactory quality of subcontracted supplies. In some circumstances where, for instance, major design and development tasks are sub-contracted, it may be necessary for a sub-contractor's quality plan to be included as part of the quality plan of the main contractor at any event to be referenced in his quality plan.

Features which may need to be considered in a quality plan:

1. Critical stages of the project or manufacturing cycle;
2. Design and development quality control procedures;
3. Manufacturing, inspection and test stages for components and main assemblies;
4. Reliability and maintainability programme;
5. Control of special processes;
6. Configuration control procedures;
7. Interchangeability control;
8. Identification of safety features and associated QA methods;

9. Sub-contract control;
10. Special calibration requirements;
11. Design review procedures;
12. Product audit procedures;
13. Material and component selection;
14. Software QA;
15. Safety critical software;
16. Qualification testing;
17. Installation and trials;
18. Special training needs;
19. Interface control;
20. Investigation of new techniques;
21. Value engineering;
22. Use of data feedback from previous designs.

Note: This list is indicative only and not exhaustive.

Design reviews

You have learned that it is cheaper to put things right early in the life cycle, and Figure 3.5 may be familiar.

Design reviews are a particularly effective **preventive** measure.

A design review is an engineering management process that is used when it is required to subject a design to a formal, systematic critical study. It is a cooperative examination.

Each review will have a different objective in the development of the configuration baseline.

At the commencement of a design, a programme of reviews will be prepared and included in the project and quality plans against significant milestones.

The overall objective of a review is to demonstrate to the senior management of both the design agency and the customer that the design will satisfy all aspects of the requirement in a cost effective manner. The detailed objectives are to ensure that:

1. All reasonable design paths have been explored;
2. All contributory factors have been considered;
3. The design meets the specification;
4. The design can be produced, inspected, tested, installed, operated and maintained in a way which is satisfactory to the customer, taking into account time and cost constraints;
5. There is adequate supporting documentation to define the design and how it is to be used and maintained.

A design review is not the same thing as a technical progress meeting. The latter is concerned with day-to-day technical achievements and problems of time and cost.

During the design process it is normal for the design project leader to hold *ad hoc* informal reviews with his or her staff to assess the manner in which the design is satisfying the requirement; in addition to these, there should be formal occasions when progress can be assessed by the customer in the presence of senior management and when potential faults can be examined by specialists.

Review procedures can differ significantly between different industries, companies and groups, but one definition of this function seems to enjoy acceptance by the majority: 'the formal, documented and systematic study of a design by the user and specialists not directly associated with it'. Depending on the nature and complexity of the product it will be usual to arrange three stages of review in the design and development cycle (shown in Figure 3.6).

Preliminary review

A preliminary review is held at the time of the product concept studies or at the planning proposal stage. It is intended to cover such points as design philosophy, specification requirements, and broad estimates of time and cost. It sorts out ambiguities in the marketing or customer specification.

Intermediate reviews

A number of reviews are held during what is known as the project definition phase. They will consider functional diagrams, energy flow design, mechanical design, quality plans, reliability programmes and so on, and will usually concern sub-assemblies or components rather than the complete equipment.

Final review

A final design review is held when the material lists and drawings are complete, the assessment of reliability and tests on pre-production units have been analysed, and the preparation of handbooks has been completed in draft.

The objectives of the design review are threefold:

1. To confirm that the design meets the requirement;
2. To ensure that the design is producible and testable against the requirement; and
3. To uncover hidden snags.

Those participating in the review should include top management, designers, production, purchasing and, when the company is using techniques, processes or materials with which it is unfamiliar, outside specialists. You should always remember that the designer is responsible for design, and though the design review may give advice, which is on record, the designer must nevertheless retain the right to make the final decisions. It is also important to avoid an atmosphere of conflict and to recognize that the intention is to criticize the design and not the designer: the aim is for a cooperative effort to achieve the best possible result.

The standard of the design is set during the design and development cycle and the design reviews bring together the various plans and programmes necessary to

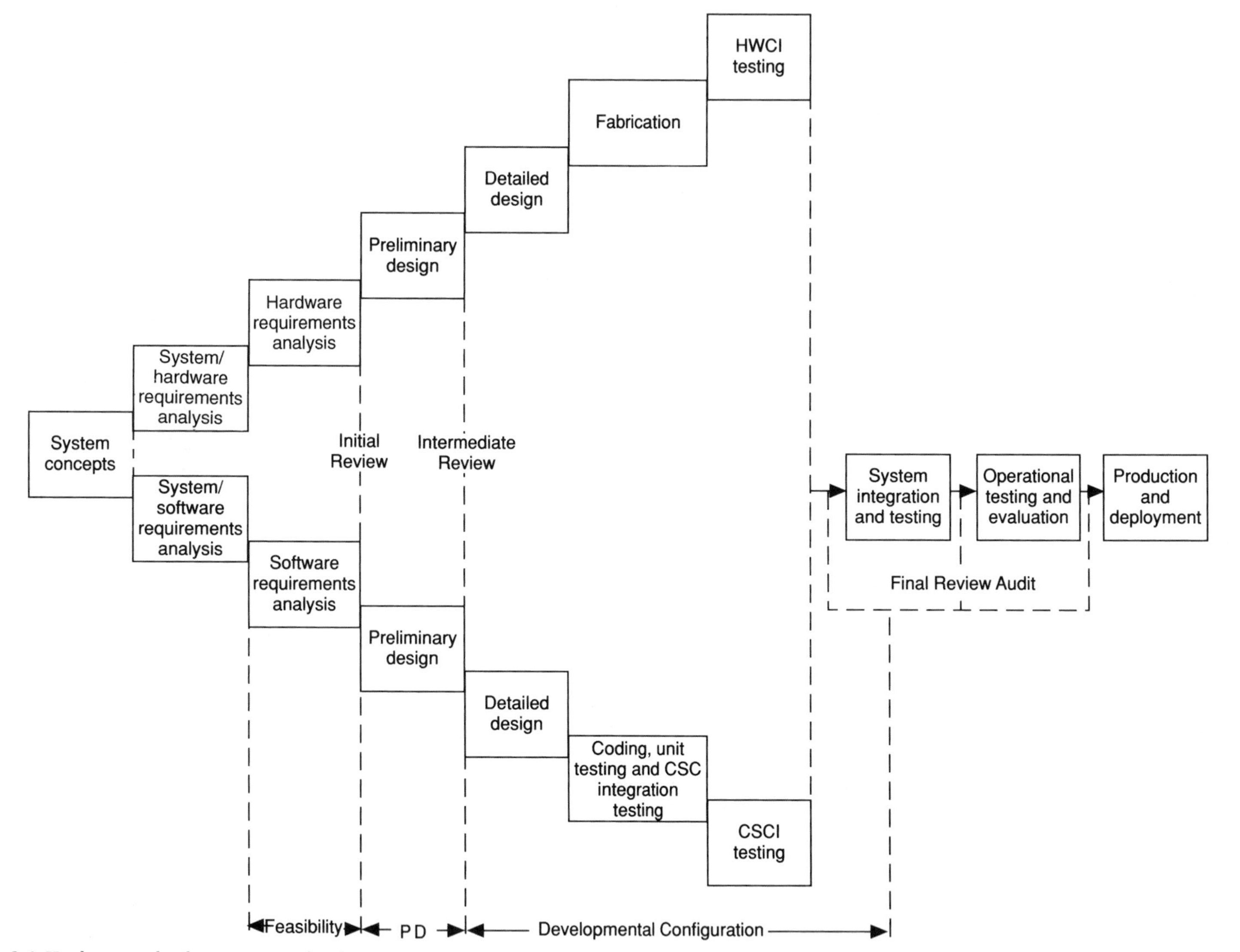

Fig. 3.6 Hardware and software system development.

produce a successful design. It gives the customer an opportunity to monitor the progress being made to satisfy the contract requirements, and, perhaps more important, permit the customers to acquaint themselves with the new equipment at an early stage.

Some important aspects of design

Interfaces

Wherever there are interfaces, there are quality risks. Interfaces may cover technical and procedural requirements and can include:

1. Hardware/software interfaces;
2. Contractor/sub-contractor interfaces;
3. Works/trials environment interfaces.

In the first case, who hasn't suffered a computer problem and found the hardware engineers' first reaction to be: 'It's a software problem!', and vice versa!

Control of sub-contractors is notoriously difficult but this will be covered in more detail in a later unit.

When something goes wrong with the equipment at a critical time, the reaction is usually to patch up the problem and forget the well rehearsed and used procedures that would have been used back in the laboratory!

Role of the quality department

We have already mentioned the 'quality manual'. It should include an organization chart showing how responsibilities for various aspects of the work are divided between the staff, but it should also define the liaison and coordination channels which are often more important. The document should contain technical procedures covering such topics (where appropriate) as design review, reliability and maintainability programmes, value engineering, interchangeability and standardization control of tolerances, quality information, change control, and the preparation and maintenance of drawings and lists of British Standards which it is company policy to implement.

It is important for the company to have the correct relationship between quality assurance (which may include reliability and safety sections) and other departments. The quality department should have three roles:

1. To coordinate quality activities of other departments, especially design and production;
2. To provide specialist advice and expertise on certain topics concerned with the measurement and prediction of quality such as metrology, calibration, reliability engineering and statistical quality control;
3. To audit the quality management of the company body by measurement and test of a sample of products and by a periodical review of systems and procedures both in theory and practice.

Responsibility for the quality of the final product should rest with the design and production departments. It should not normally be the task of the quality departments to test all output from production nor should the work of following up and rectifying quality failures be the sole responsibility of the quality department, though they should check that somebody does this work.

Feedback of information

It is only when an equipment is used that the quality of the design team's work is authoritatively assessed. Unfortunately, there are problems in feeding comments and information on the good and bad features of their work back to the designers in a form which is useful to them for future work. There is a vast flow of information on defects, and ADP is increasingly used to store and analyse this information; it is of great value to the maintenance section but is of limited use to the designer since it is often, understandably, deficient in accurate detail concerning length of use, environmental stresses at time of failure, and diagnosis.

However, the designer should be aware of its existence and availability, and may be expected to seek from it data on good features of previous designs which have shown low failure rates.

Standardization and specification

There is some controversy about standardization in techniques and variety reduction in components and sub-assemblies. Some say it is of great value in reducing cost and improving reliability; others regard it as placing undue constraint on the designer, limiting performance and bringing about technological stagnation. Each proposal for the use of a standard technique or component must be judged on its merits.

It is best to avoid generalized arguments as to whether the slowing of technological growth which standardization may cause outweighs possible advantages of improved availability and lower cost; each case has to be considered on its merits. Documented standards of preferred practice are introduced only after careful consideration; the more rigorously they are applied, the more useful they become.

Variety prevention requires an evolutionary approach to design which allows standard modules to be used in a number of equipments. Implementation of standardization requires a determined management effort and committed resources both at company level and on each project. Complex systems require a sequence of specifications of increasing refinement, including requirement specifications, defining specifications and production specifications. All specifications need to be comprehensive, accurate and realistic.

Reliability

The definition of reliability in ISO 8402 is 'the ability of an item to perform the required function under stated conditions for a stated period of time'. The reliability of an equipment cannot be established by a single measurement at a single point in time; in fact the true figure can only be measured in retrospect when all equipments have failed. Therefore, much effort must be directed at predicting the reliability of a design to discover where it needs to be changed.

Ultimate responsibility for reliability must lie with the person responsible for design, but in a large organization or project certain reliability activities may be delegated to a person or team; there could be specialists in reliability engineering in an integrated quality and reliability section, available to give specialized advice and services to the design team. However, reliability will depend on development testing and quality control during manufacture as well as design, and will be influenced by data fed back from use.

As with many aspects of quality, reliability can be considered in terms of a specification, a plan, and a programme. Taken together, these can be used as a tool with which management can measure and control the progress of a design.

Safety

With the advent of recent consumer protection legislation, safety is now a very serious topic. Because safety is important it should be considered both during development and manufacture of equipment and during use. Where appropriate there should be a safety plan and a formal safety programme, including prediction, control and verification. Safety analysis takes account of failures arising from unreliability, and credible accidental damage. Safety hazards should be eliminated by re-design, but if this is not possible they must be controlled by the use of safety devices.

Trials and testing

In the ideal world design must be finalized, proven and put under configuration control before manufacture begins. Unfortunately this is not always the case.

Feasibility, development and evaluation testing are important sources of information for the designer, and he should be deeply involved in deciding the aims of trials and in the detailed planning of them. He should be especially concerned with predicting the results to be expected from trials, working out how they are to be recorded and analysed, and considering what conclusions are likely to be drawn from them.

Quality assurance should ensure that appropriate test facilities are available, appropriate procedures for planning and conducting trials exist and effective use is made of the information obtained from the trials.

Products of the design process

At the conclusion of the design phase there will be a number of products. These may include:

1. Detailed design specifications;
2. Complete set of drawings;
3. Mathematical models;
4. Bread boards (electrical/electronic circuits not laid out as in production);
5. Trials equipments;
6. Prototypes (basic models covering aesthetic and ergonomic design);
7. Manufacturing procedures;
8. Automatic test equipment;
9. Installation and maintenance date.

All this information must be controlled, particularly after the final design review has accepted the information as satisfactory and meeting the requirements of the customer.

All the information must be archived to give the project history (and could be used for the defence in any product liability case).

The drawings, specifications and general documentation must from this stage come under configuration management (CM).

Management of change

After the design has been 'frozen', i.e. accepted by the final design review, a decision to change the design in one area will have repercussions in other areas. It is important that changes in the current state of the design should be known to all who are affected by them. Procedures and techniques for controlling changes should be detailed in either the codes of procedures or the quality plan, the implementation of which will ensure that proposed modifications are costed and assessed for operational and engineering effect. The designer should be fully aware of the role he or she has to play in these procedures and the importance of configuration management which requires tight control of the issue and amendment of documentation with some form of master record system.

Value engineering

Figure 3.1 showed a reference to value engineering – an area which has been given scant attention in recent years in the UK.

Value engineering (VE) is a systematic method of examining products, systems or services aimed at improving cost-effectiveness. VE developed as a technique in the 10 years after World War II. In those days it merely involved the examination of a product to see if it could be made less expensive by substituting cheaper materials or by altering the design so that the process of manufacture was easier. VE today is not limited to engineering and the cost of physical items, but involves the value in the widest sense. VE is frequently called value analysis as though the two terms are synonymous. However, strictly VE relates to the development of new products, and value analysis (VA) to the analysis of existing products. VE principles and techniques are applicable throughout both the development and production stages in the life of a product, but early application has obviously the greatest potential for overall economy.

Many leading firms have special VE staff or employ VE consultants. A definition of VE is:

> an organised effort to obtain optimum value in a product, system or service, by ensuring that the necessary function is obtained at the lowest life cycle cost.

As with many modern management techniques, VE is the formal use of principles and techniques which have always been applied by good designers and engineers using common sense. Common sense can be a 'hit or miss' affair; the purpose of VE is to force those examining a problem to conduct to truly exhaustive enquiry into all possible solutions. Essential to the process is not only the examination of all known possible answers but also the use of creative thinking to devise as many new alternatives as possible. There are two fundamental principles to be observed when using VE:

1. The effort employed must be organized;
2. The function of what is under examination must be defined and analysed.

The first principle is best observed by constituting teams for each study and making them work to a disciplined routine or job plan. The second is the fundamental principle of VE. The functions and sub-functions involved must first be clearly defined, and the requirement for each firmly established, before examination of and speculation on methods of achieving each function can profitably begin. The concepts of robust design and manufacturability, which will be examined in Chapter 25, could be considered as aspects of value engineering.

Summary of conclusions

This has been a very important chapter. There is little doubt that it was not until about the mid-1970s that people began to appreciate that quality in design was important. Until then design was a matter of aesthetics only and was not concerned with ergonomics, reliability, availability, durability, maintainability, etc. In the media, design is still often considered from a largely aesthetic viewpoint; consider the Design Council Awards in the UK, which appear to equate design to styling and packaging.

Now manufacturing industry appreciates quality in design and realizes that no matter how good the manufacturing processes involved if a complete package of drawings and specifications of fully trialled models are not available there will be no chance of producing a quality product.

4

Quality in purchasing

Introduction

Very few organizations are totally self-sufficient in the sense that all their products or services are generated in-house. Some materials, components or services are generally purchased from outside organizations and the primary objective of the purchasing department is to obtain the correct supplies or service from the vendors at the correct time, to the current specification at the right price. Note that I did not say 'cheapest' price. Remember our motto: 'Price is negotiable, quality is not.'

Once again, as in Chapters 2 and 3 we must refer to our diagram of the basic structure of the organization (Figure 4.1), which shows the area which we will discuss in this unit.

Evaluating suppliers

The importance of the purchasing department can be visualized when you realize what proportion of items, products, equipment services, etc. are bought-in. This proportion will vary of course between factories, industries and organizations but according to Oakland it averages 60% turnover in all industries. With this sort of percentage involved it is easy to see that if the purchasing department is not well organized and has no clear aims and procedures, problems will ensue, but before we go any further let us consider a few definitions which you will need to know about (all from BS 4778).

Definitions

1. Vendor Appraisal – Assessment of a potential supplier's capability of controlling quality, carried out before placing orders.

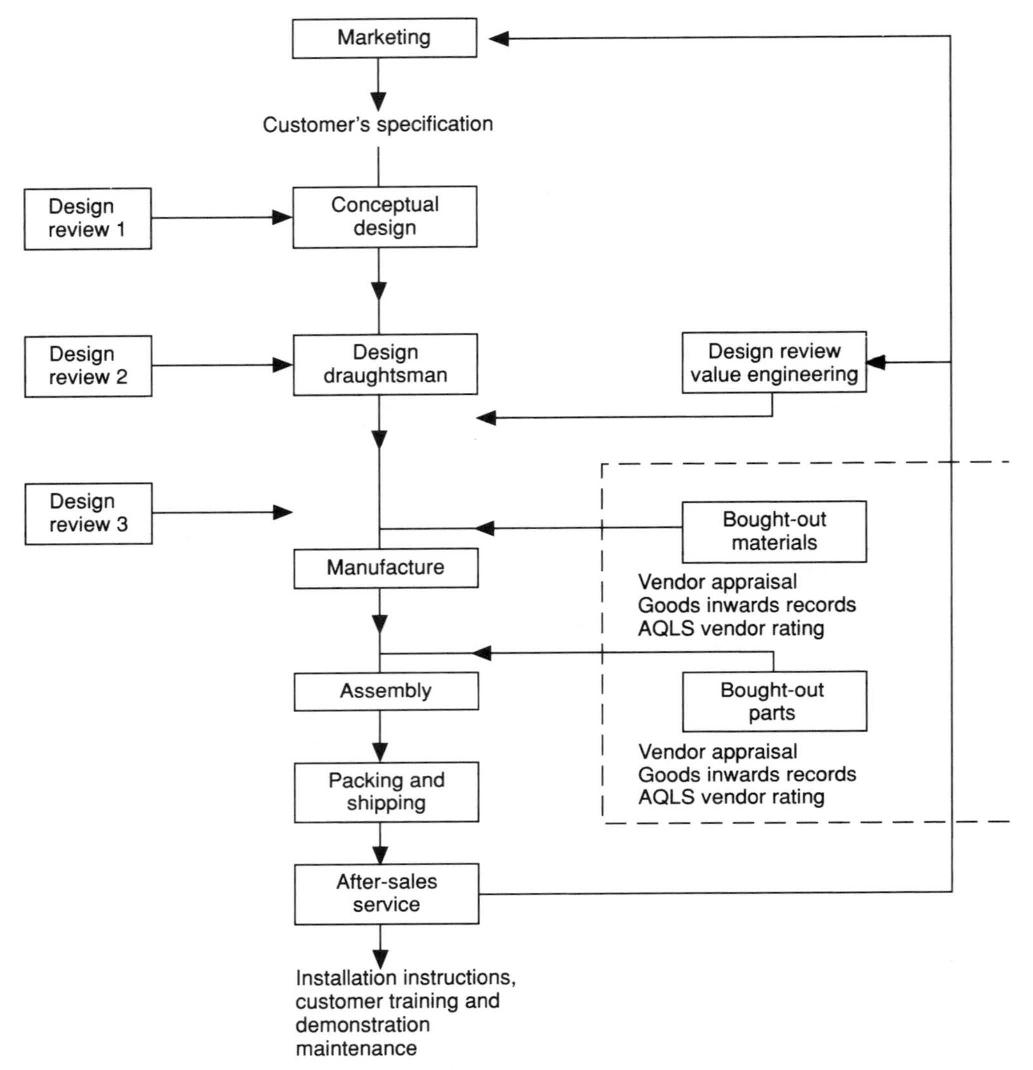

Fig. 4.1 Basic structure of an organization.

2. Supplier Evaluation – Assessment of a supplier's control of quality, carried out after placing orders.
3. Supplier Rating – An index of actual performance of a supplier. This is sometimes called vendor rating.

These are all shown in more detail in Figure 4.2.

The concept of quality audit

Quality auditing is a non-executive function, as distinct from operations such as inspection and surveillance (performed for the sole purpose of product acceptance

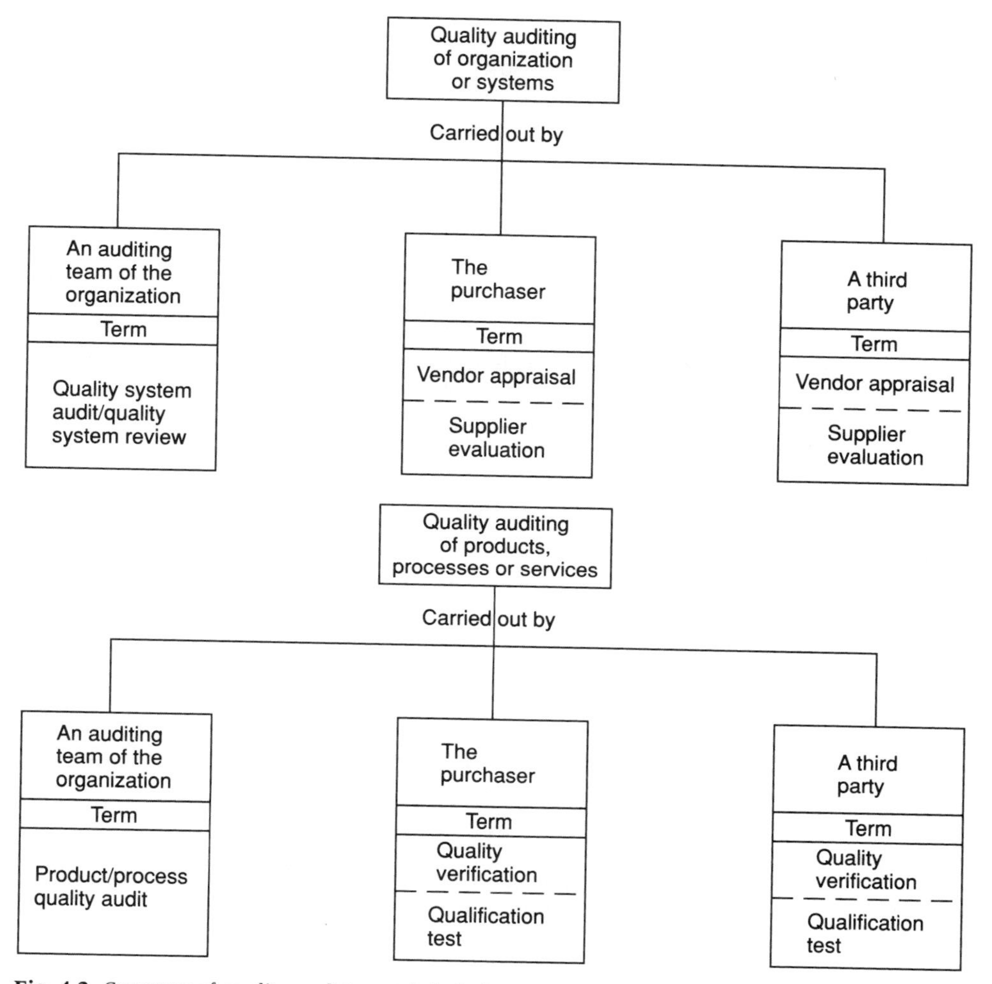

Fig. 4.2 Concepts of quality auditing and their descriptive names. Source: BS 4778.

or process control) which involve making decisions for action, although it can involve the review of such functions. We shall examine the use of auditing techniques in quality assurance in greater depth in Part Two of this book. Here, we shall simply point out its use in assessing vendors.

Auditing can relate to the quality of a product, a process or a system. Quality auditing is usually carried out on a periodic basis and involves the independent and systematic examination of actions that influence quality. The object is to ascertain compliance with the implementation of the quality system, programme, plan, specification or contract requirements and, where necessary, their suitability.

Thus it can be seen at this early stage, that the purchasing department probably controls large sums of money and of that department cannot get the quality right when making purchases then the quality of the final product will not be right.

Responsibility of the purchasing department

The purchaser is responsible for ensuring that all purchased material and services conform to the requirements specified. For this he or she must have complete purchasing data and these data must be accurately transferred to the selected sub-contractor.

The selection of sources and the type and extent of control exercised by the purchasing organization will be dependent upon the type of material and the sub-contractor's demonstrated capability – mainly based on 'supplier rating'. The organization must of course ensure that the purchaser's controls, whichever ones are chosen, are effective.

Thus the purchasing officer must conduct the affairs of his or her department efficiently, and see to it that the suppliers take advantage of every practice and technique which will enable them to achieve maximum efficiency. Details of the organization of the purchasing department, will not be given here but what we are concerned with are those functions and responsibilities which have a direct bearing upon the quality and reliability of the product.

These responsibilities should be laid down by senior management. Typically, the purchasing department should be charged with the responsibilities of:

1. Familiarizing itself with the technical characteristics of the company's products;
2. Studying and understanding the technical characteristics which are requirements of the materials and components procured from outside sources;
3. Selecting only those suppliers who are considered to be capable of meeting the required standards of quality, delivery, and cost;
4. Cooperating with quality and production engineering departments to confirm that the requisite understanding, organization and technical ability does in fact exist or can be developed;
5. Ensuring that the suppliers are provided with all necessary information;
6. Cooperating with the quality department to ensure that the ability to produce parts to the required standard is confirmed by the acceptance of first-off samples for examination;
7. Obtaining deliveries of the requisite quality and quantity, at the times required;
8. Holding discussions with the supplier to ensure that they fully understand their responsibilities.
9. Drawing up terms of business, which will define:
   (a) The responsibilities accepted by the customer;
   (b) Everything which the supplier is required to do to meet his or her obligations to the customer;
   (c) Incorporating such of these requirements as are legally binding into a formal contract between the two parties.

It is worth repeating that purchasing is essentially a two-party business. It is a primary concern of the buyer to ensure that each party is fully aware of what is expected of one other.

To help them provide the service required of them the purchasing department should have links with the following:

1. Design – for 'design or buy' decisions;
2. Legal Services – drawing up meaningful contracts;

3. Production – for 'make or buy' decisions;
4. Marketing – future demands and feedback from customers and on competitors;
5. Finance – terms of payment required, etc.;
6. Quality – feedback on quality problems concerning purchased items and their manufacturers.

The purchasing manual

These links would normally be formalized in a written purchasing manual or a set of purchasing procedures which would:

1. Assign responsibilities for and within the purchasing function;
2. Define the manner in which suppliers are selected to ensure that they continue to be capable of supplying the requirements in terms of material and services;
3. Give the appropriate purchasing documentation required by the organization – such as the purchasing order;
4. Show the organizational links of the department.

Procuring supplies

The purchase order

It is worth stressing once again the importance of unambiguous procurement specifications – successful procurement of supplies begins with a clear definition of the requirements.

These requirements, found on the purchasing documents should contain data clearly describing the product or service ordered. Elements that may be included are:

1. Precise identification of style and grade;
2. Inspection instructions and applicable specifications;
3. Quality system standard to be applied.

Purchasing documents should be reviewed for accuracy and completeness before release. This is necessary, as it is cheaper to put the problem right at this early stage, than to find out and resolve the matter at a later stage when some further value-adding activity may have already taken place.

In more detail the purchase order should carry the following information.

1. Name and address of the originating organization;
2. Name and address of receiving company;
3. Identifying number;
4. Quantity of product or amount of service required;
5. A full description of the type, style, grade, or other means of precise identification of the product or service;
6. The applicable issue of the product or service specification and any other relevant technical data (a reference to a published, current specification may be used);

7. Reference to any certification of conformity to requirements, which must accompany the delivered product (see later);
8. Price agreed between purchaser and vendor;
9. Delivery agreed between purchaser and vendor;
10. Cost allocation – this for internal use;
11. Delivery instructions;
12. Buyer's signature and standing in the organization;
13. Purchaser's conditions of business;
14. Quality control assurance requirements.

The authority to sign purchase orders is usually restricted to one or two persons within the organization and limitations may be imposed as to the amount of expenditure which may be incurred by a signatory.

The purchasing organization should also develop appropriate methods to ensure that the requirements for the supplies are clearly defined, communicated and, most importantly, are completely understood by the supplier. These methods may include vendor/purchaser conferences prior to purchase order release, and other methods appropriate for the supplies being procured.

Ensuring contractor quality

Just before we move off this important subject we must say something about the quality aspects of the purchase/sub-contract order. Always remember the main contractor is totally responsible for the sub-contractors and suppliers, hence the quality conditions appropriate to each order must be defined. A policy decision may be made to deal only with firms who meet the quality requirements of ISO 9000, or it may be decided to invoke the contractual requirements of the appropriate allied quality assurance publication in the case of defence contracts. In every case the contractor should develop a clear understanding with the supplier on quality assurance for which the supplier is responsible. The assurance to be provided by the supplier may vary as follows:

1. The purchaser relies on supplier's quality assurance system;
2. Submission of specified inspection/test data or process control records with shipments;
3. 100% inspection/testing by the supplier;
4. Lot acceptance inspection/testing by sampling by the supplier (usually based on AQL (Accepted Quality Levels);
5. Implementation of a formal quality assurance system as specified by the purchaser;
6. None – the purchaser relies on receiving inspection or in-house sorting – this will be covered in more detail later.

The assurance provisions should be commensurate with the needs of the purchaser's business and should avoid unnecessary costs. In certain cases, formal quality assurance systems may be involved (ISO 9000, ISO 9001, ISO 9002 and ISO 9003). This may include periodic assessment of supplier quality system assurance by the purchaser – 'Second Party' Assessment.

In any case a clear agreement should be developed with the supplier on the methods by which conformance to purchaser's requirements will be verified. Such agreements may also include the exchange of inspection and test data with the aim of furthering quality improvements. Reaching agreement can minimize difficulties in the interpretation of requirements as well as inspection, test or sampling methods.

Good purchasing data – now what?

We are on the right track – we have clear, complete, purchasing data, all set out on the purchasing order – but from whom shall we purchase?

Procurement – traditional method

It is still common practice in many organizations and particularly in government agencies to contact as many potential suppliers/sub-contractors as possible, issue them with the procurement specification and ask for a response to the tender document. In many cases the eventual order will be given to the lowest tender. If the procurement specification is ideal this may be satisfactory, but over-emphasis on 'price alone' is short sighted and fraught with difficulties. Price is only one element of the cost – what about:

- Reliability of delivery? (on time);
- Quality level? (to correct quality).

If the stores do not arrive on time, what difficulties this could cause. It's the same in service industries, if someone for instance is employed to clean hotel rooms between 1100 and 1200 hours, what happens if they are late? The occupant arrives to find the cleaners in the room – this does not create a good impression!

Let us return to the manufacturing environment. If purchasing take up the cheapest price in the response to tender and they have no other information on the firm available, it may be too late to arrange a vendor assessment, and so the purchaser will have to rely heavily on the goods inward inspection organization ensuring the requirements of the sub-contract/purchasing order have been met. In the type of organization using this rather traditional method of procurement the goods inward inspection department would be expected to employ a large number of personnel, as the amount and extent of inspection performed on receipt varies with the effectiveness of the supplier's delivery and quality systems – and we will know little of these if 'price' is the driving force. It should always be recognized that the quality of supplied material or service can be controlled only at the point of production. Inspection at receipt for acceptance or rejection is an inefficient and wasteful exercise, minimized in most quality conscious organizations.

Over-reliance on the goods inward department can be disastrous and should not be encouraged – it should complement and supplement source quality control rather than ignore it or duplicate it unnecessarily.

In general, however, there will always be some form of goods inward department who will establish appropriate measures to ensure that supplies which have been received are properly controlled. These procedures should include quarantine areas or other appropriate methods to prevent unqualified supplies from being inadvertently used.

The extent to which receiving inspection will be performed should be carefully planned. The level of inspection, when inspection is deemed necessary, should be selected with overall cost being borne in mind.

In addition, when the decision has been made to perform an inspection, it is necessary to select with care the characteristics to be inspected.

It is also necessary to ensure, before the supplies arrive, that all the necessary tools, gauges, meters, instruments and equipment are available and properly calibrated, along with adequately trained personnel.

Once the goods have been received appropriate receiving quality records should be maintained to ensure the availability of historical data to assess supplier performance and quality trends.

In addition, it may be useful and, in certain instances, essential to maintain records of lot identification for the purposes of traceability.

Procurement – modern method

The modern method of procurement is to use only 'approved' sub-contractors or suppliers. If price is the main driving force you will tend to deal with anyone but imagine an ideal supplier, and imagine what his or her virtues would be. A short list might include the following:

1. Products that are always 100% correct and reliable;
2. Deliveries that are always on time;
3. Quantities that are delivered are always correct: there are never too few or too many items;
4. Deliveries that occur daily to minimize the stock carried by the user;
5. The supplier is willing to accept changes in the quantities ordered, even up to a day before delivery;
6. If something does go wrong, there is total commitment to righting it again as rapidly as possible;
7. Competitive product pricing.

This is the supplier you should want to deal with – the type you must deal with if you are working with the just-in-time method and the way 'Total Quality' will drive you. It is possible to work with suppliers like this but much work will be necessary to reach this ideal. So what can you do before the sub-contract order/purchase order is placed? Well, from our early definition in this chapter we can arrange for a vendor appraisal.

Vendor appraisal

Effective procurement demands the selection of suitable sub-contractors. There are several ways of choosing satisfactory sub-contractors:

1. Review of previous performance in supplying to similar specifications;

2. A satisfactory assessment (for similar material/services) to an appropriate quality system standard by a body considered by the supplier to be competent for the purpose;
   (Note: This in no way diminishes the supplier's responsibility for ensuring that the purchased material conforms to specified requirements.)
3. Assessment of the sub-contractor by the contractor to an appropriate quality system standard, this is vendor appraisal.

The first task of the supplier is to draw up a vendor appraisal programme from a list of suppliers who are known to be capable of delivering products which can meet all the requirements of the specifications, drawings and purchase order.

The methods of establishing this capability may include any combination of the following:

1. On-site assessment and evaluation of supplier's capability and/or quality system;
2. Evaluation of product samples;
3. Past history with similar supplies;
4. Test results of similar supplies;
5. Published experience of other users.

Such a programme will normally start with a review of existing suppliers if there are any. In these circumstances the only practical approach will be to grant satisfactory suppliers provisional approval so that purchasing can continue. After that, all new suppliers should be subject to vendor appraisal, while existing suppliers should be subjected to supplier evaluation and a vendor rating system evolved to ensure adequate control is maintained. These topics are discussed in more detail later.

Vendor audit

Once the vendor appraisal programme has been established, trained, mature, experienced auditors, with a knowledge of the technology involved will be required to carry out the work. Auditing is covered in Chapter 9 and vendor assessment is one version of this. What is important is that the auditor convinces the supplier's management that conforming products really are required and that the customer is prepared to go to much trouble to get them. The supplier will also judge the customer to some extent by the competence and professionalism of the auditor.

The auditor will normally complete standard a pre-printed vendor appraisal form after the survey. This will form the objective evidence of the result and depth of the audit as well as providing a check list ensuring that all required points are covered. It will also make easier cross-comparison of different suppliers of similar products.

This report will cover the following factors:

1. Management competence/experience/attitude;
2. Understanding of contract and purchasing specifications;
3. Plant and facilities;
4. Processes;

5. Employee skill and training;
6. Quality department organization and effectiveness;
7. References from other customers/other quality approvals;
8. Financial stability.

Particularly important is 1 above. If top management is not committed to quality, the chances of obtaining quality products are diminished.

Obviously the amount of work carried out on vendor appraisal will, in many cases, depend upon the value of the purchasing order and the critically of the item requested.

Number of suppliers

The next question to ask is how many sub-contractors/suppliers do you require? If you have too many you will expend too many resources in vendor appraisals and supplier evaluations and the process will become uneconomic.

There is a thought that each contractor should have at least two suppliers per product. This allows the purchasing department to play one supplier off against another to obtain the best price and also gives a safer fall-back situation in that if the first supplier for some reason fails to deliver, there is always the opportunity to go back to the second supplier, albeit at a rather late stage, to try to recover. There is however, at this time a very important lobby of industrialists who advocate the use of a sole supplier for each item. This has the following advantages:

1. You get a better service. A supplier who has all your business for a particular product will be better able to invest in its present and future quality. You will also be in a stronger position to expect the best service from him or her;
2. You reduce the amount of monitoring of the purchased product;
3. You can develop a close collaborative relationship with your supplier, in which each party sees it as in his or her interest to support the other;
4. You can get a better price. If you are spreading your business over three suppliers, each has a relatively small proportion of your business, so that his or her unit costs are much higher than if he or she supplied all.

Single-source resource management is the objective to aim for. If you cannot find a supplier prepared to make the sort of commitment described, your risk is greater. It may still be best for you to keep to a single supplier for the product, but you will have to commit yourself to a more comprehensive programme of goods inward inspection. Goods inward inspection is necessary for any new supplier, but as you gain confidence in him or her the resources allocated to this function can be reduced.

Many firms are aiming for this ideal, but it needs a dedicated approach and can cause problems if the supplier goes out of business, but by taking this single source purchasing approach it is necessary to develop confidence in the relationship between the vendor and vendee so that surprises such as receivership should not occur.

So what are the characteristics of vendor appraisals? They are:

1. held prior to contract placement;
2. discretionary – although it should be company policy to carry them out;

3. needed to generate a qualified supplier list;
4. carried out by a broad-based team, possibly covering finance, design, production, contracts and QA;
5. carried out with agreement of the supplier and are in no way contractual;
6. systems-oriented;
7. generally not too deep (often time limited).

Supplier evaluation

Vendor appraisal will give you a good idea of the dedication, capability and stability of the firms you intend to deal with, but what happens when you give them the work or request the service.

Do you continue to monitor their performance? The answer is of course 'Yes' and this is called supplier evaluation.

Suppliers should be visited on a regular basis to see if they are performing to requirements. The number of the visits required will again vary according to:

1. Critically of item;
2. Value of the contract;
3. Evidence produced during the vendor appraisal.

This last point is important. To issue a contract to the supplier means that the basic criteria you adjudged necessary before issuing work has been met, but is this the best you can hope for? No, not really, if the relationship with the supplier can be developed and a mutual trust established, the vendor and vendee should be able to work together to ensure the product or service meets the customer's requirements in every way. The amount of upgrading necessary at the supplier and the willingness of the management to ensure this happens will be factors taken into consideration during the supplier evaluation.

So far we have concentrated mainly on getting the system right and it is true a good product can be obtained even if the system is poor, but generally it must be agreed that to obtain the best product, consistently and at reduced cost a good 'quality system' is essential.

So by all means let us concentrate on getting the system and the environment right, but we must never lose sight of the product we are buying. It is after all, the product or service we are purchasing, not the system behind it.

With this in mind it may be desirable to carry out an evaluation of a product sample. This can take the form of removing finished items from the manufacturer and subjecting them to independent test and examination, or it may be deemed satisfactory just to examine all the process, examination and test records associated with the product.

There are other factors which can be associated with supplier evaluation and these are shown in Figure 4.3.

So far, we have chosen our supplier, given a contract and carried out a number of evaluations, but what happens in the long term? It is important to see that the suppliers do not fall to an unacceptable level. Also it may be desirable to compare one contractor with another to see who should be the preferred contractor. Such a system is called vendor rating or supplier rating.

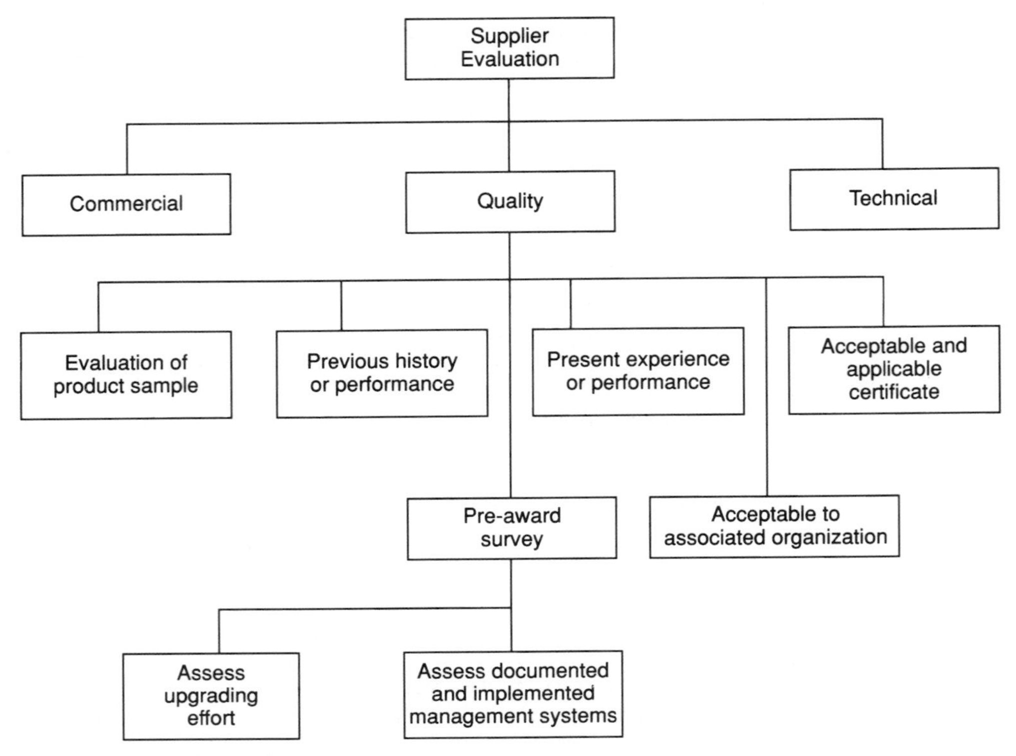

Fig. 4.3 Supplier evaluation.

Vendor rating

We have seen that vendor appraisal can help us produce an approved vendor list. If you have more than one subcontractor on that list for each procured item, then you should want to develop a system which indicates quite simply which of these subcontractors you should prefer to work with; that is vendor rating.

There are many factors which you should take into consideration when developing a system. Obviously price is one factor. All other things being equal it makes sense to deal with the firm who can provide the product or service more cheaply than equivalent competitors, but should that be the only criteria? Clearly, the answer is no.

Do they deliver on time? This is another important factor. If you have planned resources to be available at a certain place at a certain time to take delivery, and this does not occur because of the sub-contractors inefficiency, it will cost you money in some form or another. The firm has proved to be unreliable and should be penalized in some way on the vendor rating system.

Imagine the same case, but this time the subcontracted product arrives on time, however it has to be rejected because it is not to the procurement specification. Again time has been wasted and money expended due to poor subcontract work. The firm again should be penalized in the vendor rating system.

There are many of these rating systems available in industry and commerce, and some are not without shortcomings. These are often as given below:

1. Unfair systems may be evolved which do not compare like with like;
2. Some systems are based purely on 'conformance' and ignore price and service;
3. Some systems are inconsistently maintained;
4. Rating criteria may sometimes be vague and ambiguous;
5. Some systems are poorly managed.

To be successful these deficiencies must be overcome, and the system should be:

- Clearly defined;
- Restricted to necessary information;
- Economical;
- Practical;
- Maintainable;
- Auditable.

A simple system is shown below and takes into account the three factors, price, service and quality. Allocate suitable weightings to suit the company, for example:

Quality – 40 marks
Price – 40 marks
Service – 20 marks
TOTAL 100 marks

So the subcontractor who offered the cheapest price, delivered on time and with no deficiencies would be granted 100 marks. If a second subcontractor had to be used to produce the same part but the price was 5% above the first subcontractor, 2% of the items were defective and they were 1 day late in a 20-day delivery, the index would be:

Quality – 40 2 × 40 = 39.2 100
Price – 40 5 × 40 = 38 100
Service – 20 1 × 20 = 19 20
TOTAL 96.2

Obviously the first subcontractor would be the preferred one, but the second subcontractor on this basis may also prove to be acceptable.

You can of course vary the weightings and penalties as necessary to suit the industry provided they are clearly defined and well understood and applied fairly. The system can then be used to feedback information on vendor performance.

Conforming to requirements

Now we have given you some insight into vendor appraisal, supplier assessment and vendor rating. All require dedication and effort, but if these techniques are applied intelligently and religiously there should be a consistent improvement in quality and the goods inward department should be able to shed manpower to compensate. Inspection at receipt for acceptance or rejection is an inefficient and wasteful device to be replaced, as soon as possible, by the use of receiving inspection to check that the necessary systems and procedures, used by the supplier to control quality, are in fact working effectively.

Organizations which have carried out vendor appraisals or rely on audit by independent third-party assessment, often request a 'certificate of conformity' to be delivered with the purchased material. This is a simple statement by the supplier that what is being delivered actually conforms to the requirements, as laid down in a specification. A certificate of conformity should include the following minimum information:

1. The supplier's name and address;
2. The serial number and the date of the certificate;
3. The customer's name and address;
4. The customer's purchase order number;
5. A description of the product and the quantity;
6. Where appropriate, the customer's identification marks;
7. The identification of any specification to which the goods are supplied;
8. Any agreed concessions or waivers;
9. The following statement, signed by the person nominated by the firm as responsible for quality control or his deputy:

Certified that the supplies detailed hereon have been inspected and tested in accordance with the conditions and requirements of the contract or purchase order and, unless otherwise noted below, conform in all respects to the specification(s), drawing(s) relevant thereto.

Summary of conclusions

Purchasing is an important area as it controls significant sums of money. If the department does not consistently purchase 'quality' products or services, the main contractor will eventually fail.

There are many ways for the contractor to ensure conformance of purchased material or services. Good subcontracting demands:

1. An examination of the design of a product or service prior to purchasing to determine the difficulty or complexity of its manufacture;
2. Investigation of a potential subcontractor's design, manufacturing, inspection and control capabilities; and
3. The use of a subcontractor's quality histories, vendor ratings and other obtainable data.

When subcontracting the contractor decides, for the material to be purchased, the manner in which its conformance will be demonstrated, where this will be carried out and by whom. The contractor may have his/her supplier perform this function as part of the subcontract conditions, he/she may perform it himself, or he/she may hire the services of an agent for the purpose. Examples of objective evidence by which the contractor demonstrates that his/her controls and those of his/her supplier are effective are test reports, measurement and test data, etc. supplied by the subcontractor or similar documentation recorded by the contractor or the independent agent. In any case the contractor should ensure that the records provide evidence that subcontracted material and services conform to contract requirements.

5

Quality in production

Introduction

One of the shorter definitions of quality you have been introduced to is 'conformance to requirements' or 'conformance to specification'. Although these definitions are now interpreted to cover their wider fields, initially they grew out of the need to manufacture goods to 'the requirement' or 'the specification'; really quality in production.

Many of the paragraphs relate specifically to controls which can only be associated with the manufacturing industry, such as inspection and test instructions, control of inspection, measuring and test equipment, manufacturing control, in-process and final inspection, etc. It is here in manufacturing where quality control and quality management have been practised for the largest period. We shall look at the special characteristics of service organizations later, in Part Three of our book.

Once again as in previous chapters we must start with the basic structure of an organization (Figure 5.1). This time we will deal with the elements highlighted, covering 'quality in production'.

Planning – the key to quality

It should come as no surprise to realize that planning is the basis of ensuring that quality is assured during the manufacturing phase.

Figure 5.2 shows the difference between a company which relies heavily on the inspection department and has little perception of quality management, and a company which believes in planning and the use of quality management techniques.

Management should establish and maintain an adequate programme for quality assurance, which is planned from the earliest possible stage and developed in conjunction with other functions and is capable of being effectively and systematically

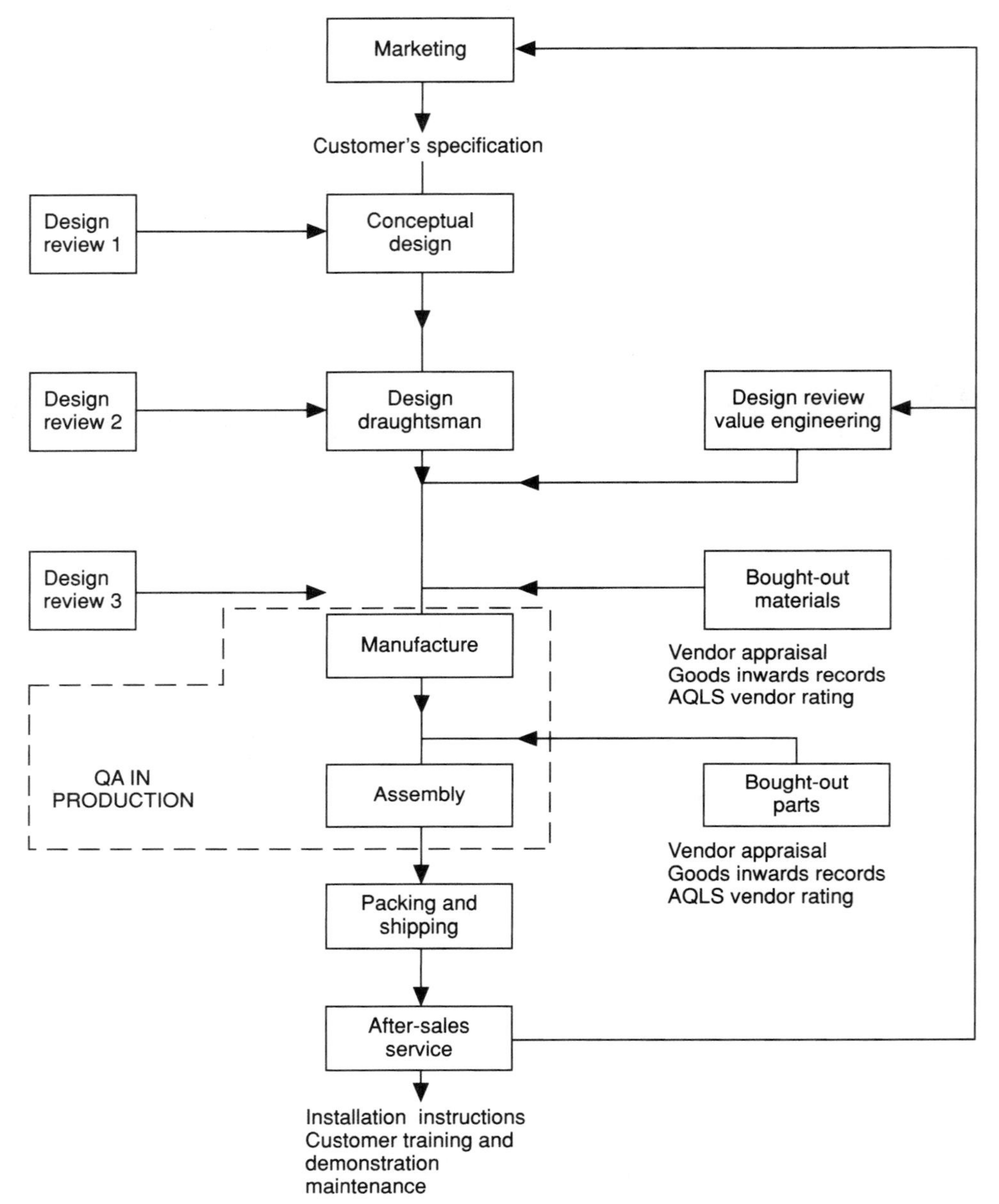

Fig. 5.1 Basic structure of an organization.

reviewed. One of the main objectives of planning is to identify any special or unusual contingencies or requirements, i.e. those requirements that are unusual by reason of newness, unfamiliarity, lack of experience, or absence of precedents. When such requirements are found there is frequently a need for study, planning and scheduling to provide appropriate operations, processes and techniques and the means for testing and proving conformance with the requirement. All this planning must be documented.

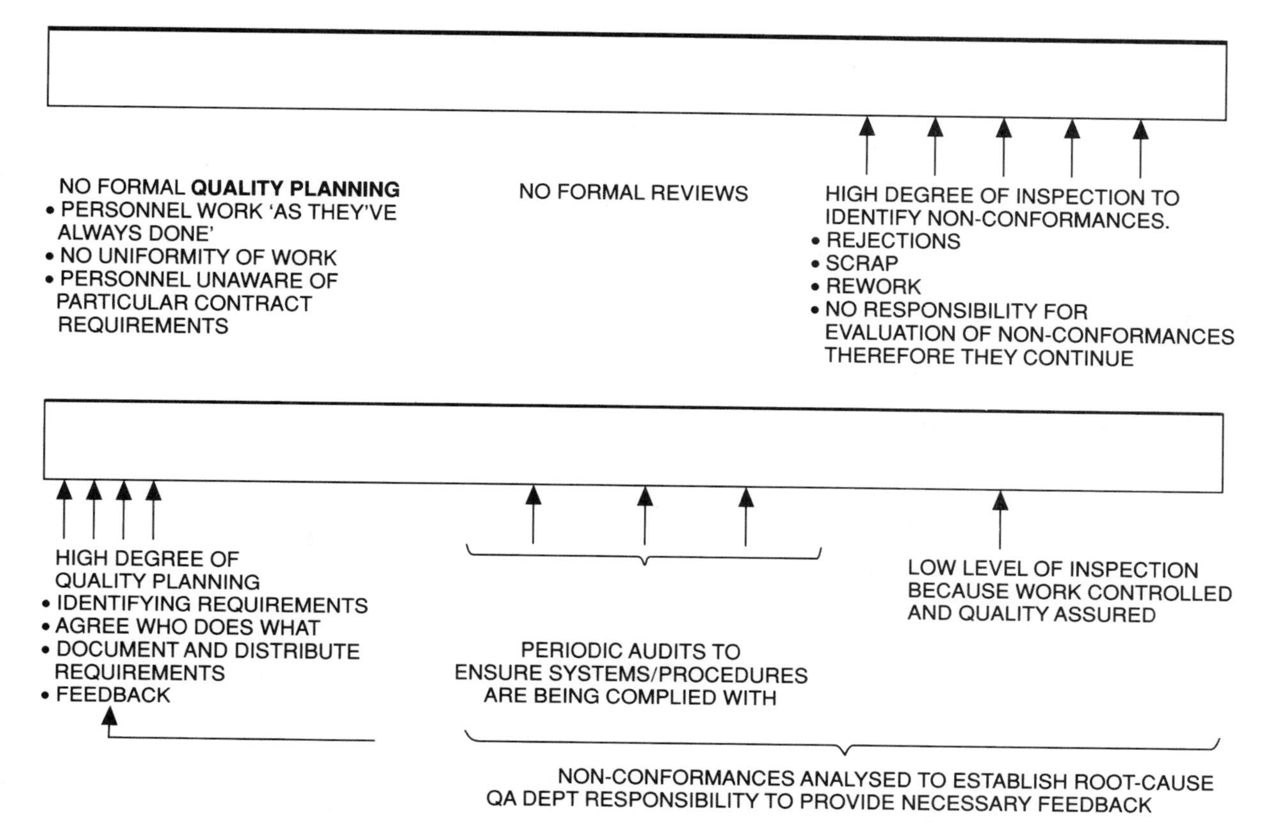

Fig. 5.2 Organization with and without QA function.

Thus in the production phase, management should conduct at the earliest stage, a sufficiently extensive review to ensure:

1. The timely identification and acquisition of any controls, processes, inspection equipment, fixtures, tooling and personal skills that may be needed to ensure product quality;
2. The up-dating of inspection and testing techniques including the development of new instrumentation (and its calibration – see later);
3. The compatibility of manufacturing and inspection and test procedures and applicable documentation before production begins;
4. The deployment of appropriate resources.

Thus throughout the whole manufacturing cycle the product and the resources used to produce that item are fully controlled.

In fact it is easy to see that for any task to be fully controlled, then each element of that task must also be controlled. If you think about it, each task consists of four elements:

1. The documentation required;
2. The item being worked upon;
3. The equipment being used to perform the task;
4. The person carrying out the task.

This can be shown diagrammatically as seen below:

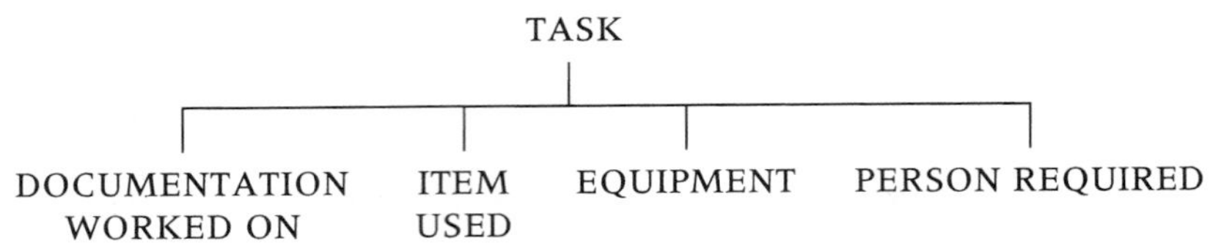

All these elements will be discussed in more detail in the following pages, with particular reference being made to the production department. The argument still generally holds if you consider any other work however.

Documentation

In Chapter 3 we came across three documents to be found in most design offices. These were:

1. Codes of practice;
2. Quality plans;
3. Quality manual.

Codes of practice are only found in design offices. Quality plans are project or contract specific documents which, although they may be found in production departments are usually more associated with research, design and development for it is here where the initial problems arise. These problems, like working with materials never used before, or working at a new frequency not encountered previously, have to be sorted out at the design/development phase before a suitable set of drawings/specifications, etc. can be issued to the production department before manufacture can begin.

The most important quality documents which you will come across in the production environment are the quality manual and work instructions/procedures.

The quality manual

Quality manuals set out the general quality policy and practices, and is the book by which the organization plans for quality. It can also be a useful aid for such purposes as training (e.g. complementary to works instructions), or marketing, especially if the consequences of modifications necessary to satisfy particular customer requirements are emphasized (e.g. costs of deviating from standards). Additionally they can provide:

1. Information from which the purchaser may derive confidence in the supplier's organization (useful for supplier/vendor appraisal);
2. An indication of the responsibilities and interrelated activities of personnel and functional groups;
3. A vehicle for auditing, reviewing and evaluating the management and/or quality control system.

Quality procedures and work instructions

These are the lower level documents used on the shop floor on a day-to-day basis. Quality procedures are descriptions of the activities that individual functional units need to implement elements of the quality system.

Work instructions should be developed and maintained to prescribe the performance of all work that would adversely be affected by lack of such instructions. These written instructions should not only be created and brought to a satisfactory state and put into use but should also be subject to continuing evaluation for effectiveness and adjusted as necessary. They provide a basis for control, evaluation and review and without them differences in policy and procedures can arise and variations in practice may occur resulting in confusion and uncertainty. Further to this, they provide means for delineating work to be done and for delegating authority and responsibility.

Documentation control

All changes to documentation should be in writing and processed in a manner which would ensure prompt action at the specified effective point, indicating whether the modifications are:

1. Retrospective;
2. Immediate; or
3. To be carried out at some later point in time.

Management should maintain a record of changes as they are made and ensure observance of the rule that notations written on copy documents do not authorize departure from the original. Documents (including drawings) should be re-issued after a practical number of changes have been made. Provision should be made for the prompt removal of obsolete documents from all points of issue and use.

Control of manufactured items and of manufacturing processes

As we have implied already, the manufacturing departments are responsible for producing goods in accordance with the drawings/specifications issued initially by the design department and these must have all acceptance/rejection criteria adequately defined.

Manufacturing operations

Management should ensure that manufacturing operations are carried out under controlled conditions. The omission of any particular operation or process from the scope of such control invites inferior quality. Ineffective, incomplete or intermittent control can lead to costly and unnecessary defects. For this purpose adequate communication is indispensable. Accordingly, manufacturing operations should be defined to the greatest practical extent by appropriate work instructions and work should be accomplished as specified in these instructions.

Process quality controls should be applied to all materials, production processes and equipment being used in the manufacture of products in order to attain specified requirements. Procedures should be provided to ensure that all materials to be used in manufacturing processes conform to the requirements of the specification.

The manufacturer should establish and maintain a system for the identification, preservation, segregation and handling of all material from the time of receipt through the entire production process. The system should include methods of handling that prevent abuse, misuse and deterioration.

Storage areas or stockrooms should be provided to isolate and protect material pending use. To detect and prevent deterioration, material in stock should be inspected periodically for condition and also to ensure that the shelf life has not been exceeded.

Every operator, machine or process has inherent variability and the extent of this variation is referred to as its process capability. Consequently, there is a need for management to:

1. Establish the process capability of work methods in order to ascertain whether a job can be accomplished satisfactorily;
2. Establish the process capability of existing plant and where necessary the need to bring it up to specified requirements;
3. Control and monitor process capability on a continuous basis to detect and eliminate potential causes of non-conformance.

This is the basis of Statistical Process Control (SPC).

Inspection and its control

The manufacturer should provide a system of inspection to ensure that specified requirements are satisfied. Many inspections are normally performed throughout a manufacturing process either for the purposes of process control or to ensure the acceptability of quality characteristics that cannot be observed or measured at a later point in the process. In addition, it is customary to perform a final inspection and a test of the completed item to obtain an overall measure of its conformance with requirements stipulated in terms of item performance.

Types of inspection

This may be accomplished in a number of ways, such as by machine operator (i.e. operator control), automatic inspection gauges, production line inspection stations, patrol inspection or by any combination of the above. Two basic methods may be used:

Inspection by attributes

Each unit of product inspected is classified as 'acceptable' or 'defective', e.g. GO or NO-GO gauging. The degree of 'acceptability' or 'defectiveness' is not taken into account.

Inspection by variables

This takes the degree of 'acceptability' or 'defectiveness' into account and operates by considering the measurements made. This is shown in Figure 5.3. (The statistics of the 'normal' and 'binomial' distributions are explained in Chapter 11, Part Two.)

The manufacturer should establish and maintain a system for identifying the inspection status of material during all stages of manufacture and should be able to distinguish inspected and uninspected material by using some identification system such as stamps, route labels, tags, etc. If stamps are used each individual stamp must be traceable to its owner.

Now let us consider the equipment used on the items produced.

Control of machinery and calibration

In the manufacturing industry the equipment used generally consists of machinery and test equipment. Both have to be maintained to ensure they meet the planned performance, so routine maintenance is important. Once again records must be

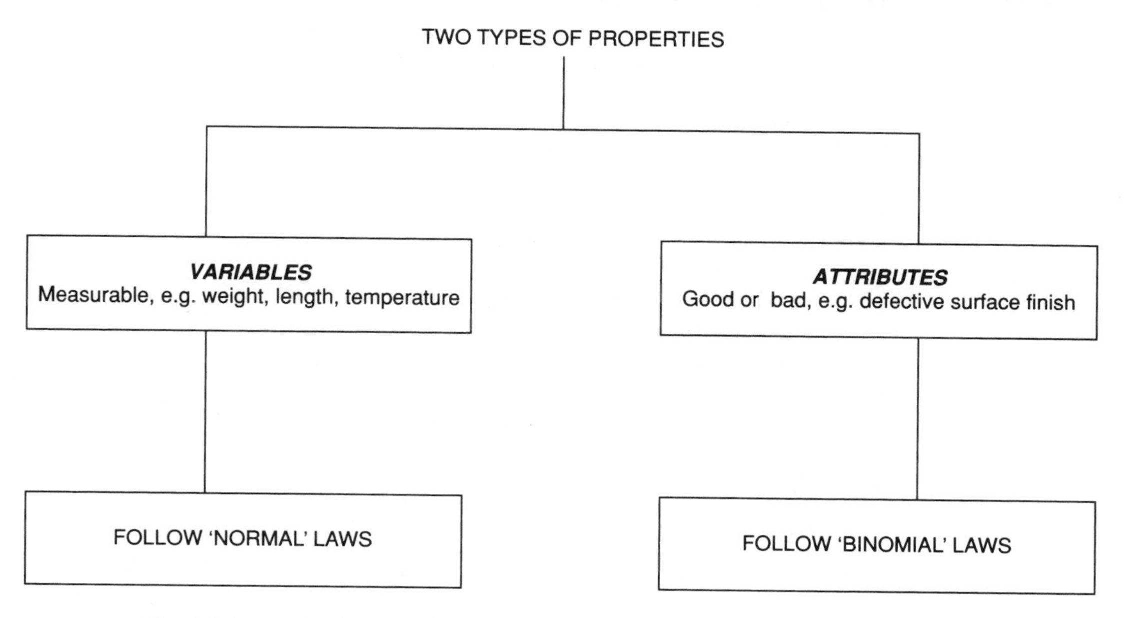

Fig. 5.3 Inspection by variables and by attributes.

kept of the procedure and the results must be regularly analysed to ensure that the routine maintenance is being carried out economically (not too frequently) and at the right periodicity (to ensure that large adjustments are necessary each time the equipment is offered for maintenance).

Management must recognize the influence that jig and tool design can have on the cost, design and feasibility of production. All production jigs, fixtures, tooling, templates, patterns, gauges and test equipment should be proved for accuracy prior to release and thereafter should be adequate between use and checked at stated regular intervals to ensure their continuing accuracy.

The accuracy of measuring instruments and gauges must be verified at appropriate intervals by comparison with certified reference standards.

The control outlined above is necessary for all measuring equipment used in the factory or laboratory. It is a very important topic the full significance of which is often overlooked. Organizations must establish the extent and frequency of calibration and keep records of the adjustments necessary as evidence of control.

Control of the people involved

It is also of course necessary to ensure identified, competent people, with the correct training, and the correct attributes, are in the right place at the right time to carry out the work. This is all part of the planning exercise.

Motivation is also important and the importance of top management commitment to quality cannot be overstressed.

Adjustment and corrective action

What we have described so far amounts to the ideal situation, but even with meticulous planning things do go wrong and unforeseen problems arise. The strength of quality management is that it demands that systems must be operating to ensure that problems which do arise are dealt with effectively in the shortest time. The shorter the feedback loop the quicker and easier it is to put right; this is also the cheapest method.

Whenever problems arise they should be analysed, and corrective action taken to ensure that the same problem does not arise again. By analysing the available data, trends are revealed, and if the trend is undesirable it may be possible to reverse the trend before any serious problems arise.

Poor work methods or non-compliance with work instructions are frequent causes of defects and failures in service. Often poor design or inadequate specifications are the cause. As the need indicates, remedial action requires:

1. Changing unsatisfactory designs, specifications and work methods; or
2. Enforcing compliance with satisfactory designs, specification and work methods.

Prompt, effective, remedial action is essential to a sound quality programme and the segregation of unacceptable material from acceptable material is rarely sufficient.

Control of non-conforming material

As we mentioned above human beings, being what they are – fallible – will occasionally produce items which do not conform to the technical data provided. This is non-conforming material. Under these circumstances the manufacturer should have in place a system which provides for the identification, segregation and acceptance, rework or disposal. All non-conforming material must be clearly identified to prevent unauthorized use, delivery, or mixing with conforming material. The customer may agree that certain non-conforming material may be used on his/her contract and if this agreement is reached then a 'waiver' or 'concession' report is generated. These again form an essential link in the chain. These reports should be frequently analysed to ensure the problem does not arise again.

It is best that concessions should not be considered. The firm producing the non-conforming item is producing 'second rate goods' and is not 'getting it right the first time'. They will try all the harder next time if concessions are not agreed.

Avoidance of accidental damage

It would be very sad if an item had been well designed and then meticulously manufactured and assembled only to be damaged in moving from the assembly shop to store.

Planning must ensure that at all stages the manufacturer has in place a system for identifying, handling, storing and packing of all material from time of receipt through the manufacturing process until delivery.

Manufacturing control

All control requires the establishment of a standard for the means of comparison, the assessment of conformity with the standard and the application of suitable corrective action as necessary. This is usually carried out by means of an audit or review – and independent examination to provide information.

Audits may be carried out on the quality system, on a process or on a product. The former – quality system review or audit is possibly the most important.

Quality system review

In this, actual practice is compared with the control systems laid down in the quality manual or other quality documents. Reviews should be carried out periodically and systematically and evaluated to ensure continued effectiveness. Such a review is conducted by, or on behalf of, top management to ensure that delegations and methods are achieving the required results; to reveal deficiencies or irregularities in any of the elements examined and to indicate possible improvements.

The review serves as a check on management at all levels and is designed to uncover deficiencies and to eliminate avoidable waste and loss.

The reviews of the system are 'periodic' in that the entire system is reviewed at planned time intervals which are established by management taking into account the effectiveness of the system as measured by previous review and operating experience.

The reviews are 'systematic' in that they are planned so that all elements of the system are examined for effectiveness over a given period of time. To be effective the reviews must be carried out by personnel who are:

1. Trained in audit techniques;
2. Independent of the activity being reviewed;
3. Aware of the technologies involved;
4. Used to the procedures involved.

Product or process audit

These are more examinations to provide information. It is all very well ensuring the system is always in operation, but we must never lose sight of the product. It is after all, the product the customer will pay for!

Product verification is used to ensure that the requirements for quality have been met. It may consist of taking the final product, stripping it and having each component tested to ensure the requirements have been met, but may also take less stringent forms providing there is adequate evidence or proof that the product meets the quality requirements. For further information and definitions refer to Figure 4.2.

Records

Throughout this chapter, we have constantly referred to records. These form the objective evidence that the system is functioning as it should, and system audits will evaluate these records to ensure the system is working as it should. In fact, management should develop and maintain records necessary to demonstrate the effective operation of the quality control system employed and to provide objective evidence of product or service quality.

Although the list is not exhaustive, such records may include:

1. Design reviews (Chapter 3);
2. System reviews;
3. Test results;
4. Calibration records;
5. Concessions or waivers;
6. Analysis of process control data;
7. Corrective actions;
8. Inspection records;
9. Vendor performance (Chapter 4);
10. Customer complaints.

Role of quality assurance in manufacturing

A typical organization chart for a large firm is shown in Figure 5.4. Note that there appears to be a management representative independent of other functions who has the necessary authority and the responsibility for ensuring that quality standards are defined and maintained. He is designated the 'Quality Manager'.

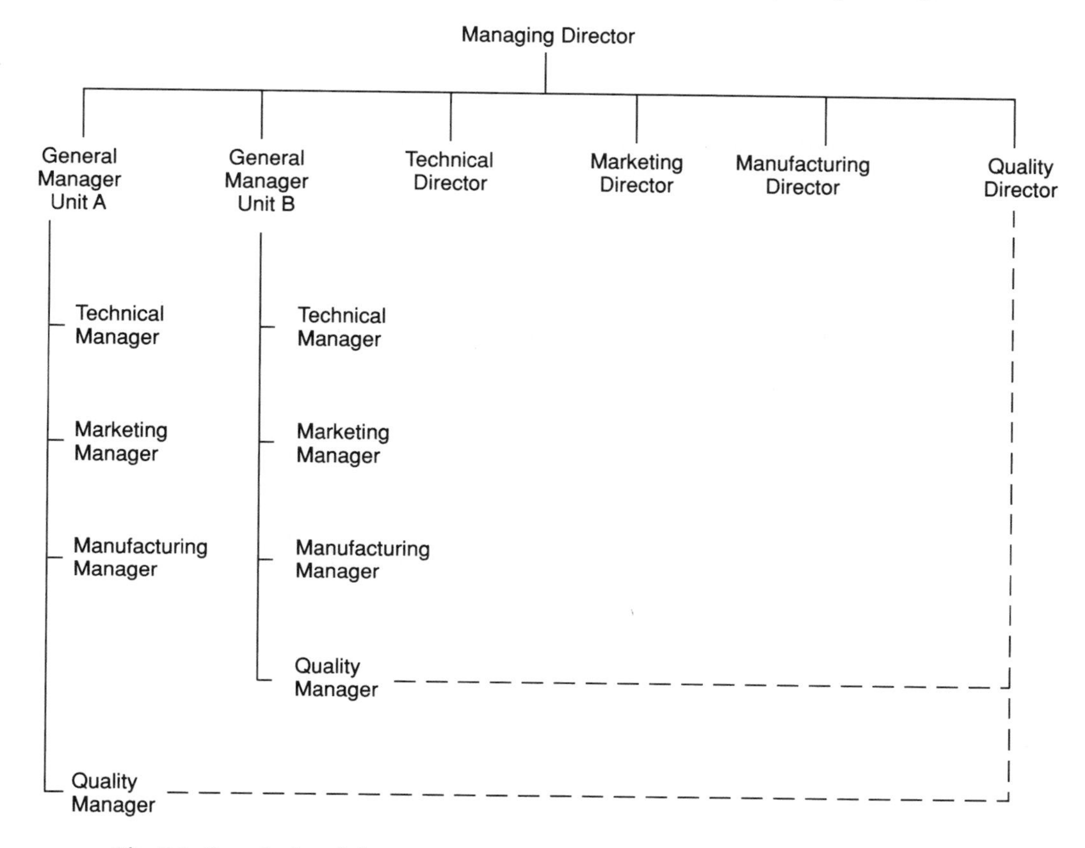

Fig. 5.4 Organizational chart.

The Quality Managers in the units may be administered on a day-to-day basis by the General Managers concerned, but if there is any conflict the Quality Managers have professional or technical access to the Quality Director, so their independence of action appears safeguarded. Once again their actual 'job descriptions' or 'terms of reference' should be scrutinized to ensure this is so.

Quality manager's department

There is no absolute right or wrong way of organizing a quality department. Much depends on the size and maturity of the organization concerned, and the nature of its business. We shall discuss contemporary thoughts on this issue, which reflect the desire to inculcate a 'total quality management approach', in Part Three of the book. The approach in this chapter represents a more traditional approach, applied to a large manufacturing organization.

It might be well to consider the role of the two main areas of the quality manager's department:

1. Inspection and test planning;
2. Quality engineering.

Inspection and test planning

Inspection and test planning deals mainly with appraisal and their main activities are:

1. Determining test and inspection equipment requirements;
2. Designing and developing test and inspection equipment;
3. Procuring test and inspection equipment;
4. Planning the calibration and maintenance of test and inspection equipment;
5. Determining test and inspection points (i.e. will final inspection be adequate or should some form of in-process inspection be incorporated on selected characteristics which cannot be inspected or tested at a later stage?);
6. Determining inspection levels – 100% or sampling;
7. Writing test and inspection specifications and instructions;
8. Developing test and inspection reporting systems;
9. Determining test and inspection manning levels.

Remember inspection and test may well be carried out in:

1. Goods inwards (Chapter 4);
2. Fabrication;
3. Assembly;
4. Final assembly;
5. Installation.

Quality engineering

Quality engineering may well be involved in what we may call the preventive measures. These include:

1. Training;
2. Ensuring quality standards are defined and understood;
3. Co-ordination of quality plans, procedures and the manual;
4. Analysis of reports;
5. Planning and participation in all types of audits/reviews;
6. Monitoring
   (a) requirement specifications,
   (b) contract reviews,
   (c) configuration, specification and drawing control,

(d) design reviews,
(e) process and product qualification,
(f) Calibration system,
(g) handling storage and transportation,
(h) quality cost recording,
(i) corrective actions,
(j) quality improvement programmes,
(k) documentation.

It must be stressed that the quality assurance department generally monitors events and receives information late. Hence everyone is responsible for the quality of the product they produce or the service they offer. The quality assurance department is not responsible for creating quality, but for monitoring and reporting on quality, and maintaining the system for supporting it.

Summary of conclusions

The 'critical' function is performed by inspectors. What might be called the 'reflective' function is performed by the quality engineering department. They collect and analyse many reports to determine quality trends and to pinpoint those processes which produce the most defectives. Remember the shorter the feedback loop the cheaper it is to take corrective action.

A good quality management system relies on just that – production of information which is analysed and action taken to put things right and to minimize the chance of a similar occurrence happening again in the future.

6

After-sales quality

Introduction

Now we come to the final chapter in Part One of this book. From what has been stated earlier you may well believe that if all the actions and procedures mentioned in previous units have been adhered to, why is there a need for after-sales quality? Sadly no matter how good the system, defective designs, products or services may well be delivered and they will be costly to the producer – both in direct costs and loss of reputation, but a system will have to be evolved which minimizes the problems to the customer after delivery.

Once again we will turn to our original organizational chart (Figure 6.1) as we have in all previous units and consider the final link in the chain. You will note that before going to after-sales service we must also consider that area which immediately follows manufacture – namely packing, preservation and storage. Then we have delivery followed by after-sales considerations.

Quality assurance extends beyond the point of sale of a product or service; efficient after-sales service and the development of good customer relationships are powerful ways of gaining a competitive edge – influencing new and repeat purchase decisions, thereby increasing market share.

Increasing attention is now being paid to what customers actually want, rather than what it is thought they might be prepared to accept. For products, this means quality which extends to installation and service, spares, the speed and competence of repair – in other words, the cost of ownership from delivery to disposal. For the service sectors, it means individual attention, flexibility and intelligent anticipation. In both cases, it adds up to the customer receiving value for money, care, and consideration thereby coming away from the transaction with a favourable impression which will enhance the reputation of the business concerned.

Storage, packing and delivery

Following manufacture, the supplier should establish and maintain a system to control packing, preservation and marking processes to ensure conformance to

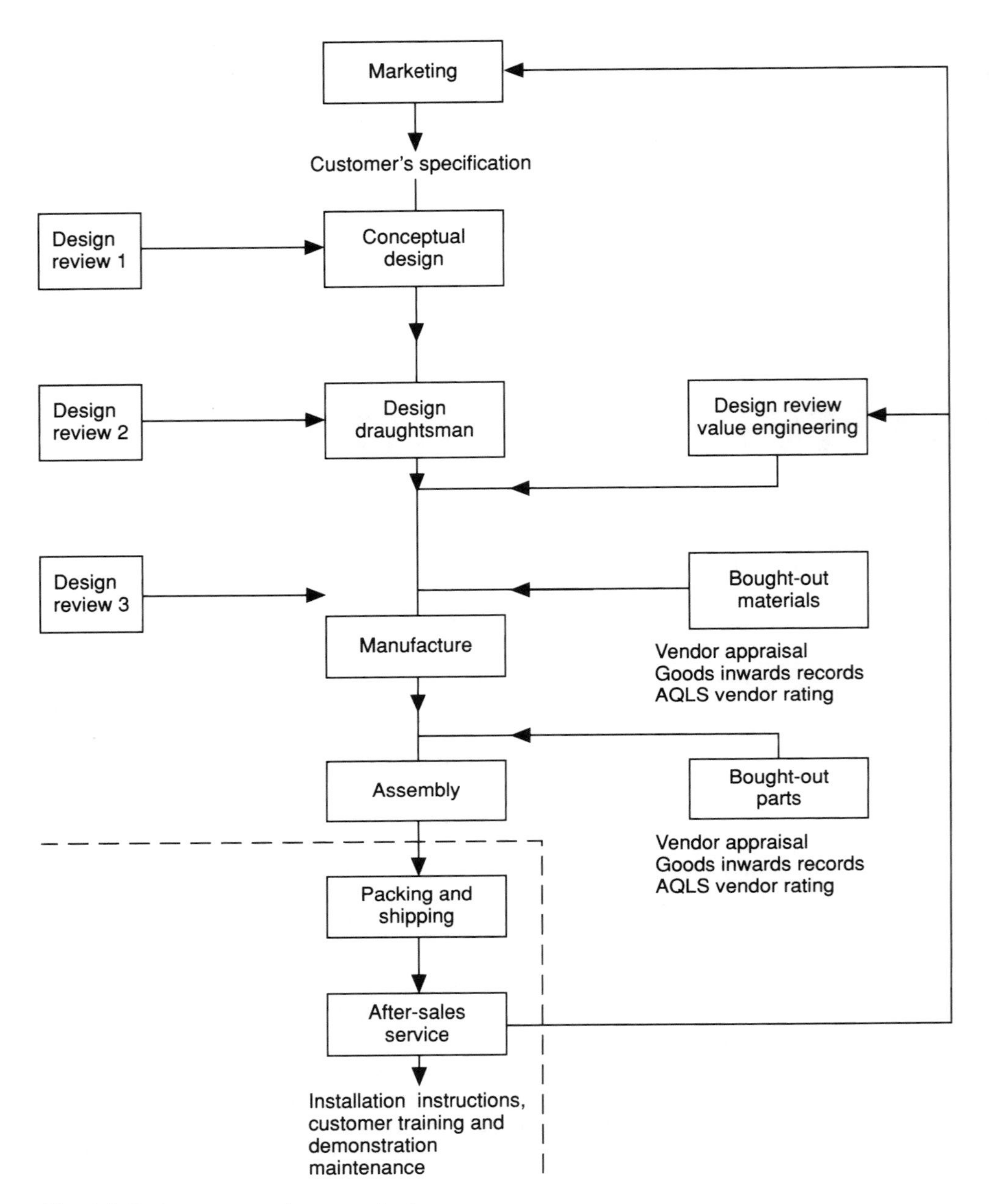

Fig. 6.1 Basic structure of an organization.

contractual or specified requirements. Such a system will include adequate written instructions or procedures for material handling and storage. These will be established to ensure that:

1. Material identity is maintained;
2. Suitable crates, boxes, containers and means of transport are properly used;
3. Adequate protection is provided against damage and contamination.

Care in storage

Manufactured items should be stored in clean, tidy and environmentally adequate storage areas with their identity maintained. Materials which deteriorate with time such as paints, rubbers, etc. should be clearly identified and controlled and for these you would expect special instructions. In these storage areas security is important and access will be limited to authorized personnel only.

You should expect a supplier to ensure that:

1. Quarantine and/or bonded storage areas are provided if necessary;
2. Material identity is maintained during storage;
3. Storage conditions are in a secure and correct environment as required;
4. Material only leaves storage when properly authorized;
5. Procedures for rotation of stock when necessary are adequate and effective.

The customer expects to receive goods in the right condition suitably marked, preserved, packed. Correct transportation is also important. Failure to control these operations adequately will negate all those effective controls you have exercised during the manufacturing process. The supplier must ensure that arrangements for delivery include protection against degradation of his/her product.

Another factor which should not be forgotten of course is, when is delivery complete? Arrangements should include protection of the material until it is put to use. In this case I am thinking particularly of war stock. Bombs, etc. produced now, may of course never be used, but they may be required in 10 years time and they must work first time. Packing and preservation in this instance is very important.

Supplier's responsibility

From the above you can see that the handling of materials, including storage, packing and delivery requires proper planning, control and documentation. This applies not only during delivery but also up to the time of being put into use. It is the supplier's responsibility to ensure this is taken care of and he/she should be able to demonstrate with actual objective evidence that he/she has these necessary controls.

Marketing and servicing

The marketing and servicing functions form the direct link with the customer and as such provide the main channel for the feedback of market reaction to goods and services supplied. Advantage should be taken of information available from these sources but care must be taken not to accumulate superfluous data and thus create analysis problems which can result from ill-conceived surveys and questionaires. It is prudent to seek professional advice regarding the confidence expectations prior to submitting anything other than the simplest type of request for information to the market, otherwise much time and goodwill may be dissipated at considerable cost.

We have discussed quality in marketing in some detail in Chapter 2, so now let us concentrate more on that other interface with the customers – after-sales service.

I am sure that you would wish to see a servicing organization which offered the following:

1. Prompt, pleasant, efficient telephone manner;
2. Prompt replies to either requests for information or visit;
3. Service conducted pleasantly, efficiently and on time;
4. Service conducted by personnel who have:
   (a) the proper training,
   (b) correct documentation,
   (c) adequate, calibrated test equipment,
   (d) correct spares;
5. The provision of skilled and prompt attention to all complaints.

These would be provided by a quality conscious servicing organization – but this does not happen accidentally, it has to be worked on, always remembering that here is an opportunity to use customer contact to gain information and, with a little luck, the edge over the other competitors. Now let us look into how we can set up the right servicing organization.

Staffing the servicing department

To ensure prompt attention to service requirements there must be an adequate number of trained personnel. This number will depend very much on the reliability of the delivered product – high reliability will mean less servicing/repair will be required, low reliability will mean more repairs are required.

Quality is a property which may change with the age of the product or service. Clearly, part of the acceptability of a product will depend on its ability to function satisfactorily over a period of time. This aspect of performance has been given the name reliability, which is the ability to continue to be fit for the purpose or meet the customer requirements.

Reliability ranks with quality in importance, since it is a key factor in many purchasing decisions where alternatives are being compared, and many of the general management issues related to achieving quality, are also applicable to reliability.

So what is reliability? In ISO 8402 it is defined as 'the ability of an item to perform a required function under stated conditions for a stated period of time'.

With this in mind the reliability should have been clearly defined at the marketing stage and every effort made during design to ensure the product would meet that requirement. Reliability testing, which will not be covered in great detail in this chapter is both costly and time consuming and, sadly, is often neglected, but it is only by knowing the reliability that we can hope to have the requisite number of servicing engineers, suitably trained and in position ready for market launch.

As reliability is an exceedingly important aspect of competitiveness, there is a need to plan and design reliability into products and services. Unfortunately the testing of a design to assess its reliability is difficult, sometimes impossible, and the designer must therefore invest in any insurance which is practicable. Some methods of attempting to assure reliability are as follows:

1. Adopt proven designs;
2. Use the simplest possible design – the fewer the components and the simpler their designs, the lower the total probability of failure;

3. Use components of known or likely high probability of survival. It is usually easier to carry out reliability tests by over-stressing components than by over-stressing the complete product or service;
4. Employ redundant parts where there is a likelihood of failure. It may be that a component or part of a system must be used which has a finite probability of failure (F). Placing two of these parts in parallel will reduce the probability of both failing to F2. Three in parallel will all fail with a probability of F3, and so on. Clearly the costs of redundancy must be weighed against the value of reliability;
5. Design to 'fail-safe';
6. Specify proven production or operational methods.

Designs should be studied and analysed in order to determine possible modes of failure and their effects on system operation. The main object of such an analysis is to discover CRITICAL failure areas and design features. It can be carried out using 'Top-down' or 'bottom-up' techniques. Within the analysis each potential failure will be considered in the light of probability of occurrence and categorized as to its probable effect on the successful operation of the equipment. This type of analysis is called Failure Mode and Effect and Criticality Analysis (FMECA). Such work will give some indication of the type and quantity of spares required by the servicing organization to give a satisfactory service.

Training the servicing department staff

By having some idea of the eventual reliability of the product we are selling – we will at least have some idea of the number of people we will require in the servicing department, but field repair personnel are only as good as their training, documentation, test equipment and availability of spare parts allow. So let's look first at the problem of training. Once again as in all aspects of quality – planning is essential.

At the earliest possible stage in product development attention should be paid to the problem of training the people who will be involved once the product has left the factory gates and will include the customer's personnel who may well be installers, operators or maintainers, and your own field service staff. This training should be clearly shown in the overall quality plan and must be carried out well before the product is issued to the customer. Imagine the loss of reputation involved if the product was delivered and there was no one capable of installing, operating or maintaining the equipment. Such a situation should never arise in a quality conscious manufacturing organization.

Stores availability

It is sad to relate that frequently organizations are so intent on producing the product on time that they pay scant attention to the problem of producing spares. This must not be allowed to happen. Before market launch, stores depots must be adequately stocked with an initial set of spares drawn up by staff who are well aware of the reliability trials which have been carried out during design and development.

Information from these trials must be circulated to staff involved with spares procurement. They can then plan and calculate the items required and in what numbers. The days of ordering a percentage of each item made should now be well passed, being superseded by this more scientific approach.

The right test equipment in the servicing department

Early in the product life cycle decisions must be made on how the equipment will be serviced. What servicing should be done by a visiting technician? What servicing can only be carried out at a centralized workshop where certain equipments (which are not easily portable) are available? What items can be economically serviced? In this latter case it can often be proved that it is uneconomical to split down and repair a component, it is easier and cheaper to replace the whole sub-assembly. With all this in mind a suitable maintenance policy must be evolved. Figure 6.2 shows the breakdown of a simple multimeter. By using different symbols an idea of the repair policy can be seen:

1. Items marked (△): can be repaired by the 'first line' repair organization – the mobile, visiting tradesman;
2. Items marked (□): can only be replaced in the specialist workshop nominated;
3. Items shown (○): may be repaired/replaced in a less specialist workshop. (Hatching indicates that the item represented is a consumable item and the old part should be discarded.)

For all these repairs suitable test equipment of some sort is required. Portable, robust equipment is required by the mobile service engineer, while more complex equipment would normally be used in the centralized facilities. Whatever servicing test equipment is used it must be calibrated and checked regularly and be of the type nominated in the repair schedules. In other words, as usual, the test equipment must be controlled in just the same way as the equipment in the manufacturing plant itself. Sufficient control must be maintained to provide confidence in the accuracy of any measurements taken.

The control of test equipment and methods employed should include the following (if appropriate):

1. Correct specification and acquisition, including range, bias, precision and robustness under specified environmental conditions for the intended use;
2. Initial calibration prior to first use;
3. Periodic recall for adjustment, repair and recalibration, considering manufacturers specification, the results of prior calibration, the method and extent of use, to maintain the required accuracy in use;
4. Objective evidence to show identification of instruments, frequency of re-calibration, calibration status and procedures for recall, handling, storage, adjustment, repair, calibration installation and use;
5. Traceability to reference standards of known accuracy and stability, preferably to national or international standards.

When measuring equipment is found to be outside the required calibration limits, corrective action will be necessary. Each occurrence should be investigated in order to avoid recurrence. This may of course result in a review of calibration methods and frequency, training or adequacy of the test equipment specified.

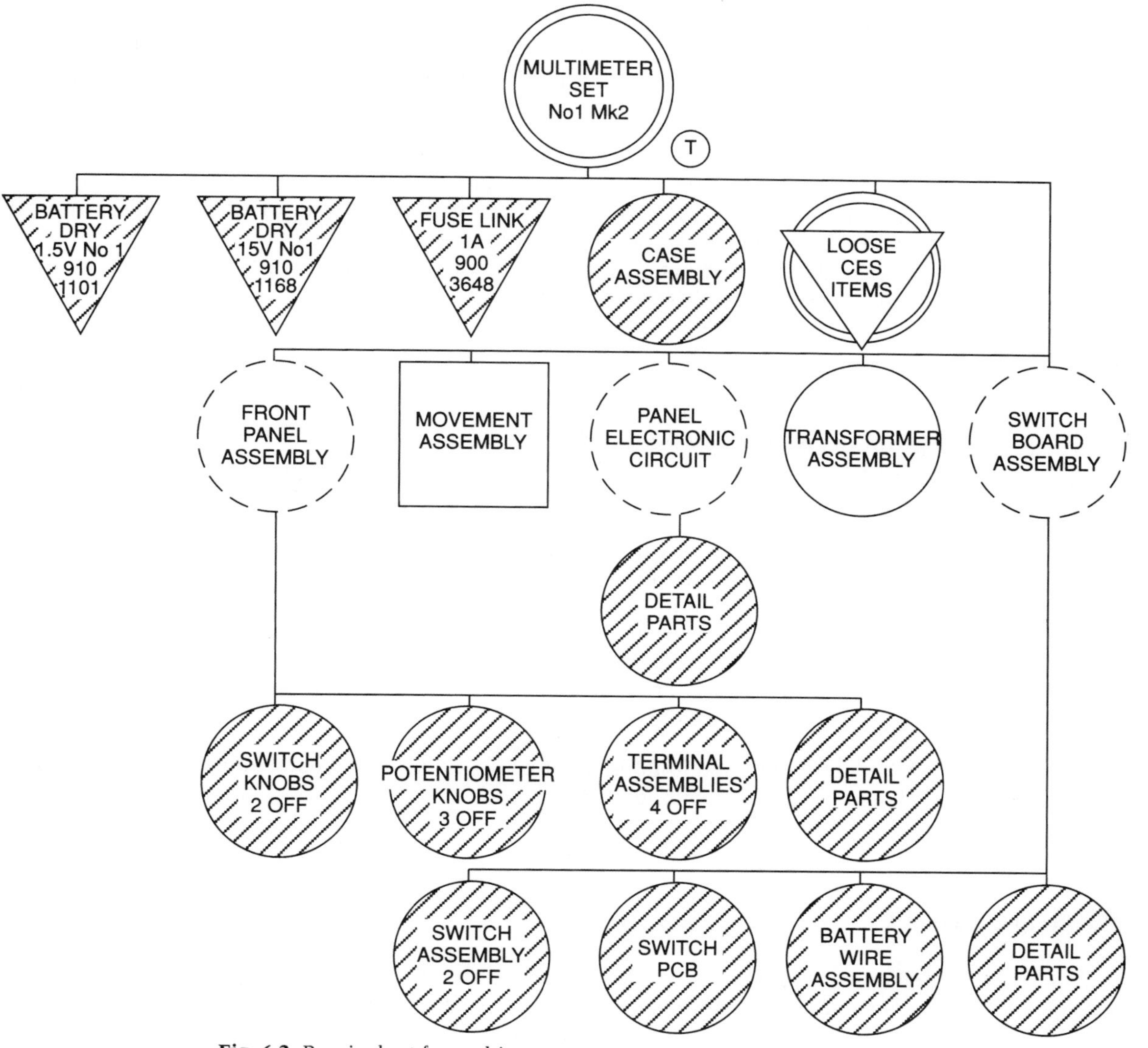

Fig. 6.2 Repair chart for multimeter set.

Calibration of measuring and test equipment is a very important issue both in the manufacturing and servicing environment and is an area which still needs more attention in industry.

The right instruction manuals

Anyone associated with the procurement of defence equipment may find it difficult to remember equipment which was delivered to the customer complete with all the necessary manuals, test equipment and spares.

Project managers tend to concentrate so heavily on the prime equipment – which is usually glamorous and worthy of wide publicity, that they tend to allocate lower priorities to the manuals, test equipment and spares. In the domestic field this approach would be disastrous. All these items must be planned and the targets achieved before market launch. Any variation from this could mean commercial disaster.

1. The correct documentation (correct subject and edition) must be available at the nominated areas in the correct quantity and in the correct readable condition. Anything other than this should not be acceptable.

   Installation instructions, user manuals and maintenance documentation all come under this heading and are an important part of the planning exercise.

   Before leaving the subject of instructions and manuals mention must be made of safety. User manuals and correct labelling is very important to ensure safety of the operation or user, failure to meet the criteria may well in the future give rise to a product liability claim. There is little evidence available in Great Britain at the moment showing what can happen in such cases, but there are some alarming case histories available in the USA which should have a salutary effect on manufacturers in other countries.
2. One other aspect is 'disposal'. Disposal of products or by-products should also be planned from an early stage. In the present-day climate when there is increasing concern for the environment, planned disposal must be considered. Would spoil tips associated with coal mines be encouraged today? Remember the problems we have had and continue to have, with the disposal of radioactive substances. All these problems should have been thought of and planned for from the beginning.

All documentation should be verified by the intended user, and a control system set up to ensure this should happen.

Maintenance

Reliability has been mentioned as the prime information required for allocating the human resources necessary to provide the necessary servicing/repair staff, but maintainability must also be considered. Again, it is essential from the earliest possible stage to obtain a maintenance specification from the customer. How does the customer organize his/her resources? What is the training policy, etc.? How will the customer validate that you (the manufacturer) have met the maintenance criteria?

A number of definitions will be useful here:

1. *Maintainability*: the ability of an item, under stated conditions of use, to be retained in, or restored to, a state in which it can perform its required functions, when maintenance is performed under stated conditions and using prescribed procedures and resources;
2. *Up time*: the period of time during which an item is in a condition to perform its required function;

3. *Down time*: the period of time during which an item is **not** in a condition to perform its required function;
4. *Mean Time Between Failures (MTBF)*: for a stated period in the life of an item, the ratio of the cumulative time for a sample to the total number of failures in the sample during the period under stated conditions;
5. *Mean Time To Repair (MTTR)*: the average time to repair an equipment.

Maintenance planning

Maintenance specifications are not too difficult to write but in many ways they are difficult to interpret and to agree if the conditions have been met. After all what is the average time to repair a complex equipment? Unless the time is taken over a very long period – not the sort of time which can be quoted contractually – the interpretation can become very difficult.

However, using a combination of the definitions shown above it is possible to give a meaningful specification from which the maintenance policy can be evolved. The use of a system as we saw for the multimeter can then be used to break down the design into repairable/throw-away items. This, in conjunction, with the correct training, documentation, test equipment and spares will allow the basis for a good maintenance system.

Product recall

The worst problem which can be encountered in the after-sales phase of a product is product recall. A few instances are described below.

The February 1990 issue of Product Liability International gave details of recent car recalls. Saab had to recall 70 000 of the 900 series cars and GM to recall 16 800 mediums and compacts. Ford may even have to introduce fuel system changes to cars made from the mid-1960s to date! A car driver was jump starting his car when the battery blew up and he lost an eye. Despite the fact that there was a warning notice on the battery which said 'Danger. Explosion can cause blindness', a total of almost $10 million was awarded against the battery maker.

Just think of the cost of each of the operations above. Someone has to pay in each case. It must of course, be the manufacturer and this cost must be considered as a 'quality failure' cost. We are always trying to minimize these costs so a product recall will be disastrous – the cost will be enormous, and every effort must be made to ensure such an episode will not occur again.

Product recalls are luckily rare – usually only necessary if safety is involved, but if a manufacturer is in doubt about the use of quality management systems, then such a huge quality failure cost should soon impress upon him/her the value of the use of preventive measures. These preventive measures, the records they produce and the actions taken as a result are what quality is about – the shorter the feedback loop the cheaper it is to put a problem right and if the records are consistently studied to highlight trends then the problems which give rise to product recall should be minimized.

Resolving customer queries and complaints

We have so far mentioned what is required in the way of resources to ensure prompt efficient maintenance and repair, but as you are aware this is not what quality is all about. 'Quality' is to ensure you gain from the experience to endeavour to minimize similar problems in the future. A feedback system must be evolved which records customer queries and complaints.

Customer information

Manufacturers who are quality conscious use the information provided by customers to gain a competitive edge. Jaguars are an example, there they have teams of people who regularly ring new owners of their cars to find if there are any problems. An obsession with satisfying the customer is really what quality is about. Use is made of any information gleaned to improve both the design of their cars and the after-sales service, their main idea being that they want repeat orders from satisfied customers and they want the satisfied customers to sing the praises of their cars at the golf club, at cocktail parties and everywhere they go. They believe they sell more cars this way than through advertising!

Customers, however, are not always satisfied, they sometimes have queries and complaints, and these must be resolved expeditiously and in a courteous manner. This will not happen automatically (as I am sure you all realize from personal experience!) but has to be planned and trained for.

British Airways are quality conscious and in an attempt to obtain a market edge, they asked their operatives what they thought their customers wanted to see most from BA. The majority of employees believed that prompt, efficient resolution of queries and booking was the answer – short queues and prompt (within three rings) answering of the telephone. Reasonable you might think – but when the customers were questioned they wanted a smiling, pleasant person to do the booking in, or answer their query. They were less perturbed by the length of the queue!

So customer complaints and queries must be answered as quickly as possible in the most courteous manner. This will require training and organization. This is interesting – there is another cliché which comes to mind 'Total quality begins with training and ends with training'. I have recently become aware that a local departmental store has become interested in quality, the staff have suddenly changed their attitude to customers. There is little doubt that they have been trained in 'customer care'.

Feedback services

It was mentioned earlier that we would address feedback services later and now is the time. It is all right resolving problems like customer complaints but the idea of quality is to minimize complaints and problems and this can only be done by highlighting these problems and feeding back the information to people in the organization who can learn from the experience and resolve the immediate problem as quickly as possible.

In the case of banks, building societies, supermarkets, etc. if a queue builds up there should be an arrangement whereby staff can be called up immediately from the office to open up another facility to minimize the problem.

In manufacturing industry it takes a little longer. Customer complaints and information from repair visits or service requests should be fed back as shown in Figure 6.1 both to the designers and marketing. The marketing organization should be interested, they will find it useful to have this information when the time comes for them to determine and define customer needs, expectations and product requirements.

As far as the designers are concerned, they will receive the information – particularly that derived from repairs and will soon realize which spares are being used the most. They should ask themselves did they order the correct number of spares in their initial calculations? If not, they must have made some wrong calculations or assumptions. Was the reliability adequate? Is the usage different to that indicated in the initial specification? Is the maintenance philosophy sound? Or is there a better way? Is the information provided satisfactory? Questions, questions, questions! If the feedback system is working as it should, it must make designers think and start to make incremental improvements which should ensure the product has a long and successful life. Incremental improvements are often better and cheaper than a radical new design which brings in renewed problems, re-tooling, etc. – always hazardous.

One final word of caution – incremental design improvements bring with them up-issued drawings, specifications, etc. A good documentation control and configuration management system is essential in this case, to ensure everyone in the chain is aware of the changes and know the build standard of all items (configuration status accounting).

The cycle begins again

Having now fed the relevant information from the after sales organization back to marketing, the cycle is now complete. At an appropriate stage a new customer requirements' specification will have to be generated. The timing will be important and this is always difficult to make. Consider the Mini motor car – should it be taken out of production? Should it be replaced? Should it continue to be incrementally improved each year? Will a Metro fill the gap? Only adequate market research can help make that decision. In the electronics field decisions may be easier to make – electronic components have a limited commercial life, so a manufacturer may not be able to buy suitable components after a time making a re-design essential.

So the cycle is completed as shown in Figure 6.3.

The quality loop

The quality system typically applies to, and interacts with, all activities pertinent to the quality of a product or service. It involves all phases from initial identification

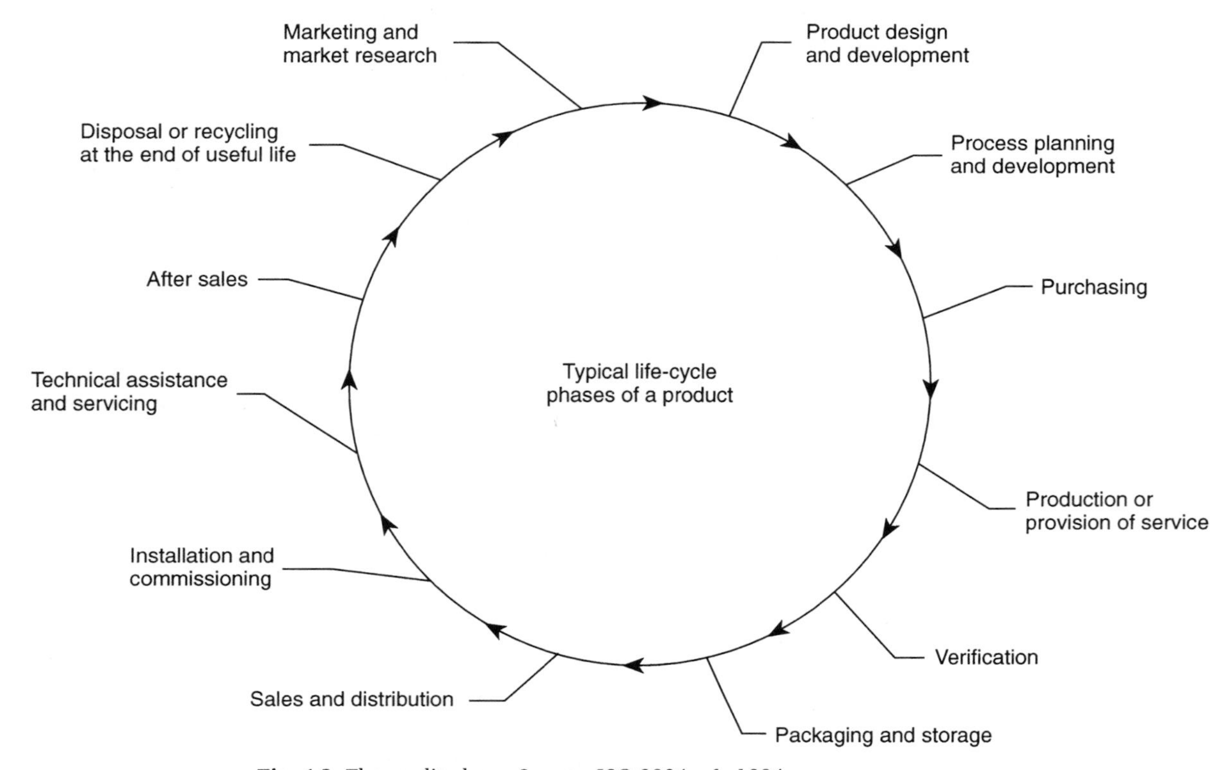

Fig. 6.3 The quality loop. Source: ISO 9004 – 1: 1994.

to final satisfaction of requirements and customer expectations. These phases and activities may include the following:

1. Marketing and market research;
2. Design/specification engineering and product development;
3. Procurement;
4. Process planning and development;
5. Production;
6. Inspection, testing and examination;
7. Packaging and storage;
8. Sales and distribution;
9. Installation and operation;
10. Technical assistance and maintenance;
11. Disposal after use.

Summary of conclusions

In this chapter we have shown the importance of efficient, prompt and courteous after sales service. It is at this stage that the customer is more in evidence and it is here the customer is involved with the product or service. Now is the time when the questions will be answered.

1. Was the customer requirements' specification generated by the marketing department correct?
2. Was the design satisfactory?
3. Was the product manufactured correctly?
4. Was the product packed and transported satisfactorily?
5. Was the accompanying documentation available and to the right standard?
6. Was the customer really satisfied?

At this stage possibly millions of pounds have been spent before market launch, now the customer sees the product and will use it.

Quality conscious firms must grasp this opportunity to feed back information which will now become available to the appropriate departments to ensure they build up a data bank which can be used effectively to improve the current equipment and will be used later to ensure that any new design offered will be a significant improvement on the present model.

Part Two

Quality Assurance Techniques

Introduction

Part One explored the nature of quality in its industrial context of supplier and customer. It stressed that concern for quality, and the contribution to achieving quality had to be 'total', i.e. company-wide. The different departments within the company which have an especially critical role to play were then examined in turn.

Part Two will examine how the diverse activities influencing quality can be coordinated, and managed effectively in order to achieve two goals: assuring the customer of dependable quality, and achieving this with the maximum economy of effort on the part of the supplier.

The management and control of quality demands: a quality system; a management representative given clear responsibilities in the pursuit of quality; and the effective use of a number of specialized QA and QC techniques. All these things and people are the topic of this module.

7

Creating the quality system

Introduction

This chapter will look in more detail at the requirements for a quality system, a concept introduced in Chapter 1 of Part One.

The quality of a product or service is a characteristic which is sometimes obvious by its presence, but always conspicuous when it is absent. Whether we define quality as 'customer satisfaction', 'fitness for purpose' or 'compliance with specification' it can only be achieved if it is planned and managed to be achieved.

An effective quality system will emphasize fault correction and improvement. For this reason I shall also include in this chapter some advice on how to gather and use data concerned with eliminating known problems.

Planning for quality

Less than total quality can be achieved with little planning, but true quality implies confidence, and it is demonstrated in regular repeat business. In one of Phil Crosby's phrases it is 'going back to the same shop twice'.

The expectation of achieving total quality depends very much on the complexity of the product or service, the skill and familiarity with the job of the people involved, and the environmental influences affecting the activities involved.

Quality and complexity

The probability of achieving satisfaction with a very simple task should be high. For example, the task may be to provide a customer with 1 kg of garden fertilizer from the local ironmonger or garden shop. A typical thought (or planning) process for this task could be as follows:

1. Set objective (providing fertilizer);
2. Set parameters (weight required);
3. Set criteria (1 kg);
4. Perform task (select fertilizer);
5. Verify task (weigh fertilizer).

This is, of course, a very simple process. However, hundreds of times each day people perform analytical, planning tasks like this without being conscious of how they are managing their needs. Complicated tasks are still composed of the same five steps I have listed as 1–5, but much more detail is involved at each stage. For example, suppose we were not considering garden fertilizer, but a substance with a much higher value or safety-critical feature, such as a medical drug. In this case the thought (or planning) process could be expanded into:

1. *Set objective*:
   (a) provision of drug XYZ;
2. *Set parameters*:
   (a) weight,
   (b) to specification ABC 12345,
   (c) shelf life used,
   (d) shelf life remaining,
   (e) packaging,
   (f) delivery,
   (g) source traceable and certified,
   (h) cleanliness/contamination level;
3. *Set criteria*:
   (a) 1 kg + 10 g – zero,
   (b) specification xyz issue C,
   (c) minimum shelf life 6 months,
   (d) double polythene sealed pouch,
   (e) standard trade pack to destination A by date dd/mm/yy,
   (f) certified source from approved supplier,
   (g) Irradiated to level 666;
4. *Perform task*:
   (a) issue material from bonded store,
   (b) weigh using sterilized equipment, ensuring all equipment is calibrated,
   (c) Fill drug in packaging,
   (d) seal packaging,
   (e) irradiate total package;
5. *Verify task*:
   (a) check order
   (b) check item issued
   (c) check correctness of drug packaged as follows:
      (i) remaining shelf life
      (ii) packaging
      (iii) delivery details,
   (d) calibration of equipment,
   (e) cleanliness of process area,

(f) effectiveness of irradiation,
(g) weight of drug,
(h) etc.

Effectiveness of planning

You may feel that advocating the planning of all tasks will lead to unnecessary activity. This is not true, because whatever the size of the task, a degree of planning is required. Moreover, it can be shown that planning tasks properly will contribute to performing them correctly, in the minimum time scale.

Ideally, a model system should be flexible enough to cope with both simple and complex tasks within its framework, and that is what guidance documents such as the ISO 9000 series of quality system standards have to accomplish. A quality system designed within the rules of ISO 9001 or 9002 would identify and document **appropriate** level of planning to perform the particular task.

Ideally one should strive to plan once, and use the plan many times for different tasks. Such an approach constitutes a quality system. Figure 7.1 indicates a hierarchy of planning activities and documents.

One aim should be to plan for product or service quality by utilizing a general or undifferentiated level of detail in the documents at the top of the hierarchy. The risk in this approach is that documents are so general that they are practically meaningless, and serve no practical purpose, and care should be taken to guard against this.

The quality manual can be looked at as the head of a pyramid of documents dealing with quality at different levels of detail and in different departments. The manual does not include all these documents, but it should show that they exist, when they are relevant, and where they are to be found.

Company-wide planning

If a company with several departments is able to coordinate their activities and achieve total fitness for purpose and total compliance with specification, it has achieved a signal success. Can it hope to do this without a quality assurance department? The answer is 'very unlikely', except in a very small, or extremely quality-aware and quality-professional company.

In an ideal environment, where each department or functional activity was perfectly discharged, quality activities would be 'part and parcel' of everything that was going on. If as examples:

1. Marketing always accurately identified the customer's requirements;
2. Engineering always designed in these requirements;
3. Manufacturing always produced to the defined specification.

Then in this case the achievement of quality would be intrinsic to the company's activities. From the point of view of cost-efficiency this would be a very desirable state of affairs. However, it is a goal that can only be approached if it is positively planned for. This means prescribing the intentions for quality achievement, and appointing a manager responsible for establishing the means to achieve them. The

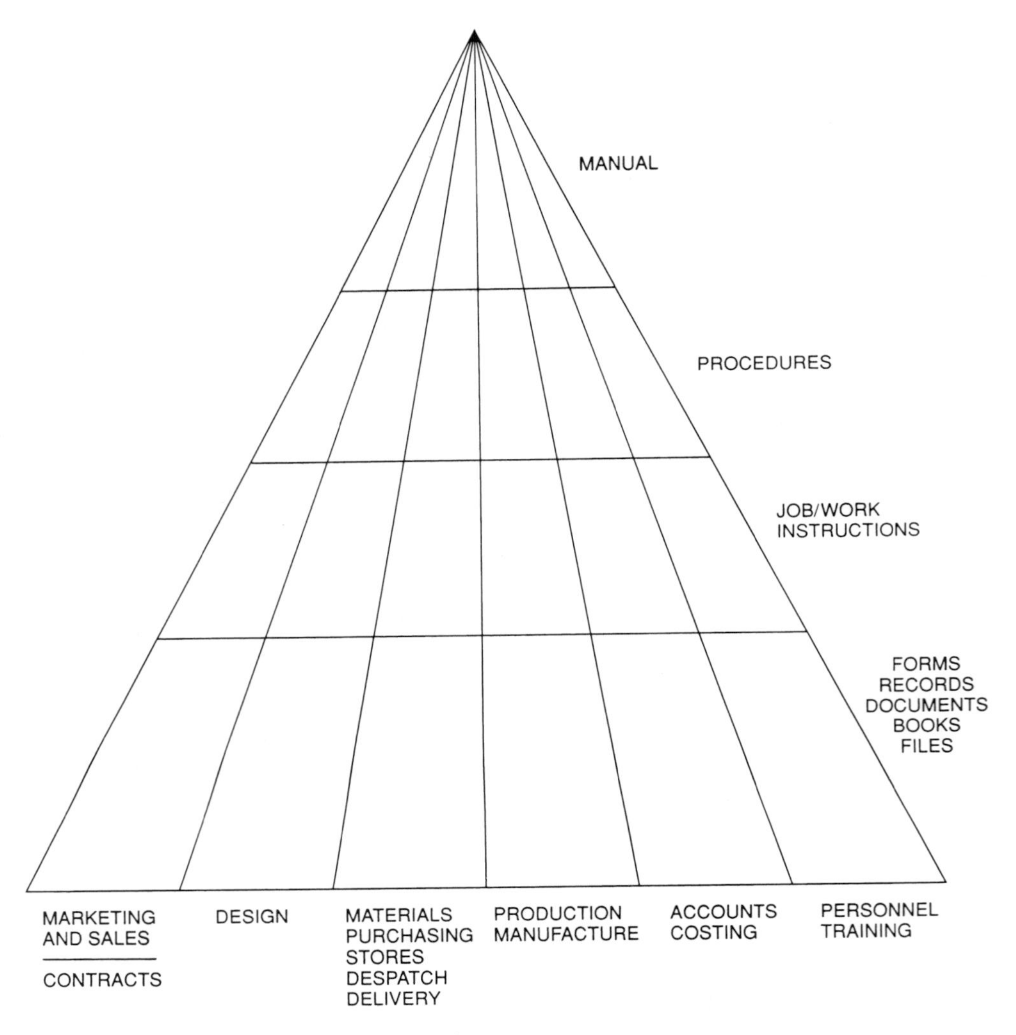

Fig. 7.1 Documentation structure for quality systems.

task of this individual, whom ISO 9000 refers to as the 'Management representative responsible for quality', is to ensure that quality goals are reached through the active involvement of all departments.

The 'ins and outs' of quality

In order to plan for quality it is helpful to look at each operation in a sequence as a process, with inputs from 'suppliers' and outputs to 'customers'. This approach is fundamental to 'Total Quality Management' (Figure 7.2).

1. *Inputs* are what are given to the person in charge of the process, to work with. They include materials, tools, and work instructions.

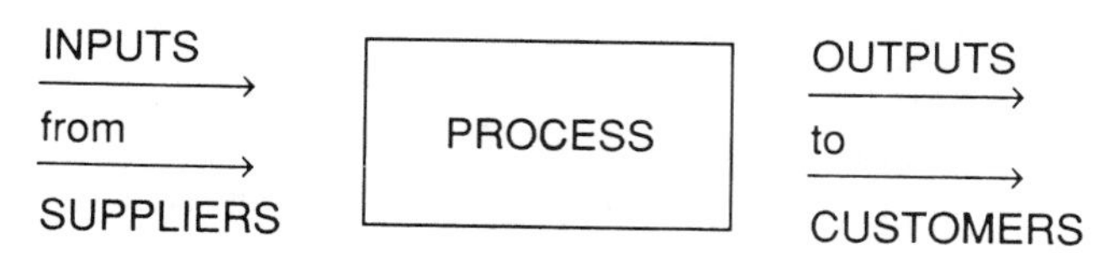

Fig. 7.2 Processes, inputs and outputs.

2. *The Process* is any operation performed. It can be a paperwork function, or a physical activity such as performing a machining operation to a product, or equally the indirect activity of maintaining a piece of equipment.
3. *Outputs* are what is passed on as a result of the process. They can include products, information and records.

The output from one process is the input to some other process, and each customer is someone else's supplier. Thus every person in the organization has immediate suppliers and customers with whom to establish communication, regardless of how divorced they may be from the initial design or raw materials, and from the company's ultimate customer. This model is explicit in TQM, as we shall see again in Part Three, Chapter 20. Whether or not it is expressed so forcefully, it has to be recognized implicitly when planning for the achievement of quality in any organization, large or small.

The quality system

Industry is characterized by advancing technology; and currently increasing customer demands and expectations, in terms of ease of use, safety and environmental protection.

The supply of goods or services has to meet varied demands: on choice, price, performance, prompt delivery, safety, ease of maintenance and many other factors. The concepts and disciplines which have been developed to ensure compliance with the current requirements expressed by customers are referred to as quality assurance.

Quality assurance is achieved through the establishment and adherence to a quality system, sometimes described as a quality '**management**' system. In other words, the quality system is the set of disciplines and procedures for 'making quality certain'.

Quality system requirements

Most quality systems are based on, and demonstrate certain basic concepts:

1. The quality of products and services depends on the supplier's control of design, manufacture, inspection/test and all other operations which affect quality.
2. Suppliers should not only be able to deliver products and services on schedule at an agreed price but also be able to substantiate by objective evidence that

they have maintained control over those aspects which affect quality, and verified the acceptability of the product or service.

Customers also have obligations which they need to perform in order to ensure the quality of the products or services they receive. They must make sure that their statement of requirements is full and clear. They should stipulate the degree of assurance required, in order to ascertain that the supplier has control over his own activities and outputs.

Quality system structures

There is no unique quality system structure, ideal for all organizations. Nor would it be true to say that for any given company, there is only one structure that would suit it. What is true is that any viable quality system structure must address and provide procedures for certain considerations affecting quality; how each is implemented, and how elaborate the procedures and their rules, is a matter for the individual company.

The quality system structure should take account of the nature of relationships between the various functions and disciplines contributing to product or service quality. It does not have to be rigid but to ensure effective, accurate communication and actions it has to be documented. So far as is possible, the system should be flexible, in order to accommodate varying degrees of product and service complexity in the most cost-effective manner.

Commitment to quality

Before quality can be present in a product or service a 'quality culture' must be absorbed by the organization producing it. Any organization which is only partially committed to the achievement of quality or which imagines that quality is secondary to price or delivery (in fact, quality is central to achieving competitive price and on-time delivery) is not going to be capable of fully satisfying its customers.

Quality will be achieved if the message of quality commitment is acted unmistakably by the Chief Executive Officer, and training is provided to enable staff at all levels to play their part.

The demonstration of commitment by the CEO is considered essential in enlightened companies. The establishment as an unmistakable fact, that the CEO is 100% **for** quality, is seen as tipping the balance towards a new respect for quality in many companies.

The management representative

Except in the smallest companies, it is unreasonable to expect the CEO to overview and control every aspect of quality. Hence the need to appoint a management representative for quality. In a small company he might double as Technical

Manager, in the larger one it would be a specialist post with some such title as Quality Assurance Manager or Director of Quality.

Most quality system standards require the quality representative to report directly to the MD or CEO on quality matters. In some companies the quality representative may have a different manager or superior for other day-to-day matters. For example, the Quality Manager might report to the MD/CEO on quality matters but to the Technical Director for salary and disciplinary issues. In a larger, multi-site or international organization the site Quality Manager might have a dual responsibility, to the site General Manager and to the corporate Director of Quality. Clumsy and less than ideal as these arrangements may seem, they can work out well enough in practice in a quality-aware organization.

The management representative must have defined responsibility and authority for quality-related matters. The job definition or terms of reference should ideally list primary responsibilities and duties associated with each. This is essential for the effectiveness of the quality planning activity.

The quality policy

The quality policy is at the heart of any organization seeking to gain or retain a reputation for good quality. It should define the objectives for quality in line with its general trading practices.

You will probably meet more poor than good company quality policies. If it is longer than the Lord's Prayer or Apostle's Creed it must be unnecessarily verbose for the importance of the message it is trying to put across. It is essential that any member of the organization can give a sensible paraphrase of the quality policy, and understand its meaning, even if they can't quote it word for word.

A quality policy should be written by the CEO, with the assistance of the quality representative. Why? Because if it is to have any credibility within the workplace, it is essential that the CEO believes the quality policy.

The quality manual

Quality manuals do not read themselves! They need to be written and presented in an attractive fashion which will encourage people to make use of them. Sadly, many quality manuals are seldom read by most of the holders, even when a comprehensive distribution is maintained. Why is this?

A quality manual can serve several purposes:

1. As a general overview of quality activities;
2. As a statement of quality activities;
3. As a set of mandatory requirements;
4. As an index of quality and management procedures.

The authors of many quality manuals are unclear of the precise purpose for which they are writing it. As a consequence, its readability and utility are impaired. There are some basic guidelines which can prevent the writing of a quality manual proving to be a wasted effort.

Structure

A useful structure is one that follows the format of an existing quality standard such as ISO 9001. This format will be readily recognized by assessors, and so make it easier for them to check if the quality system being described meets the requirements of the standard.

Purpose

This should be precisely stated in the text and referred to frequently while composing the manual to make sure that it is 'on course'. Typical purposes are:

1. To state the company mandatory quality requirements;
2. To provide a guidance statement;
3. To provide information to third parties.

Scope and application

An organization may be large and cover many diverse operations. Thus the scope of the quality manual must be clearly defined, e.g. which sites and products or activities it applies to.

Style

The manual should be written in a consistent style, which can be assimilated by the intended readers. It should reflect the purpose of the document, in that any mandatory statement should be phrased positively and forcefully; conversely a guidance statement would be less emphatic and leave scope for flexibility.

Content

If the manual is intended as a direct reference document for company staff its content must be meaningful. It is not sufficient for the manual to say 'such and such shall be done'; it must give guidance as to who (in relation to circumstances or functional position) must do it.

The contents of the manual must address all the 'elements' of the quality system (which we shall examine in detail in Chapter 8) and all the quality-related activities, company-wide. It is a **total-management** document, both in its generation and its use. No valid manual is written purely by and for the use of the quality department.

Corrective actions to improve quality

The purpose of the quality system is to ensure quality is maintained and where necessary improved. Indeed corrective action is one of the elements of the quality system. In this section we will look at the formal procedures, of which the Quality Manager is custodian, for ensuring that corrective actions are taken and pursued until they are successful, once a problem has been noted.

Maintaining quality usually turns out to be one of those situations where you have to run in order to stand still. No situation is wholly static. Even if your product range or the service you offer is established, you are likely to be subject to personnel changes and the gradual wearing out of equipment on which you rely. Quite apart from staff turnover, the training you have given your personnel will also 'wear out' in time, unless it is regularly refreshed and recertified. The effect of these factors is that even if your quality control has become routine, its effectiveness will diminish gradually over time.

In any case, most organizations find that the simple **maintenance** of quality is not sufficient. Price competition, competition on performance specification, increased customer expectations, higher safety requirements and expanding liability legislation make quality **improvement** essential. Notice I said that price competition was a reason for enhancing quality, not reducing it. True quality improvement reduces production costs it does not increase them. We shall deal with this matter of quality costs in Chapters 23 and 24 of Part Four.

The corrective action loop

The way to improve quality is through the corrective action loop. This illustrates the stages that must be passed through before before a problem can be considered solved. The loop is shown in Figure 7.3, and represents the fact that correction depends on the following sequence of steps:

1. Detecting the deficiency – what are the symptoms?
2. Identifying the cause – do we know what's the matter?
3. Proposing a solution – do we know how to fix it?
4. Applying the agreed corrective action – fix the problem!
5. Verifying that the corrective action has been carried out according to instructions – did we fix it properly?
6. Monitor if the corrective action was successful – did the fix work?
7.
   (a) *If so*: Ensure the corrective action is documented and continues in force. Carry on!
   (b) *If not*: Review the problem again in the light of the failure of the originally proposed solution – go round the loop again!

The corrective action request

The means for ensuring that a corrective action is progressed throughout the whole loop is a document which is often called the corrective action request, or

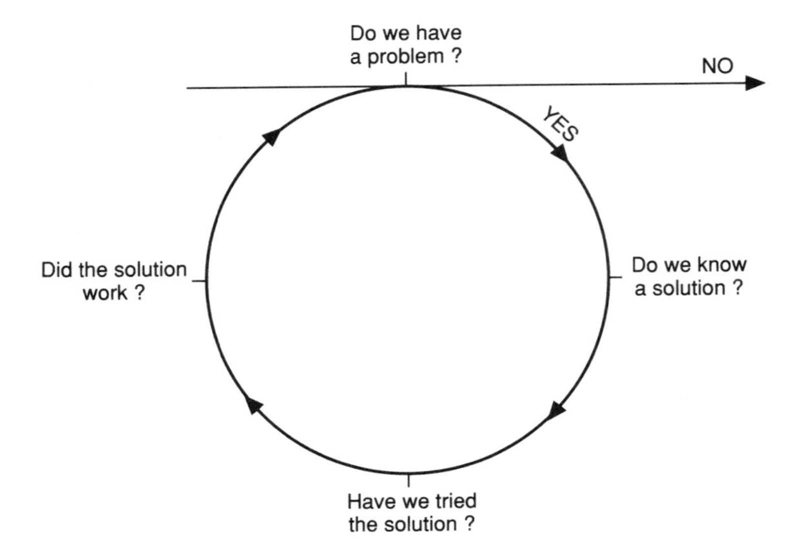

Fig. 7.3 The corrective action loop.

CAR. In other companies it can go by other names, such as Non-conformance Report (NCR) Notification of Variance (NOV) or Error-Cause Removal Request (ECRR); or more colloquially, the 'squawk'.

The CAR is characteristically employed to follow up any non-compliance found during quality system audit, but it can be used whenever a deficiency, fault or error has been identified. The CAR ensures that a cycle of the 'corrective action loop' is initiated, and thus it is the bridge between appraisal and prevention.

The key features of the CAR are:

1. It is a formal document initiated by the observer of the deficiency.
2. It describes the deficiency found, and what document it is deficient against.
3. It contains provision for quarantining a product or stopping a process until the deficiency had been investigated.
4. It requires a response within a stated time scale, from someone with responsibility for the process or product concerned, proposing.
   (a) what is to be done with any affected product; and
   (b) what steps will be taken to prevent recurrence, and by what date.
5. The proposed actions are reviewed for acceptability by QA.
6. The application of the proposed actions and their effectiveness are verified by QA before the CAR is fully signed off ('closed out').
7. In the event that no response to the CAR is received within the time laid down in the company's procedure, or that the proposed action is not implemented within the agreed time scale, the issue is escalated to a higher level of management for attention and action.

A typical Corrective Action Request form is shown in Figure 7.4.

REPORT NUMBER 123

Audit Team		
QA Dept.	Dept. Manager	Others
Requirement		Deficiency Category 1 2 3
Observation QA Signature		
Recommendations Target date for completion .... Signatures Dept. Manager .... QA		
Action taken QA signature .... Date completed		
DEFICIENCY CATEGORIES 1. Significant non-compliance with procedure 2. Significant number of minor non-compliances with a procedure 3. Minor problem area that warrants attention		

Fig. 7.4 Typical CAR form. 'Requirement' is the requirement stated in whatever document is being cited, which was observed not to be met. 'Observation' is the nature of the non-compliance with the document. Source: British Ministry of Defence form QA 157 used to report on deficiencies in contractor's quality systems.

Fact-gathering and problem-solving techniques

The previous section rather begs the question 'How do we know what the problem is?'. A difficulty which I believe besets most industries is that people want to cure the problem as soon as possible. No-one wants to stand by and let it continue for a moment longer than necessary, so 'fire-fighting' becomes the order of the day. However, you can't define an effective solution until the problem is fully understood.

Experienced, professional problem solvers will tell you that the true explanation is the one that explains ALL the facts, no matter how perplexing these may seem at first sight. How often a 'cause' and then a 'cure' are proposed while facts are still coming to light, or in the face of one particular awkward piece of information which is conveniently ignored! Time and again I have found that if the facts of the problem are accurately stated there is SOMEONE in the organization who knows the cause, and often the solution – they just hadn't realized that what was happening presented a problem to anyone!

I shall mention some useful tools for gathering and assimilating data; 'check sheets' or 'tally charts' for recording raw data, 'histograms' and 'scatter diagrams' for establishing the patterns behind the raw data.

Two techniques which can aid successful problem solving will also be mentioned here: 'brainstorming' to ensure that all possible causes are identified and considered, and the creation of 'cause-and-effect diagrams' to record the influences of possible causes on the problem seen, through their intermediate effects. A brainstorming session is typically the precursor to creating a cause-and-effect diagram, and use of the diagram aids the investigation and identification of the true cause or causes, and formulation of a cure.

Check sheets

A check sheet acts as a reminder of the classess of data we are looking for, and how they are to be classified. If fully filled in it gives all the information we are trying to collect, with a minimum of thought, time or effort. An example is given in Figure 7.5.

Non-conforming copies					
	Missing pages	Muddy copies	Showthrough	Pages out of sequence	Totals
Machine jams					
Paper weight					
Humidity					
Toner					
Condition of originals					
Other (specify)					
				Total	
Collected by: Date: Place: Formula:					

Fig. 7.5 Check sheet (data collection form). Source: BS 7850:part 2:1992.

Tally sheets

Tally sheets such as that shown in Figure 7.6 have the advantage that the information already appears in something like the form it would take when plotted as a histogram or other form of graphical presentation (compare with Figure 7.7).

Histograms and Pareto diagrams

A histogram presents the information in a more organized and graphical form, as illustrated in Figure 7.7. A Pareto diagram, as in Figure 7.8 arranges the phenomena or values in order of frequency of occurrence, in order that the most prevalent

Truck turn round time (minutes – rounded to nearest 5)	Tally	Number of trucks (frequency)
10	|	1
15	|||	3
20	𝍸 |	6
25	𝍸 ||||	9
30	𝍸 𝍸 𝍸 𝍸 𝍸 𝍸 𝍸 𝍸 ||	42
35	𝍸 ||	107
40	𝍸 𝍸	170
45	𝍸 𝍸 𝍸 𝍸 𝍸 𝍸 𝍸 𝍸 𝍸 𝍸 𝍸 𝍸 𝍸 𝍸 𝍸 𝍸 𝍸 𝍸 𝍸 𝍸	100
50	𝍸 𝍸 𝍸 𝍸 𝍸 𝍸 𝍸 |||	38
55	𝍸 𝍸 𝍸 |	16
60	𝍸	5
65	||	2
70	|	1
	Total	500

Fig. 7.6 Tally sheet.

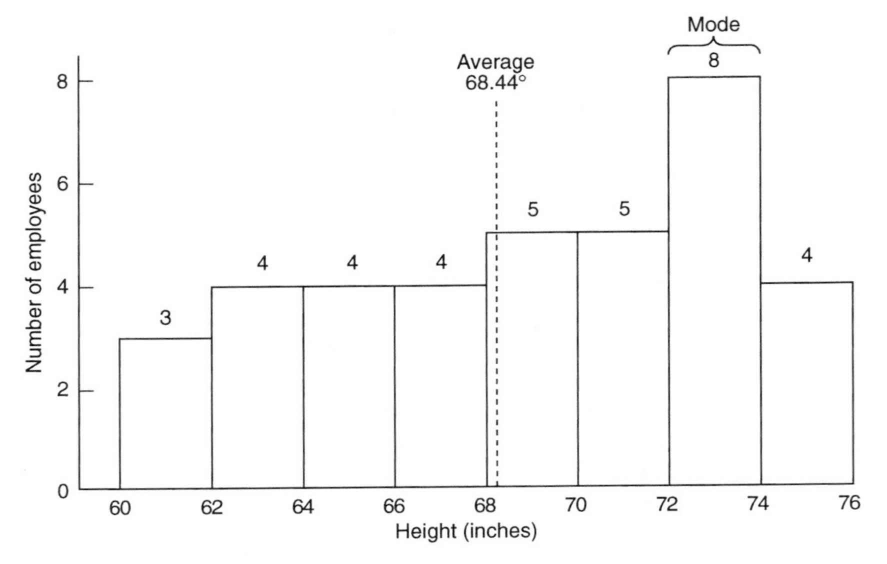

Fig. 7.7 Histogram. Source: BS 7850:part 2:1992.

can be quickly identified. This enables the 'important few' to be quickly distinguished from the 'trivial many' and be subjected to close atention.

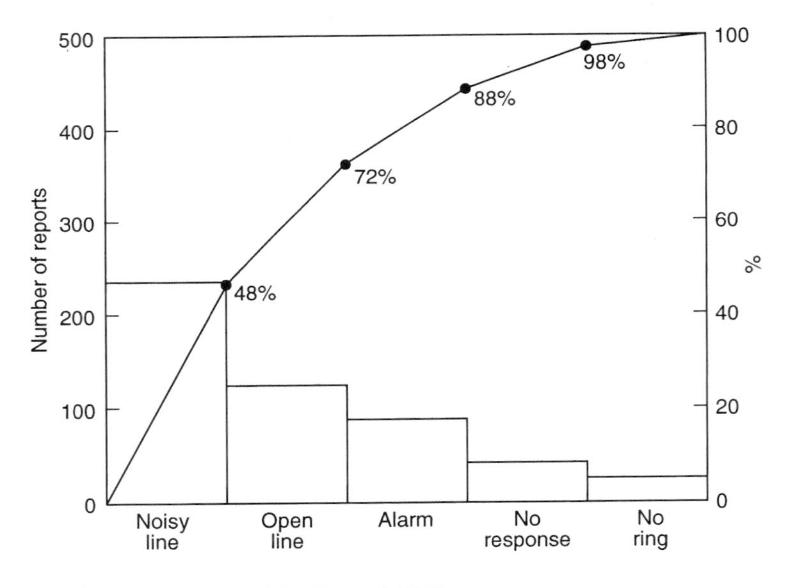

Fig. 7.8 Pareto diagram. Source: BS 7850:part 2:1992.

Scatter diagrams

Scatter diagrams plot two possibly interrelated characteristics on their two axes, so as to provide a pictorial presentation of the degree of interaction between them. Figure 7.9 is an example, showing a clear if weak 'positive' correlation.

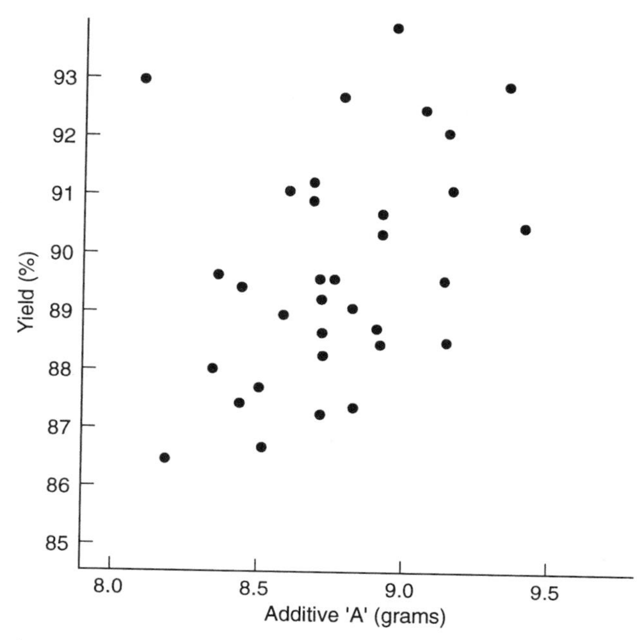

Fig. 7.9 Scatter diagram. Source: BS 7850:part 2:1992.

Brainstorming

Brainstorming is a technique aimed at utilizing the creative power of a team to generate, clarify and evaluate a rich range of ideas, in this case ideas as to possible sources of a problem.

There are three phases:

- A 'generation phase' where the objective is quickly to identify as many ideas as possible. There is no criticism value-judgement or analysis at this stage.
- A 'clarification phase' to ensure that everyone understands all the items that have been put forward. New ideas may be built on those already offered.
- An 'evaluation phase' when duplications, irrelevancies or ambiguities can be removed.

When this activity is complete it is possible to incorporate the collected ideas into a cause-and-effect diagram.

Cause-and-effect diagram

This is also widely known as an 'Ishikawa diagram', after its popularizer, or as a 'fishbone diagram', from its shape. Any process involves the deployment of the 'four Ms'; methods, materials, machinery and manpower, to which 'environment' is sometimes added. Thus the basic division of the diagram is into these fundamental factors. Each potential cause can be assigned to one of these areas and its

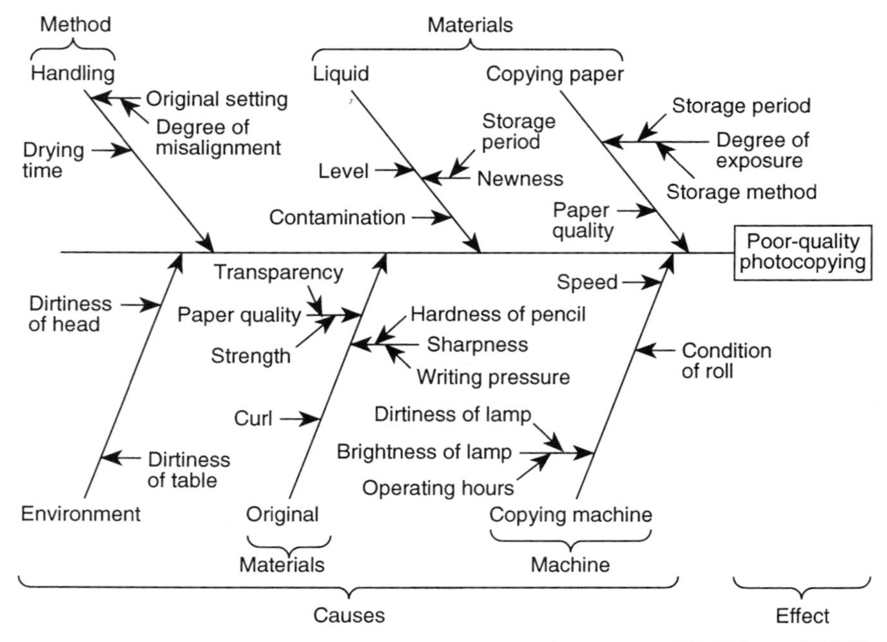

Fig. 7.10 Cause and effect diagram for poor photocopy quality Source: BS 7850:part 2:1992.

underlying cause considered. The motor manufacturers Toyota initiated the idea of the 'Five whys' where each 'cause' prompts the question 'but why was that so?'. Only after the question has been repeated five times is it considered safe to assume that the underlying cause has been uncovered. By that time the original 'backbone' of the fish will have several levels of branches, as in the case of the example depicted in Figure 7.10.

The importance of a quality system standard

We have already stressed elsewhere in this book, even before we reached this chapter, that for an organization consistently to supply a product or service of acceptable quality it must operate a quality system. This is still not the whole story; the quality system as laid down must be effective and appropriate to the size and scope of the company, and the goods or services it supplies.

Also relevant to the scope of the quality system is the extent to which the company is involved in designing the product or service at one end of the quality chain, and engaged in installation and servicing at the other. Finally, no matter how well-designed the system, it is valueless unless it is adhered to by all personnel, and at all times.

There is obviously an advantage in specifying uniform minimum requirements for a quality system which any company can adopt, and which a customer or customer's representative can evaluate. To be so flexible, yet still uniform enough to have meaning, the system has to be very carefully defined and administered.

Attributes of an effective quality system standard

In order to be credible and gain industry-wide support and recognition, a standard must have the following attributes; it must:

1. *Be comprehensive*: All the necessary elements of an effective system must be included.
2. *Be realistic*: It must leave the individual company the final responsibility for working out how it will manage its quality.
3. *Be general*: It must be equally applicable to different kinds and sizes of business.
4. *Be organized*: It must be backed up by an organization which is capable of acting as a forum for proposed changes, operate an independent assessment scheme, maintain a register of all currently approved firms, etc.
5. *Be fair and impartial*: Assessment of compliance must be seen to be confidential, professional, fair, and consistent in the standards it applies to different firms.

Such schemes have evolved from those set up by the US Department of Defense, and by the National Aeronautics and Space Agency (NASA), developing through NATO and national schemes to ISO 9000 which has much the widest scope. To achieve the criteria set out above within a scheme designed to embrace any industry (manufacturing or service) and any nation wishing to adopt the scheme, is the most severe test of the workability of such a scheme. In Chapter 8 we will examine how these requirements have been implemented within the ISO 9000 standards, and what use suppliers and customers can make of the scheme.

Risks, costs and benefits

Before going on to examine the ISO 9000 series of standards in detail, we can refer to ISO 9004–1 to highlight again the value of a recognized quality system, both to the supplier and to the customer, in terms of minimizing risks and costs, and maximizing benefits.

Risk considerations

For the company, there are risks related to deficient products or services which can lead to loss of image or reputation, loss of market, complaints, claims, liability and waste of human and financial resources.

For the customer risks include hazards to the health and safety of people, dissatisfaction and loss of confidence, uncertain availability, and unsubstantiated marketing claims.

Cost considerations

The company has to consider the costs due to marketing and design deficiencies, including unsatisfactory materials and rework, repair, replacement, re-processing, loss of production, warranties and field repair.

The customer has to consider safety, and the costs of acquisition, operating, downtime repair and disposal.

Benefit considerations

The company has to consider the opportunities for increased profitability and market share. The customer has to bear in mind the opportunities for reduced cost, improved fitness for use, increased satisfaction and growth in confidence.

Summary of conclusions

A successful business depends on planning and organizing its effective management. It follows that quality can only be achieved if it is subject to managerial planning, organization, measurement and control, in order to make it happen. How elaborate this system needs to be depends on the executives' assessment of the risks, costs and benefits which hinge on the quality needs of their company's customers.

Concepts introduced in this chapter which will be revisited regularly later in the book include:

- Company-wide responsibility for quality.
- The company's activities seen as a network of processes converting inputs from suppliers into outputs for customers.
- A documented quality system defining quality creation, improvement and preservation methods; based on the company's evaluation of its exposure to quality risks, costs and benefits. The source of this information is normally referred to as the company's 'quality manual'.
- A management representative charged with maintaining the quality system, with or without a supporting 'quality assurance' department.
- The need to have methods in place for collecting data, recognizing problems and systematically formulating, implementing, evaluating and sustaining cures for them.

8

ISO 9000 – the international quality system standard

Introduction

Customers of industries which include space research, aeronatics, medicines, telecommunications and nuclear power have sought to impose standards for their contractors' quality management systems. Initially this was on a purely national basis, then it widened through international procurement and through multinational organizations (such as NATO in the case of defence procurement).

The international standard for quality systems

BS 5750 was first published in 1979, one of the earliest attempts to produce a formal national standard applicable to any industry. This work has been expanded by the International Organization for Standardization to produce standard equally non-specific to a particular industry, but internationally agreed and adopted, and first published in 1987.

The purpose of ISO 9000–9004

This section is quoted from the introductory section of ISO 9000–1:1994, and requires no amplification since it effectively reiterates many of the points already stressed in this course of this book:

Organisations – industrial, commercial or governmental – supply products intended to satisfy customers' needs and/or requirements. Increased global competition has led to increasingly more stringent customer expectations with regard to quality. To be competitive and to maintain good economic performance, organisations/suppliers need to employ increasingly effective and efficient systems. Such systems should result in continual improvements in quality and increased satisfaction of the organisations' customers and other stakeholders (employees, owners, subsuppliers, society).

Customer requirements often are incorporated in 'specifications'. However, specifications may not in themselves guarantee that a customer's requirements will be met consistently, if there are any deficiencies in the organisational system to supply and support the product. Consequently, these concerns have led to the development of quality system standards and guidelines that complement relevant product requirements given in the technical specifications. The International Standards in the ISO 9000 family are intended to provide a generic core of quality system standards applicable to a broad range of industry and economic sectors.

Structure of the ISO 9000–9004 series of standards

The ISO 9000 series actually consists of several documents, numbered ISO 9000 to ISO 9004. Two of them are guidance documents, each comprising a set of parts covering various topics, whereas the other three are standards which can be contractually enforced, applying to different scopes of organization. Full numbers and titles are tabulated in Table 8.1.

ISO 9001–9003 are the documents which provide a standard against which an organization's quality system can be assessed. In their ISO titles they are referred to as 'models', and in the guise of British Standards (see Chapter 19) as 'specifications'. They can either be used internally by a company on which to voluntarily model its quality system, or else compliance can be stipulated by a customer.

1. ISO 9001: *Quality Systems – Model for Quality Assurance in Design/Development, Production, Installation and Servicing.*
2. ISO 9002: *Quality System – Model for Quality Assurance in Production Installation and Servicing.*
   This applies where the organization is producing a product or service designed and developed elsewhere, or where quality considerations during the design phase are not considered critical by the customer.
3. ISO 9003: *Quality Systems – Model for Quality Assurance in Final Inspection and Test.*
   This applies where an organization is inspecting and/or testing a finished product or service, or where the customer considers that inspection alone is a sufficient safeguard of quality to be cited in a contract.

The first parts of each of the two guidance documents are general, the others are more specific.

4. ISO 9000 *Quality Management and Quality Assurance Standards* ISO 9000–1 *Guidelines for selection and use.*

This document is intended to:
(a) Clarify the distinctions and interrelationships between the principal quality concepts (i.e. establishing definitions).
(b) Provide guidelines for the selection and use of the series of international standards on quality systems that can be used for internal quality management or external quality assurance purposes.

The wording in the cited paragraph means that ISO 9001–9003 can be used internally by a company on which to voluntarily model its quality system, or that compliance can be stipulated by a customer.

5. ISO 9004: *Quality Management and Quality System Elements* ISO 9004–1 *Guidelines.* This explains the need for a quality system if an organization is to achieve business success, without aligning the text with any contractual situation.

You will see from Figure 8.1 that the elements of the quality system discussed in ISO 9004–1 correspond well with those in ISO 9001 (except that the latter omits economic and safety considerations). They also provide a sound basis on which to construct a TQM programme (see Chapter 20), being:

- Management responsibility
- Quality system elements
- Financial considerations of quality system elements
- Quality in marketing
- Quality in specification and design
- Quality in purchasing
- Quality in processes
- Control of processes
- Product verification
- Control of measuring and test equipment
- Control of non-conforming product
- Corrective action
- Post-production activities
- Quality records
- Personnel
- Product safety
- Use of statistical methods

Contractual use of the standards ISO 9001–9003

As I have stressed, ISO 9004 in its various parts is an advisory or 'guidance' document. Its wording demonstrates this, frequently for example using the word 'should', whereas ISO 9001–9003 use the word 'shall'. ISO 9004 offers advice, ISO 9001–9003 issue instructions.

The reason for this difference is that whereas, as we have pointed out, ISO 9004 offers guidance and explanation as to the desirable scope of a comprehensive quality system, the documents ISO 9001–9003 have to be specific since they are designed to be:

1. Used by independent assessors, to determine whether a company's quality system is adequate and can be recommended for registration.

2. Invoked by customers, in their contracts or orders, to ensure that goods or services are created under the disciplines imposed by the quality system in accordance with ISO 9001–9003, whichever is appropriate.

National implementations of ISO 9000–9004

Numerous national standards bodies have not only been involved in the drafting of the ISO 9000 of ISO 9000–9004 documents, but have adopted the finally agreed text as their corresponding national standard. The United Kingdom is a case in point, where upon publication of ISO 9000–9004 in 1987, the British Standard BS 5750: 1979 was withdrawn and replaced by BS 5750: 1987 with identical text to ISO 9000–9004.

In the case of Europe CEN, whose members are BSI and other European standards bodies, it has adopted the ISO 9000 series of documents and assigned them 'EN' (European Standard) numbers such as EN 29001 for ISO 9001, etc. The EN designation ensures that they must have precedence over any national document covering the same topic, in a member country. BS 5750 remained in force with text aligned with the new ISO standards, e.g. BS 5750:part 1:1987 was technically identical with ISO 9001:1987 and EN 29001:1987, and the standard bore all three numbers on its cover. Now that the 1994 revision of ISO 9001 has appeared, the British version has been identified as BS EN ISO 9001:1994, making the relationship clearer.

Similar situations exist in other countries, where the ISO 9000 series standards are adopted nationally, translated into the local language and issued under a national number. For example ISO 9001 is DIN ISO 9001 in Germany, and ANSI/ASQC Q91 in the USA.

Administration of national registration schemes

We have referred to the fact that ISO documents such as ISO 9000 are prepared by a working committee on which interested countries are represented, drafts circulated for comment, and voted on before being issued when a consensus has been achieved. Any future amendments will be implemented in similar fashion, i.e. by a written proposal to the appropriate committee, followed by review and approval of an agreed text.

To fully exploit the ISO 9000 standards an individual nation may maintain a register of approved firms, and maintain a national certification scheme wherein skilled independent assessors can investigate an applicant company and decide whether to recommend registration.

Many industrialized nations will maintain at least one accredited Certification Body, entitled to award certification under the scheme to applicant companies. In Britain there are now many organizations recognized by the NACCB (National Accreditation Council for Certification Bodies).

Less developed countries may elect not to maintain their own national inspectorate but permit other countries' approved inspectorates, e.g. those of countries they export to, to assess their local manufacturers.

Reciprocity

I stress again that the care taken to establish similar wording of the standards, and equivalent capability of the inspection organizations is intended to facilitate reciprocal recognition of approvals granted by the different members. Thus a British firm's registration under BS 5750 part 1 would be equivalent to a competing German firm's registration under DIN ISO 9001, or an French firm's under NF X 50–131. Each in turn complies with the European Standard EN 29001; and, equally with an American firm holding ANSI/ASQC Q91, or an Australian one with AS 3901, as fulfilling the requirements of ISO 9001.

Elements of a quality system: ISO 9001

Quality system elements

Using Feigenbaum's definition of a total quality system, quality system elements can be interpreted as each company department or work function, cooperating and communicating in order to achieve quality.

If you consult ISO 9001, or ISO 9004, or many of the other quality system standards you will find a diverse list of formally identified quality system 'elements'. A summary, based on one provided in the Draft International Standard for ISO 9000–1: 1994, is presented in Figure 8.1.

The list of quality system elements listed in Figure 8.1 is diverse; some can be related to the activities of specific groups or departments, some relate to specific techniques, others are universal good practice.

For example, we have 'Quality in procurement/purchasing' covering the applications to the work of a defined department; we have 'Personnel/training', applicable to all departments; and we have 'Use of statistical methods', which in particular contexts may be essential, useful, or of no conceivable relevance.

It is very important, not only to be familiar with the elements of the quality system as listed in Figure 8.1, but also to feel that you understand the purpose behind each of them, and how this purpose can be interpreted in terms of the kind of business in which you are involved.

Naturally enough, the requirements of ISO 9001 are the most comprehensive of ISO 9001–9003, since firms approved to ISO 9001, the ones who design the products or services they offer, contain the most complete chain of supplier–customer relationships. Because of the importance of the quality system, I shall outline the elements of an effective quality system once more; but this time in more detail, and in the terms given in ISO 9001.

Some commentators have seen the 20 system elements of ISO 9001 as falling into three groups, dealing with system management, with system methods, and with system maintenance. You may find this more analytical approach makes it easier for you to remember the elements if they are grouped as follows:

1. **System management**
   - 4.1 Management responsibility
     Quality system
   - 4.14 Corrective and preventive action.

Figure 8.1 ***Cross-reference list of clause numbers for corresponding topics.***

ISO 9001	ISO 9002	ISO 9003	Clause number and title in ISO 9001:1994		QM Guidance ISO 9004–1	Road map ISO 9000–1
●	●	○	4.1	Management responsibility	4	4.1, 4.2, 4.3
●	●	○	4.2	Quality system	5	4.4, 4.5, 4.8
●	●	●	4.3	Contract review	–	8
●	–	–	4.4	Design control	8	–
●	●	●	4.5	Document and data control	5.3, 11.5	–
●	●	–	4.6	Purchasing	9	–
●	●	●	4.7	Customer supplied product	–	–
●	●	○	4.8	Product identification and traceability	11.2	5
●	●	–	4.9	Process control	10, 11	4.6, 4.7
●	●	○	4.10	Inspection and testing	12	–
●	●	●	4.11	Control of inspection, measuring and test equipment	13	–
●	●	●	4.12	Inspection and test status	11.7	–
●	●	○	4.13	Control of non-conforming product	14	–
●	●	○	4.14	Corrective and preventive action	15	–
●	●	●	4.15	Handling, storage, packaging preservation and delivery	10.4, 16.1, 16.2	–
●	●	○	4.16	Control of quality records	5.3, 17.2, 17.3	–
●	●	○	4.17	Internal quality audits	5.4	4.9
●	●	○	4.18	Training	18.1	5.4
●	●	–	4.19	Servicing	16.4	–
●	●	○	4.20	Statistical techniques	20	–
–	–	–	–	Quality economics	6	–
–	–	–	–	Product safety	19	–
–	–	–	–	Marketing	7	–

KEY: ● Comprehensive requirement
○ Less comprehensive requirement than ISO 9001 and ISO 9002
– Element not present

Fig. 8.1 Cross-reference list of clause numbers for corresponding topics.

4.17 Internal quality audit
4.18 Training

2. System methods

4.3 Contract review
4.4 Design control
4.5 Document and data control
4.6 Purchasing
4.7 Customer supplied product
4.8 Product identification and traceability
4.9 Process control
4.10 Inspection and testing
4.13 Control of non-conforming product
4.15 Handling, storage, packing, preservation and delivery
4.19 Servicing
4.20 Statistical techniques

3. System maintenance

4.11 Inspection, measuring and test equipment
4.12 Inspection and test status
4.16 Control of quality records.

The elements one by one

The stated scope of ISO 9001 is:

> Quality system requirements for use where a contract between two parties requires the demonstration of a supplier's capability to design and supply a product.

The requirements referred to are listed under 20 headings as 'elements' of the quality system. These will be considered in turn, and phrases of particular significance will be emphasized.

The clause reference for each element in ISO 9001 is quoted below.

4.1 Management responsibility

This first element reflects the need for senior management commitment, involvement and acceptance of responsibility for the achievement and maintenance of quality. Subdivisions are:

1. A **quality policy** issued and understood and observed at all levels of the organization;
2. An **organizational structure** defining:
   (a) the responsibility and authority given to personnel for the maintenance of quality;
   (b) the personnel and resources used to manage, perform work and verify quality;
   (c) the identity of the *management representative* who is given the authority and responsibility for ensuring that the requirements of ISO 9001 are met.
3. Regular executive meetings to review the quality system.

4.2 Quality system

This section demands that a **documented** quality system is established and the procedures and instructions contained or referenced in it must be implemented and maintained including a quality manual and quality plans.

4.3 Contract review

If work is undertaken under a contract between supplier and customer, then a **contract review** must be undertaken before and during the progress of the contract, to ensure that quality requirements are understood and can be met.

4.4 Design control

Procedures must be established and maintained to control and verify that the design of the product meets the specified requirements. This process is divided into the following phases:

1. **Design and development planning** to ensure that the responsibility for each design and development activity is clear;
   (a) **activity assignment** to appropriately qualified and resourced personnel;
   (b) identification of, and communication across, **organizational and technical interfaces**;
2. **Design input** requirements to be identified and documented, reviewed and any ambiguity resolved;
3. **Design reviews** to be performed at appropriate stages.
4. A **design output** to be documented and quantified, to;
   (a) meet all the design input requirements;
   (b) define acceptance criteria;
   (c) conform to regulatory requirements (e.g. consumer safety) whether explicit in the input information or not;
5. A **design verification** must be undertaken to establish that the design output is achieved, by such means as:
   (a) conducting design reviews;
   (b) tests and demonstrations;
   (c) carrying out alternative calculations;
   (d) comparison with similar, proven designs;
6. **A design validation** to ensure that the resulting product meets the defined user needs;
7. **Design changes** must be recognized, recorded, reviewed and approved.

4.5 Document and data control

All documents relating to the requirements of ISO 9001 must be reviewed and approved for adequacy by approved personnel before use.

Changes or modifications to documents must go through the **same review** and approval procedures as the originals.

4.6 Purchasing

The supplier must ensure that purchased products conform to specification. This includes:

1. Assurance that **subcontractors** fulfil the quality requirements of the subcontracted operations, and are chosen on the basis of suitability and if possible of demonstrated capability and performance;
2. The supplier must ensure that the **subcontractor's quality system** controls are effective;
3. **Purchasing documents** must identify all the quality requirements of the product being purchased;
4. Notwithstanding that the **supplier is responsible** for providing acceptable product, the purchaser has the **right to verify** this at source or on receipt and to reject product then or subsequently.

4.7 Control of customer-supplied product

The supplier must verify, store, maintain and account for materials supplied by the purchaser for incorporation into the supplier's product.

4.8 Product identification and traceability

Product must be identifiable at all stages of production, delivery and installation. Where specified, individual batches or units must be traceable.

4.9 Process control

The general requirements for process control lay down that production and installation processes affecting quality must be done under controlled conditions. These controls are taken to include process qualification, documented work and inspection instructions, adequate environment and appropriate monitoring of workmanship.

Special processes, those which cannot be verified retrospectively, are subject to special safeguards.

4.10 Inspection and testing

1. Receiving inspection and test:
   Incoming material is **not to be used** or processed unless **previously inspected or certified**; or else in emergency, if released without testing, to remain fully traceable.

2. In-process inspection and test:
   (a) Product must be inspected, tested and identified **according to the quality plan** or documented procedures;
   (b) Processes must be monitored and controlled;
   (c) Product must be held until inspection is completed unless release under positive recall procedures;
   (d) Non-conforming product must be identified;
2. Final inspection and testing:
   (a) All preceding tests and inspections must have been satisfactory;
   (b) No product must be dispatched until final inspection and test according to the quality plan or documented procedures are successfully completed;
3. Inspection and test records – these must be **retained**.

4.11 Control of inspection, measuring and test equipment

The supplier is responsible for **maintenance, calibration and control** of all such equipment used; whether owned, hired, or customer-supplied. Equipment measurement uncertainty must be known and taken into account.

All aspects of control, recording and verifying calibration are recorded in this section.

Test hardware and software are required to be checked as suitable for use, records kept, and information made available to the customer when requested.

4.12 Inspection and test status

Status of product must be identified by appropriate means, to show conformance/non-conformance. Records must be kept showing who was responsible for releasing conforming product.

4.13 Control of non-conforming product

There must be procedures which prevent inadvertent use or installation of non-conforming product.

4.14 Corrective and preventive action

The supplier must:

1. Have procedures for **investigating causes** and potential causes of non-conformance, and actions needed to prevent recurrence;
2. Implement **corrective actions** and **verify** that corrective actions were taken and were effective;

3. **Update procedures** to include the amended practices.
4. **Investigate the cause**, and take steps to prevent recurrence.

4.15 Handling, storage, packing, preservation and delivery

Procedures must be established and maintained which prevent damage or deterioration during handling, storage, packing or delivery. In respect of packing, marking and identification must additionally maintain the identity and segregation of products.

4.16 Control of quality records

Records must be retained and accessible, including records on subcontractors. They must be capable of demonstrating that the quality system functioned, and that the requisite product quality was achieved. They may have to be made available to the customer, if contractually agreed.

4.17 Internal quality audits

The supplier must comprehensively audit his activities to verify that his quality system is being implemented effectively. Audit procedures and results must be documented, and corrective actions taken on deficiencies.

4.18 Training

Training needs must be identified and training provided. Critical tasks must be undertaken by qualified personnel, and records of training kept.

4.19 Servicing

Where servicing is part of a contract, procedures must be established for performance and verification that servicing is properly carried out.

4.20 Statistical techniques

Statistical techniques must be employed where appropriate for verifying process capability and acceptable product characteristics, and the methods used documented.

Guidance on the use of ISO 9000 series documents

Organizations' quality systems are evaluated and registered against the appropriate model described in ISO 9001, 9002 or 9003 according to the scope of registration sought. The purpose of ISO 9001–9003 is described in ISO 9000, as we have noted. Furthermore, ISO 9004 explains the need for a quality system.

It has been felt that these guidelines alone need supplementing in order to clarify the application of ISO 9001–9003 to specific industrial contexts. Consequently certain supporting documents have been generated. For example:

ISO 9000–3 (BS 5750 part 13): *Guide to the Application of ISO 9001/BS 5750 part 1 to the Development, Supply and Maintenance of software.*

ISO 9004–2 (BS 5750 part 8): *Guide to Quality Management and Quality Systems Elements for Services.*

Table 8.1 Standards within the ISO 9000 series

Standard	Title
ISO 9000	Quality Management and Quality Assurance Standards –
ISO 9000–1:1994	Guidelines for selection and use
ISO 9000–2:1993	Generic guidelines for application of ISO 9001, ISO 9002 and ISO 9003
ISO 9000–3:1991	Guidelines for application of ISO 9001 to the development, supply and maintenance of software
ISO 9000–4:1993	Application for Dependability Management
ISO 9001:1994	Quality systems – Model for quality assurance in design/development, production, installation and servicing
ISO 9002:1994	Quality systems – Model for quality assurance in production, installation and servicing
ISO 9003:1994	Quality systems – Model for quality assurance in final inspection and test
ISO 9004	Quality management and quality system elements –
ISO 9004–1:1994	Guidelines
ISO 9004–2:1991	Guidelines for services
ISO 9004–3:1993	Guidelines for processed materials
ISO 9004–4:1993	Guidelines for quality improvement
ISO 9004–5:*	Guidelines for quality assurance plans
ISO 9004–6:*	Guidelines to quality assurance for project management
ISO 9004–7:*	Guidelines for configuration management
ISO 9004–8:*	Guidelines on quality principles and their application to management practice

*ISO standards for which no date is given are still in preparation.

Other documents are being added. These are listed in Table 8.1. Furthermore, in addition to the ISO 9000 numbered series of standards, a number of other standards are being published which bear on the management of quality systems. Some of these are specifically referenced in ISO 9000 series documents, and with them are referred to as the ISO 9000 'family' of standards. The remaining members of the ISO 9000 family so far issued or in preparation are listed in Table 8.2.

Summary of conclusions

The International Standards Organization's ISO 9000 series of standards has become accepted world-wide, over many industries and by some of the world's

Table 8.2: Other ISO standards within the ISO 9000 'family'

ISO 8402: 1986 Quality vocabulary	
ISO 10011 Guidelines for auditing quality systems –	
ISO 10011–1: 1990	Auditing
ISO 10011–2: 1991	Qualification criteria for quality system auditors
ISO 10011–3: 1991	Management of audit programmes
ISO 10012 Quality assurance requirements for measuring equipment	
ISO 10012–1: 1992	Metrology qualification systems for measuring equipment
ISO 10012–2:*	Measurement assurance
ISO 10013:*	Guidelines for developing quality manuals
ISO 10014:*	Economic effects of Total Quality Management
ISO 10015:*	Guidelines on continuing education and training

*ISO standards for which no date is given are in preparation, and ISO 8402 is undergoing revision (at the end of 1993).

most influential purchasing agencies, as a practical and legitimate model for a good quality system.

It is applicable to organizations offering either products, or services, though its wording tends to retain a product-oriented bias.

Its essence is a list of approximately 20 'elements' or features of a successful quality system. A company seeking certification has to document its system in such a way that it explains either **how** each element in turn is implemented; or else **why** it is considered inapplicable, and hence omitted.

Different documents in the ISO 9000 standard series deal with different business scopes (ISO 9001–9003). Guidance documents (ISO 9000–1 etc., 9004–1 etc.) elaborate on different aspects of the system model, and applications to different industries.

9

Quality auditing techniques

Introduction

The purpose of this chapter is to introduce the most fundamental of all the quality assurance techniques, namely quality assurance auditing. The activities to which quality assurance audit can be addressed are varied. A number of these will be mentioned, but especially audit of the quality management system. Because of its importance and topicality, systems audit will be examined in some detail.

Quality auditing

The expression 'quality auditing' is sometimes used synonymously with 'quality management system auditing' which we shall examine in some depth. However, it is possible to look at many other quality assurance activities as being audits. We shall therefore review some of them in that light before focusing on 'systems audit'.

The concept of auditing

To all non-quality specialists, 'auditing' means first and foremost financial auditing. At first sight financial audit and quality audit have little in common, but in principle rather than in application there are strong similarities.

The following analogies can be recognized.

1.

(a) A financial audit is performed by independent professionals (the auditors) in the interests of the company and on behalf of its members (the shareholders), to safeguard their investment. The auditors are not experts in the company's business, but they are experts in accounting matters and the techniques of auditing.

(b) A quality audit is performed by independent professionals (the quality auditors) in the interests of the company and on behalf of the customers, in order to safeguard the quality of their purchases. Quality auditors do not have to be experts in the processes they are observing, but they do need a thorough understanding of the objectives and techniques of quality assurance.

2\.

(a) A financial audit cannot hope to witness or examine all the company's financial records but it inspects some records, chosen at random, and looks for errors and inconsistencies.

(b) A quality audit cannot examine all the products or activities, but chooses random samples and examines them for defects or discrepancies.

3\.

(a) The financial auditors are not responsible for correcting errors or malpractices, merely for reporting them so that others can take the necessary action. They are not censured for exposing inefficiency or malpractice, but they would be criticized if it came to light that they had failed to discover and expose them.

(b) A quality auditor is similarly responsible for reporting a situation so that others can correct it. The 'Quality Auditors', or QA department, do not have to provide the resources for correcting quality problems, but must be capable of discovering them.

What constitutes a quality audit?

On the strength of the analogy given above, most activities performed by the QA department can be considered as audits, since:

1. They are performed by QA because of their independence and special skills;
2. They are based on sampling, or on checking records;
3. They are intended to report situations on which others must act.

The areas readily acknowledged as audits are:

1. Quality management review;
2. Quality system audit;
3. Process audit;
4. Product audit;
5. Service audit.

These terms will be clarified in the following sections.

Management review of the quality system

This is not exactly an auditing exercise, though it is sometimes treated as part of quality system audit, and it a requisite of a quality system as defined in ISO 9000 (see ISO 9001 or ISO 9002, section 4.1.3, or ISO 9004–1 section 5.5).

Whereas a quality system audit examines whether the system is being operated correctly, the purpose of a management review is to examine whether the system, as currently designed, is appropriate and effective.

You could say that whereas the quality system audit examines whether the quality manual is being adhered to, the management review has to decide whether the quality manual, as written, adequately covers the company's needs.

All the relevant questions must be addressed regularly; annually as a minimum, jointly by the company's senior management. Any changes in policy, objectives and quality system found necessary must be implemented. For example:

Quality policy

Is the stated company policy on quality still in line with other company policies and customer expectations?

Quality objectives

Are new objectives or enhanced performance standards needed? Have there been failures or serious difficulties in achieving any of the stated objectives? Do objectives need to be changed in the light of changes in consumer demand, new competitors or new legal requirements?

Quality system

Does examination of past quality system audit findings expose any areas of chronic difficulty or poor performance in applying the system? Are any changes in the system desirable because of the development of new technology, or new QA techniques, etc.?

Product, process and service quality audits

These will be considered in relation to each other in this section of the chapter, with the following section devoted to quality system audit.

Before considering these audit activities, it is important to make clear the distinction between the audits and the corresponding inspection procedures. Just by watching what is being done, it could be difficult to distinguish between a product **inspection**, e.g. a sample gate inspection such as we shall discuss in Chapter 10, and a product **audit**.

The difference lies not so much in the activity, as the purpose behind it. When we come to talk about product and process audits shortly, we shall give the BS 4778 definitions of these activities. You will see that they are both 'The independent examination of (product/process) quality to provide information'. The two key words are 'independent' and 'information'.

Inspection and test activities are done on all work in progress, either 100% or according to some strict sampling rules, in order to make an immediate **decision** (this unit, or this batch can/cannot be passed on to the next customer). An audit is done by an independent person who is not the regular inspector/tester; not to sentence product, but to obtain information on which others must act if it is necessary.

Thus audit is not an appraisal activity, aimed at safeguarding the product being immediately examined; it is a preventive action which will result in action being taken to prevent the recurrence of any error discovered.

Product audit

BS 4778 defines product quality audit as 'The independent examination of product quality to provide information'. This may refer to finished products, or to materials and work-in-progress. Thus statistical process control, dealt with in Chapter 12 of Part Two, is a form of product audit, although the resulting actions will be taken on the process.

It should be noted that in some industries in-process inspection of incompleted product would be classed as process audit, whereas I have defined it as a product audit. I reserve 'process' audit for cases which do **not** involve examination of the product, such as in the illustrations given below.

Process audit

BS 4778 defines process audit as 'The independent examination of process quality to provide information'.

Hence process audit does not involve checking the product, but rather the processes used to create the product; also the environment in which the processes take place, if these can effect the efficiency of the process. Examples might be:

1. Monitoring workplace temperature, humidity and dust-count;
2. Monitoring calibration status of measuring and inspection equipment;
3. Monitoring machine speed settings, oven temperatures, gas flow rates, etc.;
4. Monitoring packing, labelling and storage arrangements.

The performance standard is compliance with the requirements (methods, setting, limits) stated in the process instructions. The form of the audit could be:

1. Direct observation or measurement;
2. An audit of the records kept by the process personnel;
3. A combination of 1 and 2 where examination of process control logs is supplemented by less frequent independent measurement.

Service audit

BS 4778 gives no definition for service audit, but it will be evident that if a company is providing a service as opposed to a product, the quality still needs to be safeguarded in the same way. Any audit of the quality of service supplied would hence be a service audit rather than a product audit.

Auditing the quality system

The quality system is described in Part Two, Chapter 7. Management review periodically examines the question of whether the quality system, as designed, can fulfil its intended purpose. Given that this is the case, quality system audit answers the question: 'It can, but does it?'

The need for system audit

The quality system cannot achieve its objectives unless it is properly implemented. Explanation of the techniques used to evaluate this need some elaboration, and this will be done in the following sections.

First we will pose a question to highlight the importance of quality system audit. Bear in mind that the QA department frequently has the task of evaluating the quality systems of current or potential suppliers, and recommending whether they should be preferred vendors.

The importance of system audit

The crucial test of a quality system is this. If, for whatever reason, a quality problem does arise, will an organization be able to quickly:

1. Recognize that a problem exists?
2. Minimize its impact on customers?
3. Accurately determine the cause?
4. Formulate and implement a solution?
5. Maintain the resulting improvement?

A company without an effective quality system will find it very difficult to achieve the goals stated in the preceding paragraph.

Quality system audit affects most companies in three different contexts.

1. They must perform self-audit activities to ensure that their own quality system is working effectively.
2. Their quality system is liable to be assessed independently by outside organizations – customers, ISO 9000 assessors, etc.
3. They will equally want to appraise the quality systems of their suppliers, especially suppliers of components or services which are critical to the safety and reliability of the completed product or service.

Arising from the second of those three points, there is another important consideration. Quality system audit is a technique which is used at all three stages in the chain supplier – manufacturer – customer. Unlike most aspects of industrial know-how, the knowledge of quality system auditing techniques is available on the same terms to yourselves, your suppliers and your customers. In your dealings with them (auditing them, or being audited by them, whichever is appropriate) they can make their own judgement of your competence in the skill of quality auditing. This judgement, be it favourable or unfavourable, will inevitably colour their view of your overall competence.

Categories of quality system audit

We have just mentioned two situations where quality system audit can be applied: audit of an actual or prospective supplier, (seen both from the supplier's and customer's angle), and internal audit of your own organization. We have also alluded to a third case, where a company seeking registration seeks recognition of its quality system from an independent assessment service.

These situations are categorized as:

1. Internal, or first-party assessment, where the organization is auditing its own capability;
2. External, or second-party assessment, where one organization is auditing another (usually a customer assessing a potential vendor);
3. Extrinsic, regulatory or third-party assessment, where an independent organization is auditing one organization on behalf of another, or on behalf of a national/international accreditation scheme.

Depth of audit

Two levels of audit can be distinguished:

1. **Shallow audit**: This is intended to ascertain whether a documented quality system exists, and whether it is adequate as designed. The judgement is primarily made through inspecting the quality manual and associated documentation.

   It is often done as a pre-assessment, a form of initial screening to decide whether to progress to a deep audit.
2. **Deep audit** or **compliance audit:** This looks for evidence of whether the procedures laid down are actually being followed. Thus it will involve examining records, observing operations, and interviewing personnel.

Conducting a formal quality system audit

In this section we are going to examine the actual conduct of a quality system audit, as it is undertaken in a formal fashion. The advice in the following section applies especially to second- and third-party audits; internal audits can be treated with more flexibility if the organization has a mature attitude to quality and will not abuse a more informal approach.

Why be formal?

If an audit is not treated formally it may well not be performed effectively, and is likely to reflect discreditably on the image of the auditing organization. An audited organization will have had experience, if not of doing its own internal audits, then of being audited by other customers and outside bodies.

Stages of a formal quality system audit

The actual face-to-face auditing activities are only the central phase of a series of preparatory and continuation activities:

Planning

It is necessary to decide the scope and objectives of the audit, and what standard it is to be conducted against, e.g. ISO 9002, the auditee's own quality manual or whatever is appropriate. It is also necessary to establish a schedule to determine what locations, or what elements of the quality system are to be audited, how often and when.

Preparation

For conducting a specific audit which has been scheduled in the plan, preparation includes:

1. Choosing the auditor, or audit team and leader;
2. Notifying to the auditee the place and time of audit, its scope, and the facilities the auditor(s) will need;
3. Identification, acquisition and study of the documentation relevant to the audit (e.g. quality manual, operating procedures and work instructions);
4. Preparation of the audit check list. This is an *aide-mémoire* prepared by each auditor, listing questions or points to be checked, based on the stipulations in the procedural instructions.
   The auditor's task will be to establish, through objective evidence, that the procedures are or are not being followed;
5. Setting out the detailed timetable for the audit.

Performance

Performing the audit comprises:

1. An '**entry meeting**' with the responsible management, at which introductions are made, the purpose of the audit explained, the auditee's acceptance of the timetable and provision of requested facilities verified;

2. Audit of compliance with the relevant procedures, done so far as possible directly at the places and with the people to which the procedures apply. Collection of objective evidence through questioning, observation and inspection of records;
3. Time to collate and compare team findings, prepare exit meeting briefing and corrective action requests;
4. An '**exit meeting**' with management, at which a verbal presentation of the most significant findings is given, CARs are presented, time scales for follow-up actions agreed and any misunderstandings resolved.

Reporting

This entails issue of a written report, the contents of which were indicated verbally at the exit meeting. The report will incorporate reference to the CARs which were raised, and have been handed over at the exit meeting.

Follow-up

The follow-up activities may take the form of a re-audit, the objective being to verify that the actions required in the CARs have been undertaken and were effective.

There is a longer term aspect to follow-up also. It is of interest to both auditor and auditee; namely, whether administration of the quality system is improving. For example, in its positive aspects:

1. Is the need to issue CARs becoming less frequent?
2. Are CARs being closed more speedily than used to be the case?

And in its negative aspects:

1. Are any old CARs still unresolved?
2. Have problems been solved and then resurfaced later?
3. Are numbers of CARs and the time needed to resolve them increasing?

The auditor

The auditor, in particular the external auditor or extrinsic assessor is often cast in the role of bogeyman.

The auditor as bogeyman

If a quality engineer or quality manager feels his/her position to be weak or unsupported by senior management, he/she will use the threat of an external/extrinsic auditor as reinforcement. The auditor becomes seen as a person with purely negative powers and attitudes. Yet the warning 'You must correct that or we shall lose our approval' is not necessarily a more effective reproof than 'You ought to stop doing that because it is impairing quality, reducing customer satisfaction, and costing you money!'

The auditor as 'guide, philosopher and friend'

This expression is taken from Stebbing's book *Quality Assurance*. The point is that an external auditor has seen inside a variety of companies, good and bad. Any hints, observations or suggestions you receive are free management consultancy – a very scarce commodity at that price.

Some words of warning must be given to the auditor, however.

1. The auditor must be even-handed in regard to auditees. All must be confident that they are being treated on the same footing;
2. Never breach commercial confidence in respect of what you have seen or been told;
3. Never give an auditee explicit advice on how to solve a problem. That is not part of the auditor's job. The auditee has to accept ownership of the problem, devise a solution that suits the circumstances, and make it work.

Surprise audits, or appointments?

Some auditors believe it is their job to catch people out, and this is known as the 'gotcha' attitude. Differences of attitude surface when the alternatives of surprise and announced audits are compared. The 'gotcha' school back surprise audits, on the grounds that people don't have time to put things right.

The opposite view is that the whole purpose of an audit is to get people to put things right, once they know they are presently wrong. On that basis a pre-announced audit saves issuing corrective action requests, administering them and following them up, if the threat of a forthcoming audit is a spur to the auditees addressing problems that they were 'going to get around to'.

If a quality system is in deep disarray, prior notification to an auditee is not going to hide the fact. Most experienced auditors recommend the pre-announced audit since too often the surprise audit is wasted because of absence of key people, important processes not running, etc. which could be avoided by prior consultation.

Qualities needed of the auditor

Ideally, an auditor needs to be:

1. Professional and ethical; to do the job confidently, confidentially, honestly and efficiently;
2. Unbiased and impartial; in adjudicating between different parties, and also setting aside his/her own preferred solutions;

3. Articulate and communicative; to impart information persuasively, and encourage people to talk;
4. Enquiring and observant; in order to find all the relevant facts;
5. Diplomatic and understanding; pointing the finger at the non-conformance, not at the person responsible;
6. Thick-skinned and with a sense of humour; because auditors are not always popular!

Auditor experience

An auditor **does not** have to:

1. Have expert knowledge of the operation being audited;
2. Be a member of the QA department.

An auditor **does** need:

1. A general appreciation of the industry concerned and its methods;
2. An understanding of the principles of quality assurance;
3. Some training in auditing skills;
4. Auditing experience, initially gained as an observer or assistant;
5. Some engineering experience;
6. Some supervisory experience.

Officially recognized auditors

It is clearly important that someone accorded the responsibility of recommending companies for registration or deregistration to, for example, ISO 9001 should be of acknowledged competence.

In the UK such a scheme already operates on a national basis, under the International Registration Scheme for assessors of Quality Systems.

The scheme recognizes three levels of competence: lead assessors competent to lead and manage an audit team, assessors competent to act independently or as team members, and provisional assessors who have the necessary qualifications but do not yet have the required body of auditing experience to progress to the other grades.

To become registered, an auditor has to fulfil certain criteria of education, professional experience and training which includes an approved cause of auditor training with examination.

Teams conducting assessments on behalf of NACCB-registered certification bodies must be led by a lead assessor and all team members must be assessors or lead assessors.

Formal guidance on auditing practice

There are only a very limited number of books on quality system auditing practice though, at least in the UK, there are very many training courses offered, some of them recognized as meeting the training requirements for Registration.

There is, however, and ISO standard; ISO 10011: *Guide to Quality System Auditing*, of which the British implementation is BS 7229. It is issued in three parts:

1. ISO 10011–1: 1990 (BS 7229: part 1: 1991) – *Auditing.*
2. ISO 10011–2: 1991 (BS 7229: part 2: 1991) – *Qualification criteria for auditors.*
3. ISO 10011–3: 1991 (BS 7229: part 3: 1991) – *Managing an Audit Programme.*

Summary of conclusions

The chapter has introduced the concept of the quality audit, distinguishing it from inspection. Auditing is a method of endorsing or instigating a change in practices, rather than a method of accepting/rejecting products.

Auditing can be applied to the quality system, to products, processes or service provision. Its objective is to find facts, not to find fault; to establish a cure not to apportion blame.

The previous chapter explained how the company's quality procedures should demonstrate that 'what the company says it's doing is what it should be doing'. The live audit tests if 'it IS doing what it says it's doing'.

An audit cannot be conducted successfully without thorough preparation and follow-up. The abilities of the auditor are crucial to a satisfactory audit.

10

Inspection techniques

Introduction

Until not many years ago, the control and management of quality was regarded in a very narrow fashion, and treated as being essentially identical with inspection. Indeed, the title of the person we now call 'Quality Manager' and the scope of his/her duties was summed up in the title 'Chief Inspector'. The main technique used in controlling quality would have been seen as the use of mathematically derived statistical sampling inspection.

Inspection

Inspection is the ancestor of quality assurance. By now, you should have accepted that quality assurance is about preventing the recurrence of known errors, and anticipating and averting possible ones, but old habits die hard. There are still companies where 'quality control' means simply inspecting products to pull out defective items, and they imagine that 'corrective action' means mending the defects.

Nowadays the contribution of inspection to the maintenance of quality is emphasized less strongly, being overshadowed by prevention, but inspection still has a key contribution to make. It is important to accept what inspection can and cannot achieve, and the costs and penalties associated with it.

The role of inspection in quality assurance

We must not undervalue the need for inspection. Inspection does protect the customer; it is also a necessary precursor to corrective actions. You can take sensible measures to anticipate possible problems but you can't actually correct a problem, much less prevent its recurrence, until you know that you have got it – and what it is!

The important thing is not to rely on inspection to solve your quality problems. That is just what it cannot do. What it can do, however, is to:

1. Limit the effects of a problem;
2. Protect the customer from a problem;
3. Supply information which can be used to find a cure for the problem;
4. Save money, through each of the above actions.

Specifications and inspection procedures

By inspection, we mean any kind of test or examination which can be used to distinguish between conforming and non-conforming product. Inspection can be done manually or automatically, mechanically, electrically or using the human senses. A wine-tasting can be an inspection. If you can make a judgement against a standard, then in the context of this book, you are doing an inspection.

There has to be a defined standard which is practical and capable of being worked to. The standard must give unambiguous guidance to the inspector. Take as an example the requirement 'No scratches'. If this is to be taken at its face value, fair enough. If it really means 'no big scratches and only a few tiny ones' then the inspector must be trained to recognize where the borderline is intended to be drawn, and have access to a reference sample which is just on the edge of acceptability.

It is also essential to give an inspector adequate tools. All instruments must be calibrated at appropriate intervals, and the calibration status must be evident to the user. There must also, of course, be an operating procedure for the inspection activity.

In summary, any inspection activity demands:

1. A specification;
2. Work instructions;
3. A trained inspector;
4. Calibrated, well-maintained instruments;
5. Means of recording and analysing findings;
6. Means of identifying rejected product;
7. Means of highlighting any need for corrective action.

Attributes and variables

A test or inspection can be performed by making a judgement based either on attributes, or on variables information.

Attributes inspection

This takes account only of a quality or attribute, e.g. is this coloured red? Is it scratch-free? Does it meet the specification? The answer is either 'it is acceptable' or 'it is not acceptable', 'yes' or 'no'. It either possesses the required attribute or it does not.

Variables inspection

This requires a measurement to be taken and compared with a specification of what values of the measurement are acceptable, e.g. 'this plank should be 1.2 m long ± 1 mm, but it is only 1.1 m'. This is a parametric or variables statement. The actual measurement is compared with the desired value and the permitted tolerance.

Alternative ways of expressing an inspection

It will be clear that any variables test can also be expressed as an attributes test. The plank of wood was 1.1 m long but it also had the attribute 'failure' since the correct length should have been 1.2 m plus or minus a millimetre.

People are sometimes less quick to accept that, in principle, all attributes tests could also have been expressed as variables. For example, the scratches in one of our other examples could have been counted, and items categorized as having no scratches, one scratch, two scratches, etc.

Choice of variables or attributes inspection

Since any inspection can be done either as a variables or as an attributes inspection, the manager or engineer responsible for defining the activity has to decide which shall be used.

An attributes inspection can often be done more quickly and cheaply than a variables measurement; a machined component can be measured with go/no-go gauges as an attributes test. The alternative would be to measure the dimensions with a micrometer, and compare the variable data with the measurements and tolerances given on the drawing for the unit. If a routine measurement has to be done frequently, it is usually more convenient to arrange it as an attributes test.

On the other hand, variables testing can give much more information, of kinds which are particularly useful for identifying and solving problems. As we shall see when we discuss statistical process control in Chapter 12, a record of actual measurements can enable us to:

1. Establish the 'capability' of the process;
2. Predict the onset of 'out-of-control' conditions;
3. Gain clues as to the causes;
4. Sample more economically.

So the choice has to be made. But, if it is not intended to **record** the measurements, or if there is no time or incentive to follow them up, it is better to use attributes. There is no virtue in collecting data which is not going to be used.

Grading of defects

In some situations, a formal grading of defects is used, and this is recognized when writing quality specifications and using sampling tables. Defects are classified as critical, major or minor according to definitions given in BS 4778.

Critical defect

A defect that analysis, judgement and experience indicates is likely to result in hazardous or unsafe conditions for individuals using, maintaining or depending upon the product, or that is likely to prevent performance of the function of a major end item.

Major defect

A defect other than a critical defect, that is likely to result in a failure, or to reduce materially the ability to use the item for its intended purpose.

Minor defect

A defect that is not likely to reduce materially the ability to use the item for its intended purpose, or that is a departure from the established specification having little bearing on the effective use or operation of the product.

So, a critical defect is one that represents a safety hazard or would render the product wholly unusable. A major defect is one that would prevent the product from working properly. Any others are minor defects.

Observe that any defect which can be a safety hazard to humans is classed as critical regardless of its effect on the equipment of which it is a part. This consideration will be important when we come to consider the issue of manufacturer's liability in Chapter 26, Part Four.

Inspection limits

When numerical limits are specified, manufacturers and customers both have the opportunity to test products. It is necessary to perform measurements accurately, and to know the limits of this accuracy. It is necessary to take into account both the random error inseparable from the measurement method, and the limit of the error introduced by uncertainty of calibration.

These considerations give rise to the practice of setting 'inset limits' or 'guardbands'. This is the practice of adjusting inspection limits to allow for these known uncertainties. In cases where a parameter will be inspected several times, e.g.

100% production inspection followed by QA sample inspection, possibly followed by customer incoming goods inspection it is usual to place the most severe inset limits on the first inspection and then ease out the tolerances towards the actual specification for successive later inspections. This is intended to minimize the number of items first accepted, then rejected at a later stage due to measurement variation.

Destructive testing

Certain tests are necessary in order to disclose important information, yet cannot be done without destroying the product. They must be performed, yet this can only be on a limited sample basis.

Accelerated testing

Other tests may be required to demonstrate the continuing quality of a product over a long period of use, or after a long period of storage. This is done by means of accelerated testing. Accelerated testing can only be done successfully when there is a valid physical model which:

1. Indicates the environmental conditions which have to be enhanced in order to accelerate the test realistically;
2. Enables the degree of acceleration of the test to be calculated.

Frequency of inspection

This section of the unit will deal with the various levels at which inspection can be applied, ranging from inspecting every item, to 'first-off' inspection with, in between, the various ways in which sample inspection can be organized.

'100%' inspection

When the object of the inspection is a batch of similar small units, or an assembly consisting of a number of similar elements, the most extensive form of inspection that can be done is 100%, i.e. inspecting every unit or element.

There are many situations in which this is done, either manually or automatically, and compared with statistical sampling it is the only method which does not involve a statistical likelihood of some defective items being overlooked. However it is not necessarily the most valuable nor even the most thorough method of inspection.

The above statement may seem unexpected, but consider the practical situation.

1. In particular, when examining complex items visually, the ability to detect a blemish depends very strongly on the amount of time taken;

2. Fatigue can cause loss of concentration, and hence errors;
3. The time taken for 100% inspection may mean that for the sake of speed, it has to be treated as an attributes' inspection, whereas measuring and recording variables would give more information on the state of the process.

For these reasons some form of statistical sampling can, in fact, be more effective in appropriate circumstances.

Statistical sampling

Providing sampling is done in such a way that it is statistically sound, it can be an effective safeguard of a specified quality level. The risks incurred in taking a sample rather than doing 100% inspection can be quantified.

The applicable statistics and the use of the tables will be discussed later in this chapter.

'Gate' sample inspection

The most rigorous approach to sample inspection is to divide up all production into 'inspection lots'. The size of the inspection 'lot' or 'batch' will be dictated in part by convenience, but it is a fundamental assumption that each lot is 'homogeneous', i.e. there is no reason to suspect abnormal variation within the batch.

Thus a known variation, such as starting to use a fresh batch of raw material, or shutting a machine down and then recommencing work on a different shift, would be reason for assigning subsequent work to a different inspection 'lot'. The way the inspection lot is determined therefore depends on:

1. The uniformity of the process;
2. The ability to draw off a truly random sample.

After each batch is completed a random sample is drawn for inspection, the sample size being determined by the lot size and the instructions given in the sampling plan. Each lot is then accepted or rejected *in toto*, according to the result of inspecting the sample and the requirements of the sampling plan (which indicates the maximum number of defects allowable).

This procedure is often referred to as a 'gate' inspection. Every piece of work forms part of an inspection lot and has an equal chance of being sampled. It has to wait at the 'gate' until its lot has been sampled and accepted before it is let through.

The way in which a gate sampling procedure works means that although only some items are inspected, each item has an **equal likelihood** of inspection. A short-term failure of process control should be detectable even if its duration is shorter than that required to complete an inspection lot. The range of product which is affected can be identified and isolated for further investigation.

The disadvantages of using gate inspection is that production flow is held up until a complete inspection lot has been accumulated and tested, and the result of the inspection known.

'Patrol' inspection

In patrol inspection, the samples are drawn at set time intervals, or in response to a particular situation. Examples are:

1. One sample drawn every hour;
2. A sample drawn at the beginning and end of each shift;
3. A sample drawn whenever a machine has been set up for a new job.

The advantages of patrol inspection is that the flow of finished work is not held up, and also the opportunity of quicker feedback if a problem is discovered.

A disadvantage is that a short-term problem which appears and disappears again between patrols will not be detected. Consequently, in practice patrol inspections are often performed on incomplete items during the production process, in order to pick up and react to any problem as quickly as possible. This is followed by a gate inspection at the end of the line to capture anything which has escaped the patrol, and to be able to guarantee a specific quality level (i.e. a quantified AQL or LTPD) to the customer.

'First-off' inspection

Some defects can be introduced simply as a result of errors in setting up a piece of machinery. For instance, a numerically controlled drill might have been programmed to drill a hole in the wrong place. If a fault like this is allowed to happen it will affect all subsequent work until it is corrected. This kind of error can be guarded against by simply inspecting the first piece off the machine, before the rest of the batch is allowed to be processed.

In other situations a 'last-off' inspection may be employed, either as an alternative to first-off or in addition; the latter case checking that no change in process settings has occurred during the production run.

Sampling 'flow processes'

Acceptance sampling is normally discussed in terms of the manufacture of large numbers of discrete solid objects, and my presentation has followed these lines. However, given safeguards on the way in which the samples are collected, sampling techniques can be applied to bulk solids, liquids or gases, i.e. to flow processes. Details are given in BS *5309*: *Sampling of Chemical Products* and in other standards.

Statistical process control

This topic will be dealt with in Chapter 12, but it is worth pointing out that data collected for lot acceptance purposes can often be used simultaneously for control chart plotting.

Who carries out the inspection?

The line between inspection done by production personnel and that done by QA personnel is often drawn in an arbitrary fashion.

Direct and indirect labour

Production personnel are classed as 'direct labour' and the cost of their time is allocated to the goods they are making as a direct manufacturing cost. A QA inspector doing the same job would be classed as 'indirect labour'; a support function whose cost is loaded on to the production department as an 'overhead'.

From the accounting point of view then, it is very difficult to compare apples with apples. From the technical standpoint you could argue as follows.

100% inspection

If anything has to be done 100% it is part of the production process and should be directly costed into the price of the product. One hundred per cent inspection is therefore clearly a production process, and should be carried out using direct labour. Two significant points arise from this.

1. If production rate increases, with the result that more inspection is required, the production manager can put in temporary people, authorize overtime, etc. If QA are doing the same inspection using personnel classed as 'indirect' procedures for responding to fluctuating workload are typically less flexible;
2. It is strongly in the production manager's interest to improve quality to the point where 100% inspections become unnecessary, so effecting a direct saving in his production costs.

Sample inspection

This is traditionally a QA function, but there is no absolute reason why this has to be so. This is especially true in the case of intermediate inline inspections, where there are good arguments for entrusting them to production personnel, ideally the process operators themselves. It gives them an interest in the quality of their work, and they are in the best position to apply immediate remedial action.

An acceptable compromise would be to give all the intermediate inspections to the production personnel, audited by QA, and to have personnel reporting to the Quality Manager doing the final acceptance inspections.

In recent years it has been quite common for an acceptance sample to be taken by QA immediately after a 100% inspection by production. The QA Manager must ask: 'Why am I doing this?' 'Am I taking the sample on behalf of, and as agent of the customer?' If this is **not** the reason, you should think very carefully about whether the sampling serves any purpose at all. All that is being found may be particular kinds of defect that the 100% inspectors have difficulty in recognizing, for one of a variety of possible reasons. They may not be typical of the kinds and proportion of defects the process is producing. To find and understand those, the quality assurance function need to look in the scrap bucket at the 100% inspection station.

Measurement and calibration

A theme we shall continue to stress in this book concerns objective judgement and reproducibility; whether it involves an inspection or an audit, the properties of a product or a service. In an earlier section of the chapter, on 'specifications and inspection procedures' I highlighted the need to provide the inspector with a meaningful standard for comparison.

This leads to concern for the adequacy of the standard itself, and a system for ensuring its continued accuracy. The standards that can be used for reference may be natural, material or subjective. A good example of a natural standard might be the use of ice and boiling water to calibrate a thermometer. A material standard might be a gauge block or slip gauge traceable to the national standard of length. A subjective standard might be the requirement that a soldered joint should be 'shiny', or a boiled egg 'soft'.

Subjective standards

Subjective standards by their nature can be judged differently by different people. One man's 'medium' steak could, to someone else, tend towards 'rare' or else be sent back to the kitchen as 'overdone'. So the process operator, no less than the inspector, needs to be provided with a clear reference sample or description to make the judgement as independent as possible of the particular person performing the task.

Material standards – the calibration chain

In the case of material standards, an agreed standard does exist. However, unlike the melting or freezing water for calibrating the thermometer, it may not be readily to hand, but locked away in the National Physical Laboratory (UK) or National Bureau of Standards (USA). Thus there may be a 'calibration chain' of many links from the ultimate standard, via calibration laboratories and the company's internal standard, to the working instrument whose accuracy has to be maintained.

Because of the accumulation of measurement error at each calibration undertaken in the calibration chain, these uncertainties must be taken into account in deciding if the calibration of the final instrument in the chain gives sufficient accuracy.

Accuracy and precision

We are used to hearing of measurements at the cosmological or sub-atomic level being established with great precision; but what is important, both there and in our own more mundane measurements, is accuracy. A tolerance band of uncertainty will be acceptable providing we know what the uncertainty is, and that the tolerance allowed on our measured product is much wider still.

What is the difference between accuracy and precision? Frank Price, in *Right First Time*, puts it nicely. He invites us to imagine two wartime snipers. One raises his rifle and fires five rounds from the magazine. All just miss his adversary's head, but embed in the tree trunk against which he was leaning, so close together that you could cover them all with a cigarette packet. The enemy spins round with a start and returns the fire. Two miss but one hits his opponent in the thigh, one in the chest, one in the head. Precision is pretty, says Price, but accuracy is deadly.

Environmental controls

Accurate calibration may only be possible if stringent environmental conditions are maintained, such as a specific temperature, or humidity, or freedom from vibration. A calibration laboratory has to be maintained and controlled in such a way that the environmental requirements for the particular kind of measurements it performs are met.

Natural standards

There is not a great deal of scope for the use of natural standards in the typical organization's calibration laboratory, but where it is possible to calibrate directly using a natural standard the opportunity should not be ignored. For example, it will be evident from the discussion of material standards that, compared with calibrating your mercury thermometer against a digital electronic thermometer, using the natural standards of freezing and boiling water offers advantages in time delay and expense; and with its much shorter calibration chain comes a more directly gaugable uncertainty.

Recall and traceability

Just because a measuring instrument has been calibrated once, perhaps by the manufacturer, does not guarantee that it will always remain correct. It may be subject to a slow drift due to wear or due to its properties changing through ageing; or it may suffer a sudden change due to an accident or misuse.

So all working instruments must be identified and recalled for calibration at defined intervals, which will have been set on the basis of manufacturer's recommendation, level of expected usage, and past experience and results.

Calibration systems

How can the requirements of a calibration system, such as those of traceability to reference standards, calculation of possible error in the calibration chain, environmental control, and periodic recall and recertification of working instruments be maintained? Historically, a common solution has been to assign responsibility for the calibration laboratory to the quality assurance department. But this is not essential, any more than any other activity must be done by them. What is needed is a management system for calibration, which can be overseen and audited, as an integral part of the overall quality system.

ISO 10012

ISO 10012, *Quality assurance requirements for measuring equipment* fulfils the function of defining a management system for calibration. Within the UK it addresses a need formerly met by BS 5781: 1979, a British Standard entitled *Measurement and Calibration Systems*. As used in the UK, ISO 10012 has inherited the BS 5781 number. It is worth referring to because it illustrates the extent to which the system for managing any activity is essentially the same as a 'quality' system. I shall make the same point later in the book in the contexts of reliability management, safety management and environmental management.

Like our quality system, the calibration system is broken down into various elements. BS 5781: part 1:1992/ISO 10012–1:1992 *Metrological confirmation system for measuring equipment* lists these under the headings:

1. General;
2. Measuring equipment;
3. Confirmation system;
4. Periodic audit and review of the confirmation system;
5. Planning;
6. Uncertainty of measurement;
7. Documented confirmation procedures;
8. Records;
9. Non-conforming measuring equipment;
10. Confirmation labelling;
11. Intervals of calibration;
12. Sealing for integrity;
13. Use of outside products or services;
14. Storage and handling;
15. Traceability;
16. Cumulative effects of uncertainties;
17. Environmental conditions;
18. Personnel.

Some of these headings warrant brief clarification.

1. **General** lays down that a Supplier must document the methods used to implement the provisions of ISO 10012–1.

2. **Measuring equipment** states that measuring equipment must have the metrological characteristics (accuracy, stability, range etc.) necessary for its intended use. These characteristics must be documented and maintained.
3. The **confirmation system** called for an effective documented system; for managing, confirming and using measuring equipment; to be established and maintained.
4. **Planning** calls for the adequacy of the measuring system to be reviewed relative to the needs of any new job or project.
5. **Uncertainty of measurement** from all sources must be taken into account; **cumulative effect of uncertainties** in the calibration chain must be allowed for, and action taken if it compromises the ability to maintain measurements within the limits of permissible error.
6. Recalibration must be performed at realistic **intervals of confirmation**.
7. **Sealing for integrity** must be applied in cases where tampering with instrument settings by unauthorized personnel is otherwise possible.

I believe that the content and intention of the other elements will be sufficiently obvious by comparison with the analogous elements of a Quality System, which are discussed in Chapter 8 of the book.

Accreditation of calibration laboratories

Just as there are national and international schemes for the accreditation and registration of firms' capability on the basis of their quality system, so it is possible for a laboratory's calibration system to be evaluated. There is provision for this in ISO Guide 25: *General Requirements for the Technical Competence of Testing Laboratories* and in the Euro-norm EN 45001: *General Requirements for the Operation of Testing Laboratories*.

In the United Kingdom the organization recognizing and accrediting calibration laboratories is NAMAS, the National Measurement Accreditation Service. Laboratories have to demonstrate that they operate a quality system complying with NAMAS requirements. They are assessed against the NAMAS accreditation standard, M10 and regulations, M11. These include requirements relating to management, staffing, facilities, equipment and procedures. After granting accreditation, NAMAS maintain a schedule of periodic surveillance visits. The parallels with ISO 9000/BS 5750 accreditation procedures will be obvious.

Summary of conclusions

Quality cannot be inspected into a quality or service, but to a limited extent bad quality can be inspected out. Inspection does not add value, but it can show where money is being wasted. It can provide the information on which prevention may be based.

Inspection can never be totally relied on; especially it is fallible where subjective judgements or fatigue are concerned. Because of these limitations a careful sample inspection can be more helpful than a hasty 100% inspection.

The use of inspection and monitoring equipment introduces the need for calibration and maintenance. This is true of subjective standards as well as tangible ones – for example how do you calibrate the examiners of educational tests? Calibration systems are dealt with in ISO Standard 10012.

11

Statistical sampling inspection

Introduction

We shall begin this chapter by introducing three significant statistical distributions which provide the mathematical basis for the topics of the remainder of this chapter, and also the following one. These distributions are known as the binomial, the Poisson and the normal. They are important when we wish to predict the likelihood or significance of finding specific numbers of defective items, defects, and measured values respectively.

We shall cover acceptance sampling applications in the remaining part of this chapter, and control charting in Chapter 12. We shall use the distinction between 'attributes' and 'variables' data which we introduced in Chapter 10 of Part Two of this book.

Statistical frequency distributions

There are many applications in quality control where we wish to take a small sample, yet still gain a reliable guide to the status of the whole group from which the sample was taken. For example:

1. What proportion of the total number of resistors, from which I have drawn a sample, is defective?
2. How many defects in total does the reel of newsprint paper, from which I have cut a sample, contain?
3. How representative are the measurements I have just made, on the output from a screw-cutting lathe, of its machining process capability overall?

An important supplementary question in each case is 'what is the risk of the statistical conclusion being misleading?'

Fortunately there are basic 'distributions' established in statistical theory, and represented by mathematical equations, which can give answers to these questions. I shall not quote the formulae in this book, but I shall indicate where tabulated data derived from these formulae have been compiled as an aid to inspection, and how they can be used.

The binomial distribution

The binomial theorem deals with a series of events, each one of which has only the same two possible outcomes. For example in tossing a coin, the possible outcomes are 'heads' and 'tails'. The formula allows the probability, in any number n of throws, of 0, 1, 3, . . . n heads occurring. The outcome must always be head or tail; the probability of each is equal, and of one or the other is complete certainty, so the probabilities for each throw can be stated:

$$P_h + P_t = 1$$
$$P_h = P_t = 0.5$$

A similar calculation could be set up for the case where, of the two outcomes, one is very likely and the other very unlikely; e.g. in the case of a computer program where an error occurs on average in one line of code per thousand, the probability of a sampled line being defective is $P_f = 0.001$, $P_p = 0.999$. Since once again there is no third alternative, $P_f + P_p = 1$.

Whereas what the binomial theory does is to start with an assumed level of defectives in the population and work out the likelihood of particular proportions being found in the sample, our quality control applications often require the information to be presented the other way round – having found a certain number of defectives in the sample, what does it tell us about the overall defective level in the remainder that was not part of our small sample?

The applications of the binomial distribution are in those cases where we are concerned how many units fail. We are not concerned whether each has one or more than one defect. One fault will be enough for it to be rejected. The applicable charts and tables are those designed for:

1. Acceptance sampling by attributes (number defective);
2. Process control 'p' charting (for proportion defective).

The Poisson distribution

The binomial distribution considered the case where a sample is taken and the number of times a particular event which could have taken place on any occasion was actually observed (e.g. failure to meet the inspection criteria, relative to the number of pieces inspected).

The Poisson distribution, on the other hand, deals with the number of times a random event occurs within a particular length, area, volume or timeframe. Unlike the binomial case, no meaning can be attached to the number of times the observation didn't occur. For example, we can say that a light bulb had to be changed three times during the space of 10 000 hours. However it makes no sense to ask 'How many times didn't a light bulb have to be changed during the same interval?'

So the Poisson case applies where the possibility of committing a mistake is continuous, but in fact only a few mistakes are made; for example, number of ink smudges per 100 pages of manuscript. The Poisson distribution measures defects, not items defective. Applications are in:

1. Acceptance sampling by attributes (number of defects);
2. Process control '*c*' charting (for number of defects).

The normal distribution

Whereas the binomial and Poisson distributions both deal with discrete facts or decisions (attributes), the normal distribution deals with continuous variables. It is also known as the 'error' distribution (since it was first studied in relation to the reproducibility of astronomical observations) and as the Gaussian distribution after the mathematician Gauss.

The normal distribution is concerned with the way in which measurements cluster about a central number which is the true or desired value. Astronomers accepted that the angular distances and positions they were trying to measure were constant, but that the measurements they obtained showed a random scatter about a central, most frequent, and presumably correct value. While the scatter is random, the effect is that the greater the deviation from the true value, the less likely is that amount of deviation to occur. A bell-shaped distribution was typical, and it could be described by: the mean value, the measure of its 'central tendency'; and the 'standard deviation', a measure of its dispersion about the mean.

In the case of quality control, a target figure is demanded by the drawing or specification, and the variability is a combination of incomplete control of the process, and measurement error. The applications of the normal distribution are in:

1. Acceptance sampling by variables;
2. Process control charting by variables.

Acceptance sampling

Whereas control charting, both of attributes and more particularly of variables, will be discussed in Chapter 12, the current one will deal with the techniques of acceptance sampling. Again the distributions mentioned, and some related ones can be employed, to provide sampling tables which can be employed to make decisions as to whether the frequency of defects, or defectives in a population is acceptable, on the basis of checking a small sample.

We shall make specific reference to two sets of tables which exist within the British Standards series. These are:

1. BS 6001 which deals with inspection by attributes – either defectives or defects;
2. BS 6002 which deals with inspection by variables.

Statistical acceptance sampling

The purpose of statistical sampling is to allow a relatively small sample of a product to be inspected, either by attributes or by variables, before acceptance for the next stage of production or shipping to the final customer.

If a known risk of the lot being unacceptable to the customer is exceeded, transfer can be prevented. The risk of rejecting, on the basis of the sample result, a lot that would satisfy the customer is also known.

Sampling tables were first designed (e.g. by Dodge and Romig of Bell Telephone Laboratories, around 1930) on the basis that if the level of rejects established by the sampling plan was at a low enough agreed level, 100% inspection would not be attempted. Only lots which failed the sample test would be 100% screened, and it was assumed that this 100% inspection would be completely effective.

Tables for sample inspection by variables and by attributes

We defined attributes and variables inspection in Chapter 10, Part Two. The tables in general use within the UK are:

1. BS 6001: *Sampling Procedures for Inspection by Attributes*. Its three parts each have ISO equivalents, and deal with:
   Part 1:1991 (ISO 2859–1:1989) AQL plans for lot-by-lot inspection.
   Part 2:1993 (ISO 2859–2:1985) LQ plans for isolated lot inspection.
   Part 3:1993 (ISO 2859–3:1991) Skip-lot procedures.
2. BS 6002: *Sampling Procedures for Inspection by Variables*.
   Part 1:1993 (ISO 3951:1989) single-sampling AQL plans for lot-by-lot inspection.

BS 6001 is the more widely used of these sets of tables. Its potential uses include the inspection of:

1. End items;
2. Components and raw materials;
3. Operations;
4. Materials in process;
5. Supplies in storage;
6. Maintenance operations;
7. Data on records;
8. Administrative procedures.

Efficient use of BS 6001

BS 6001 includes many features which enable its use to be efficient and effective. Its flexibility is not always fully appreciated, and so it is often not used in the ways for which it was designed. Some of its features are:

Inspection levels (IL)

Seven different inspection levels are offered, giving different sample sizes for inspection. These are levels I, II, III and special levels S1, S2, S3 and S4. The smaller sample sizes, especially those in the 'S' levels, have the limitation of poorer discrimination between 'good' and 'bad' lots.

Switching rules

Inspection always starts as 'normal' inspection, but it must switch to 'tightened' if frequent rejected lots are found. It may also be switched to 'reduced' inspection after a long run of accepted lots. The switching rules are clearly defined; tightened inspection demands larger samples to improve discrimination and reduce the risk of a 'bad' lot getting through, whereas reduced inspection allows the economy of smaller sample sizes.

Skip-lot

'Special reduced' inspection allows as few as one lot in five to be sampled after especially extended sequences in which no failures were found.

Double, multiple and sequential sampling

Discussion so far has assumed the taking of a single sample, but sampling plans are also included wherein smaller initial samples may be taken, then other items added to enlarge the sample only if no decision can be reached on the initial sample.

Acceptable quality level

The Acceptable Quality Level (AQL) is defined in BS 6001 as 'the maximum percent defective (or, the maximum number of defects per hundred units) that, for purposes of sampling inspection, can be considered satisfactory as a process average'.

The Process Average (PA) is 'the average percentage defective (or average number of defects per hundred items, whichever is applicable) submitted for original inspection'.

BS 6001 and BS 6002 are designed to be used to protect a chosen AQL. They are concerned more with the average quality level of production than with the delivered quality of isolated lots, and they are well adapted to:

1. Controlling a continuous process;
2. Safeguarding a customer who is receiving the majority, or an aggregate mixture, of the product made.

Producer's risk and consumer's risk

Tables can be drawn up to protect either the producer's risk (of the random sample failing an acceptable lot) or the consumer's risk (of the random sample passing an unacceptable lot).

We can state these terms more formally, by following the wording given in BS 4778:

1. Producer's risk is the probability of rejecting a lot whose proportion defective has a value stated by the given sampling plan as acceptable;
2. Consumer's risk is the probability of accepting a lot whose proportion defective has a value stated by the given sampling plan as rejectable.

Producer's risk sampling plans

Acceptable quality level plans like BS 6001 and BS 6002 provide a safeguard against the supplier's risk of rejecting a good batch, i.e. one where the proportion defective is no greater than the AQL figure agreed with the customer. The sampling plans are normally designed so that there is a 95% chance of good lots being accepted, and for this reason the 'accept number' may seem generous.

Figure 11.1 shows two operating characteristic curves for an AQL of 1.0%. One is for a sample size of 13 (accept on zero rejects, reject on 1) and the other is for a sample size of 1250 (accept on 21 rejects, reject on 22).

Notice how much better the discrimination is for the larger sample size, i.e. its ability to distinguish between process averages better and worse than the AQL. (The 'ideal' AQL would be a 'step function' with 100% acceptance of lots with PA numerically lower than the AQL, and 0% acceptance of lots with AQL higher than the AQL. Both our sample sizes are quite good at recognizing and accepting lots better than the AQL but the smaller sample size is markedly poorer in its ability to throw out lots worse than the AQL.)

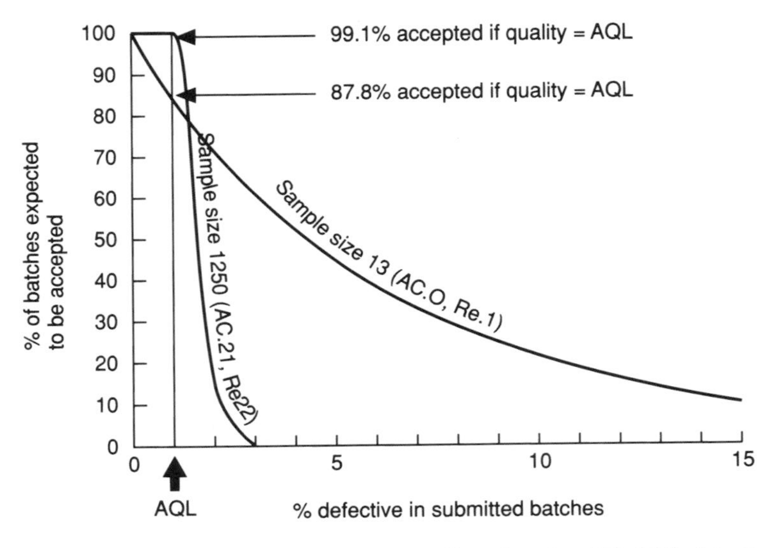

Fig. 11.1 OC curves for two 'normal inspection' sampling plans for AQL 1.0% defective.

Consumer's risk sampling plans

Although often used to protect the consumer's interests, as we have shown in Figure 11.1 AQL-related plans are not well adapted for that purpose, because they focus on the average quality over a succession of batches. It is no consolation to the customer to know that taking one batch with another, the AQL is maintained, if he/she was unfortunate enough to receive the one batch whose outgoing quality was much worse.

Instead of being based on an Acceptable Quality Level, the consumer's risk plans specify a Limiting Quality (LQ) or Lot Tolerance Percent Defective (LTPD). These measures give an assurance to the customer in relation to each individual batch in isolation.

Limiting Quality is 'the fraction defective that corresponds to a specified and relatively low probability of acceptance'.

Lot Tolerance Percent Defective is similar except that the population level defective is expressed as a percentage rather than as a fraction. Thus a Limiting Quality of 0.1 is the same as a LTPD of 10.

BS 6001 has a supplement giving LQ sampling tables. LTPD tables are commonly used in the USA military procurement sphere, where they are based on a consumer's risk of 10%.

Refer to Figure 11.2 which shows the operating characteristic for a particular sampling plan; i.e. for a given sample size and acceptance number. Notice marked on the curve the AQL for a producer's risk of 5% (i.e. probability of acceptance 95%) which is 1.2% (defective), and also the LTPD for a consumer's risk of 10% which is 5.3 (% defective).

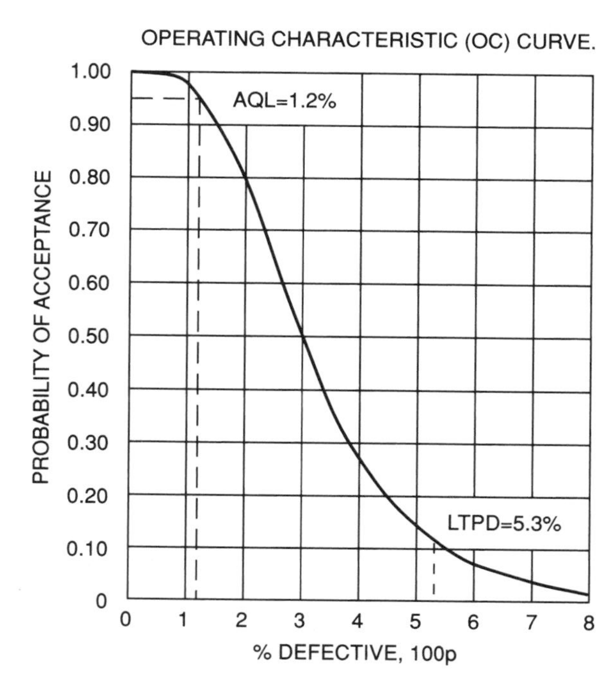

Fig. 11.2 Operating characteristic (OC) curve.

Relationship between AQL and LTPD sampling plans

Many people feel that for any given AQL there must be a universally equivalent LQ and LTPD. They read a specification requirement for an AQL of 1.5%, let us say, and want to know 'What is that as an LTPD?' There is no answer. Any given sample size and acceptance number can be related to both an AQL and an LTPD, but the equivalence AQL value x = LTPD value y only exists at that particular sample size and accept number. For a different sample size and accept number, *x* or *y* or both will be different.

Let me illustrate this in our Table 11.1. The data are taken from BS 6001,

Table 11.1 Examples of AQL and LTPD data

Sample size	*Accept number*	*Limiting quality (nominal) (%)*	*Acceptable quality Level (%)*
50	0	5	0.25
80	1	5	0.65
125	3	5	1.0
32	0	8	0.65
80	1	5	0.65
315	3	2	0.65

You will see that the figures quoted demonstrate the lack of a fixed relationship between the LQ and the AQL. You will also see that in each case the AQL appears a much more optimistic figure even though both the AQL and LQ plans are using the same sentencing criteria (sample size and accept number). This is a side-effect of the LQ plan safeguarding the consumer's risk and the AQL safeguarding the supplier's risk. Only as sample sizes increase to a larger proportion of the lot size do the AQL and LQ values converge.

Troublesome definitions

We have defined Process Average (PA), Acceptable Quality Level (AQL), Limiting Quality (LQ) and Lot Tolerance Percent Defective (LTPD) earlier. There are other terms relating to sampling for quality which can cause confusion, not least AOQ and AOQL. Figure 11.3 is an attempt to clarify these relationships.

In Figure 11.3 the level of defects (or defective items) in the population as it leaves the production line is the process average (PA). The acceptance sampling is applied in relation to a required acceptable quality level (AQL). The lots which meet the sample criteria pass through; those which fail are sorted 100%. What finally reaches the customer is the average outgoing quality (AOQ) which is better than the PA because bad lots have been rejected and rescreened. The worst possible value of AOQ is called the average outgoing quality limit (AOQL).

AOQL is often assumed to stand for Average Outgoing Quality Level, and to be synonymous with AOQ. This is, as you have learned, not the case. The AOQL is independent of the quality of product submitted (PA), it is a direct function of the sampling plan being used.

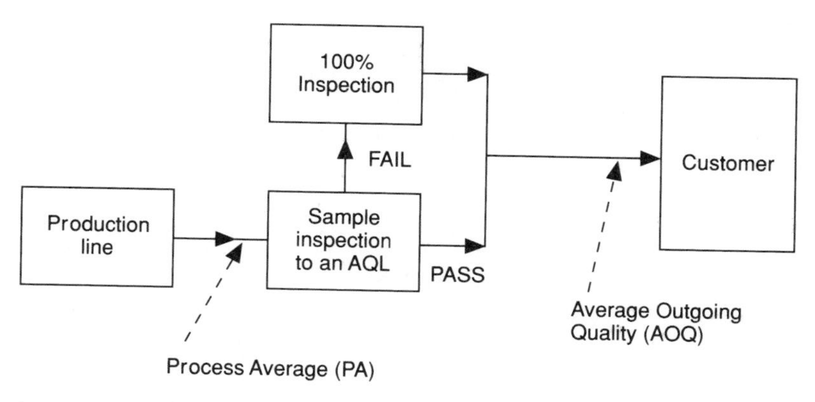

Fig. 11.3 Process average and AOQ.

This fact is illustrated in Figure 11.4. If it is still hard to accept that the AOQ cannot exceed a certain pre-determined value, regardless of how bad the PA might be, consider these two extreme cases:

1. A 0% defective PA will result in a 0% AOQ because there are no rejects to count;
2. A 100% defective PA will result in a 0% AOQ because all lots will have been rejected on first sampling, and all units will have been 100% inspected and repaired.

The peak value of escaped rejects will occur for some intermediate value of PA, for which some lots will pass and others fail and be rectified; and that peak value of AOQ is the AOQL.

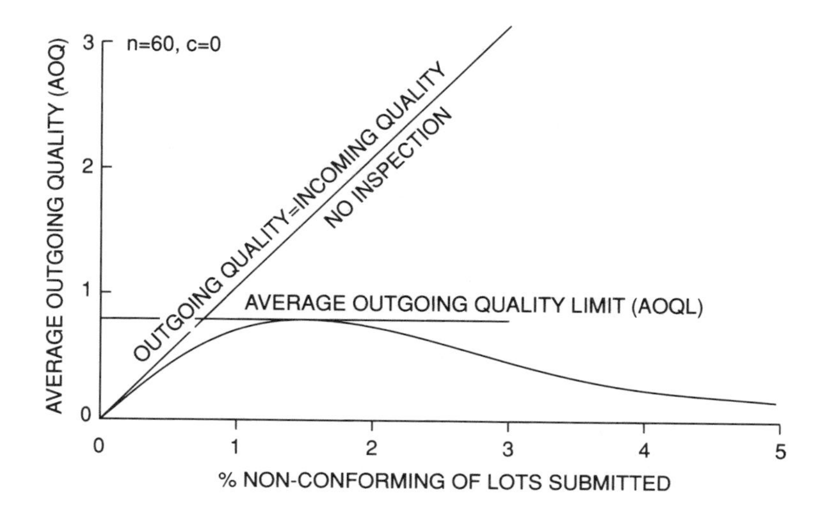

Fig. 11.4 AOQ and AOQ limit. For a given sampling plan, the AOQ can never be worse than the AOQL, the Average Outgoing Quality Limit.

Summary of conclusions

This chapter dealt with the application of statistical theory to sampling inspection. As indicated in the previous chapter, sample inspection often enables more detailed information to be recorded than a more widespread review. However, it involves a risk of false conclusions being reached. The need for a clear mathematical basis for the sampling is that it allows this risk to be measured.

Sampling plans can be organized either around the risk to the producer, or the risk to the customer. Which is appropriate depends on the circumstances. Various sampling tables are available.

Illustrations in this chapter were taken from a British Standard, BS 6001 which deals with inspection by attributes (pass/fail), the more common method, rather than by variables (measurement), but enables either producer's or consumer's risk to be held at a constant value. The tables offer alternative sampling plans which balance uncertainty against economy of sample size.

12

Statistical process control

Introduction

In Chapter 11, Part Two, we looked at the application of statistical techniques to quality appraisal, in the form of statistical acceptance sampling. In this chapter we shall be looking at more applications of statistics in aid of quality assurance, but this time at the role of statistical techniques in anticipating failures, through statistical process control (SPC); the design and interpretations of experiments to optimize process conditions and manufacturing tolerances; and the limitations of traditional SPC in assuring 'parts per million' quality.

The objective of the chapter is not to repeat all the detailed information on SPC methods given in the references, but to offer working managers a 'bird's eye view' of the field. In this way I hope I can aid you in deciding the scope for the use of SPC in your own particular business; this scope may be considerable, or it may be very small.

Statistical quality control and process control

The terms statistical quality control (SQC) and statistical process control (SPC) are sometimes used interchangeably. More correctly, SQC represents all quality control methods which rely on the use of statistical disciplines – for example statistical acceptance sampling techniques, as described in the previous chapter. Statistical process control (SPC) in contrast, is the aspect of SQC which I shall discuss in this chapter. It relates to statistical monitoring and control of the operation of a process, whether this is a manufacturing process or a process concerned with the delivery of a service.

Both terms (SQC, SPC) are apt, and stress the importance of process control to preventive quality assurance. Indeed to some people, even to some writers on quality assurance, SPC appears synonymous with quality assurance just as statistical acceptance sampling came to be seen as synonymous with quality control. SPC is indeed a very powerful tool in an appropriate context, but it is perhaps better to say that it predicts bad product rather than prevents it. It prevents only in the sense that it gives warning to stop the process and investigate, when it is drifting out of control. It does not give direct guidance where to look for the source of the problem. Hence, as one industrial statistician has said, 'SPC is 20% statistics, 10% common sense and 70% solving problems'.

Statistical process control

The correct application of SPC enables the capability of a process to be established; i.e. the limits of variability of what is produced by the process with the given methods and equipment. Following on from this, deviations in the results of the process which are due to assignable causes can, when they occur, be differentiated from fluctuations in output which are inherent limitations of the process.

A methodical and legitimate approach to SPC demands that implementation is approached in three stages:

1. Collection: collect sufficient initial data from the process.
2. Control: calculate the control limits from the data collected.
3. Capability: once the process is in statistical control, calculate the capability index(es) C_p (and C_{pk}).

Explaining these 'three Cs' in more detail:

Collection

It is necessary to decide how the performance of a process can best be expressed in numerical terms, and set up a monitoring system to collect and present the data. Information can then be collected and presented in chart form using data from perhaps 25–30 subgroups taken at agreed intervals. From this 'initial study' preliminary control limits can be estimated, and used to form tentative conclusions as to the general performance of the process.

Control

Periodic sampling is continued in order to determine with the aid of the initial control limits whether or not the process is in statistical control; i.e. if only random variations are present. If special (assignable) causes of variation are present the causes can be investigated, identified and removed from the system. Causes of assignable variation can be things like tool wear, incorrect settings, improper operations, poor materials etc.

Capability

Capability is a measure relating to the actual performance of a machine or process to a performance in respect of a specification or as otherwise required. As assignable causes of variation are removed, so capability will improve. When only common (random) causes remain further improvement in capability is not possible until the common causes are addressed.

When the process capability has been established, but only then, taking and measuring regular small samples enables the existence of an assignable cause of variation to be detected and investigated before it develops to a point where rejectable material is produced.

Graphical methods of presentation and control can be used, and the rules are simple enough for the records to be kept and interpreted by the process operator.

Process control represents action on the process to avoid the production of substandard work, rather than action on the output to segregate any substandard work that has been made. Controlling the process is clearly preferable to inspecting the output.

Factors creating process variation

Process variation can be caused by various factors, classed as:

- People, e.g. in setting up a machine, operating a machine;
- Equipment, e.g. bearing wear, tool wear;
- Materials, e.g. size, composition, properties;
- Methods, e.g. operating procedures, maintenance provided;
- Environment, e.g. temperature, power supply fluctuations.

Random and assignable variation

Variation can arise from a number of sources, and in order for it to be reduced the applicable causes of variation must be identified. Some variation is inherent in the equipment, materials, methods, environment and people performing the process.

This variation gives rise to a statistical fluctuation of performance around a central value. If performance, as measured by variation of a parameter, has a smooth variation around a central value (a 'normal distribution') it is said to be random. Irregular, unstable and intermittent changes are the result of some significant alteration occurring in one of the factors creating the variation. These point to assignable causes which, in principle, are discoverable.

Local actions and actions on the system

Because common (random) and assignable variations have different sources they require different kinds of action. Assignable causes are created locally (refer to the examples given in subsection 'factors creating purpose variation') and can be dealt with by local actions, e.g. by changing a worn cutting tool, or discarding faulty material. Limitations inherent in the process (e.g. using a diamond saw as opposed to a laser cutter) can only be controlled by managerial action or assistance from other departments. These forms of intervention are known as 'actions on the system'. They are precisely the kind of problem which are best addressed by a group such as a 'problem solving team', or 'quality circle'. We shall examine the functioning of such teams in Part Three.

Statistical control and process capability

Let us suppose that a machine is to be used to manufacture components to specific limits. The process is run under normal conditions, and a series of measurements taken on the parts produced. The distribution of measured values represents the 'process capability' of the machine. If the tolerances on the dimensions of a particular job are wider than the process variability, the job is within the process capability of the machine. If the tolerances are narrower, the job is outside the process capability and should not be attempted (Figure 12.1).

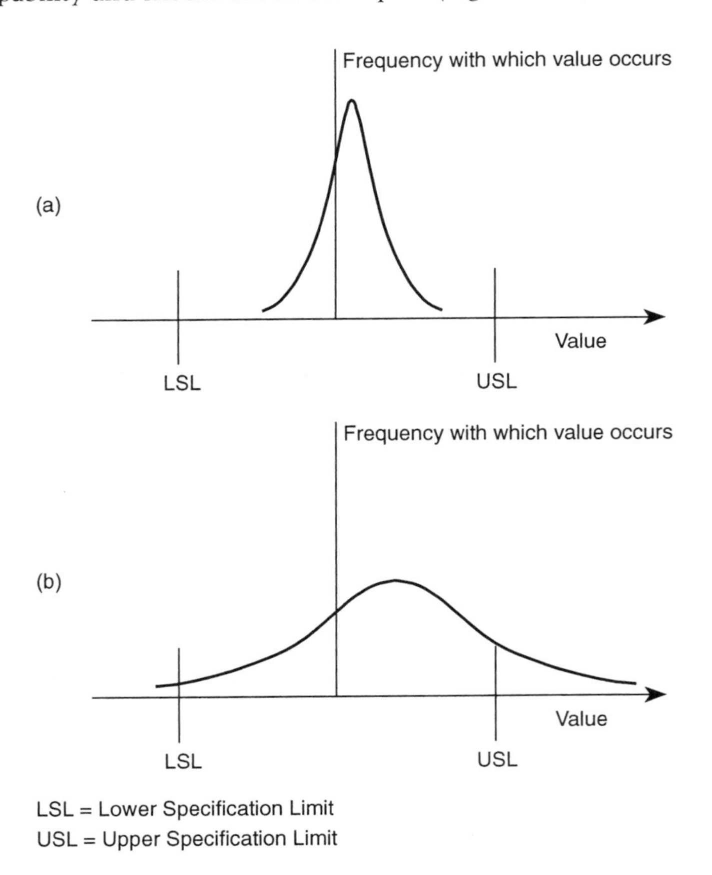

Fig. 12.1 (a) Capable process, (b) incapable process. LSL = lower specification limit, USL = upper specification limit.

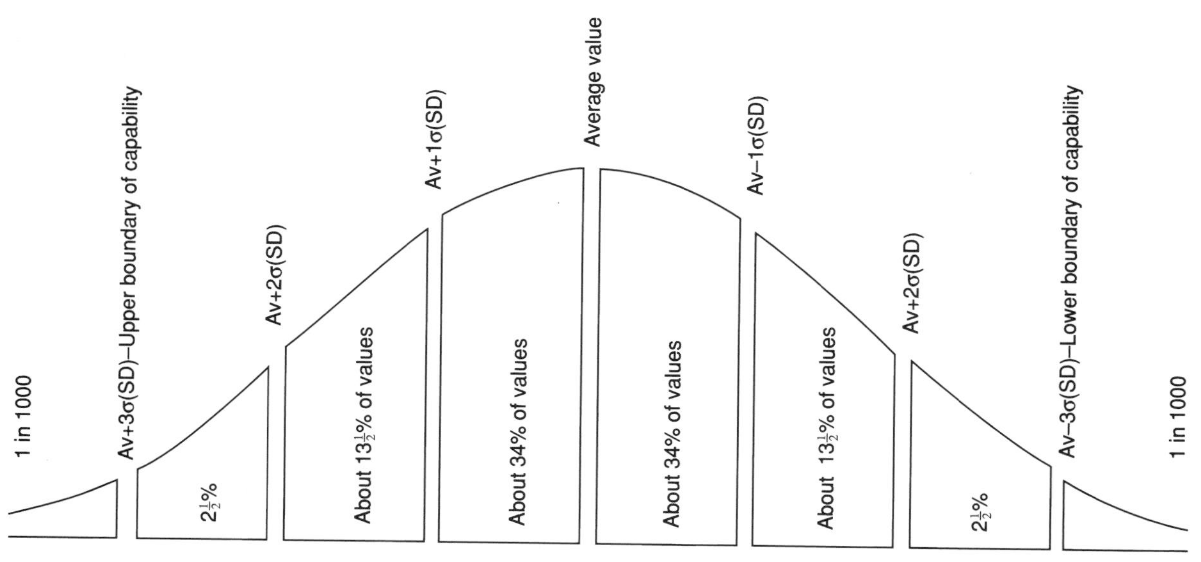

Fig. 12.2 The normal distribution.

It is important to note that the process capability is not based solely on the range of the measurements taken during the statistical investigation. Assumption of the 'normal distribution' allows the probability of occurrence of measurements outside the range actually witnessed to be expressed.

The normal distribution depicted in Figure 12.2 is also known as the 'Gaussian' distribution. All normal distributions follow this bell-shaped curve which can be fully described by two values; the mean or average of the values measured, and the 'standard deviation' denoted respectively by the Greek letters mu (μ) and sigma (σ). They measure the process's 'central tendency (μ) and its dispersion about the central value (σ)'.

By collecting a series of measurements from your process output, the normal curve corresponding to it can be established. Every true normal distribution is symmetrical, 50% of measurements will fall on either side of the central value. On either side of the mean, only 16% fall outside 1σ, 2% outside 2σ and as few as 0.1% outside 3σ from the central value.

The basis of SPC control charting is that if a measurement is found outside a 3σ limit on a previously stable process, there is only about one chance in a thousand of it happening by chance; in the other 999 cases it must be the result of some unwitting change in the process parameters which has altered the process capability. The process change may have altered the spread (i.e. the value of σ), or the value of the mean, or both. The result is that the point is not outside the 3σ limit of the new process capability (Figure 12.3).

Approximation to the normal distribution

In many practical cases measurements do not follow a normal distribution; the distribution may be much more skewed than symmetrical, or contain irregularities.

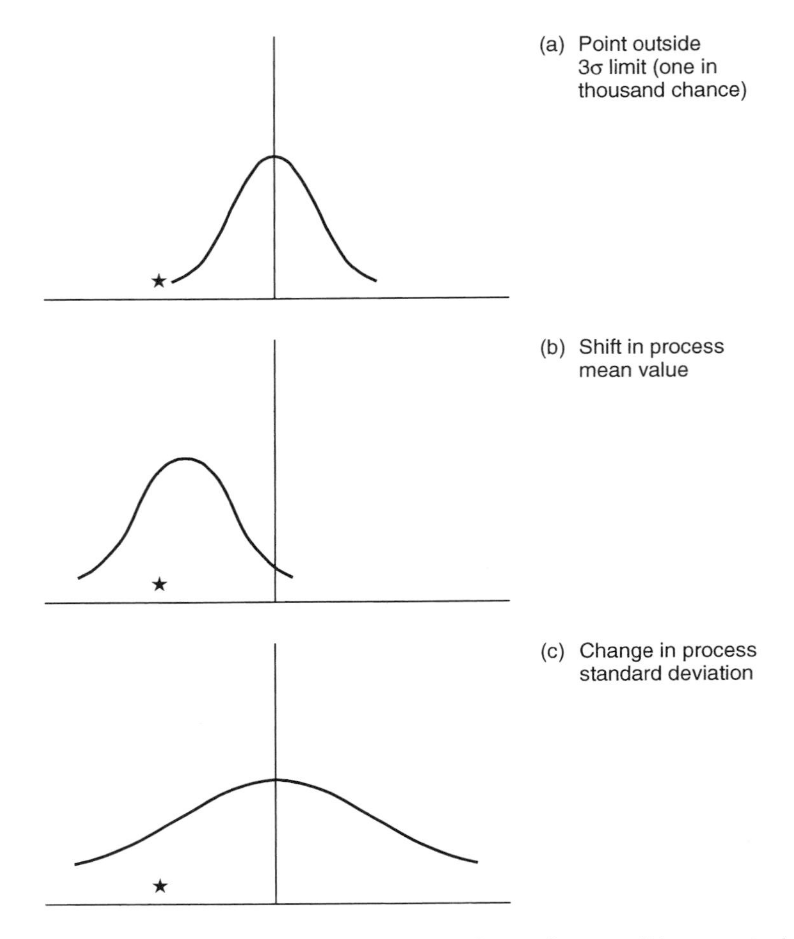

Fig. 12.3 Points outside former process capability $\pm 3\sigma$ limits; three possible causes (a, b, c).

In these cases groups of measurements can be taken and the average plotted. This will produce a distribution more normal than plotting individual measurements, which can then be more reliably manipulated according to the statistical theory of the normal distribution.

For this reason it is a much more usual practice to collect data in sets of perhaps five measurements, or measurements on five items and to average them before proceeding further, rather than take and plot individual measurements one at a time. Control charts used for SPC will then use and display limits appropriate to sample averages.

New processes and equipment

An important aspect of quality assurance in a manufacturing environment is the need to ensure that any new manufacturing equipment introduced has the process capability demanded of it. Equally, that no specification is accepted which is known to be outside the process capability of the equipment which will be used to make the item.

Often the process capability is not known with sufficient accuracy to meet this safeguard. This is the state of affairs which results in reliance on post-manufacture inspection and sorting to protect the customer, with the wastage of time and money which this implies.

We can define process capability more exactly than in our introduction, by saying that a process is capable with respect to a certain specified tolerance if both the upper and lower $\pm 3\sigma$ limits are within the upper and lower specification limits.

Thus a process may be (and often is) in statistical 'control' without being 'capable'. That is to say, if assignable causes of variation are absent it is in statistical control, but when the specification limits are taken into account the random variation may be too wide for the specification to be consistently met. Conversely, a poorly controlled process may be capable of meeting a specification which is sufficiently loose, even though the process is statistically out of control.

Process capability index C_p

There is clearly a need to establish a figure of merit which shows how well matched the capability of a process is, to the specification it has to meet. A simple measure is the index C_p.

This index is calculated as a result of assuming that a process is centred at the mid-point of the specification and comparing the spread between the $+3\sigma$ and -3σ points with the interval between the upper and lower specification limits, thus:

Capability Index (C_p) is given by:

$$C_p = \frac{USL - LSL}{6\sigma}$$

The stipulated value of 6σ is appropriate in the context of controlling the process by SPC to correct any deviation outside the $\pm 3\sigma$ limits. I will leave you to verify that any capable process must have a C_p value > 1.0.

Measuring within limits

The steps leading to establishing the process capability only represent one stage of statistical process control. Once process capability has been characterized, it is possible to draw up a chart for the required measurements with limits outside which, with statistical confidence, future measurements will not fall except under strong suspicion of there having been a change in the process capability; in other words, that an assignable change has been introduced.

Thus if sample measurements are taken and plotted, any point falling outside the control limits (the $\pm 3\sigma$ limits) are much more likely to be the result of assignable than random variation, indicating that a change has occurred in the process.

There is an important point to be stressed here – even though the measurement may be within the specification tolerance (USL–LSL) the fact that it is outside the

$\pm 3\sigma$ limits previously established as characteristic of the process, means that some action is needed to investigate the change that has occurred. It is prompt action which prevents the process control worsening to the point where the specification tolerance limits are exceeded.

Control and warning limits

Close observation of the results of data gathering can give additional warning of possible problems. Since 95% of production will fall between the $\pm 2\sigma$ limits (refer back to Figure 12.2) so as well as treating the $\pm 3\sigma$ limits as points outside which immediate action is required, the $\pm 2\sigma$ limits can be treated as giving advance warning that something may be wrong.

Further warning pointers may be received by looking out for trends in shift of successive measurements, or successive measurements grouped on one side of the mean, even if neither 3σ nor 2σ limit is exceeded. To go into further detail is outside the scope of this book, but these facts are explained in practical texts on control charting.

Upper and lower limits

If we take the case of a machining operation to produce a mechanical component, there will be an upper and a lower specification limit, and so both the upper and lower 3σ limits are of practical significance. If one limit is exceeded the fit of the component may be too tight, in the other case too loose.

In other cases only one limit may affect the utility of the goods. A bonded joint must have a certain minimum strength, but there is normally no practical need to specify a maximum strength. In such cases the control chart will tend to be treated in a one-sided fashion. This is bad practice.

If points start to appear which mark significantly stronger bonds than the process average, then the process must have been unwittingly improved; perhaps by a more effective preparatory cleaning process. The improvement should be investigated, identified and standardized before it is lost. Successful manufacturing companies are those which constantly attempt to improve the accuracy with which they can maintain the desired mean value of any parameter, and reduce its variation. This both saves costs, and allows specifications to be improved.

Leaving well alone

A measurement outside the $\pm 3\sigma$ limits of an established process indicate with a high degree of confidence that the process has changed and action is required. No less so, a measurement within the $\pm 2\sigma$ limits indicates that it is not statistically wise to assume a change requiring local action. Thus keeping control charts discourages the operator of the process from making frequent minor adjustments or 'tweaks' to the process in a misguided response to what are most likely to be purely random variations beyond their control and, by trying to intervene, increasing the variability.

Control charting techniques

Control charting methods of statistical process control are based on the theory I have sketched out, but with simplifying approximations where these are of practical assistance. For instance, the standard deviation of a sample is calculated using a mathematical formula. This may be a minor inconvenience to an engineer collecting data to estimate initially the process capability, but it is less acceptable if a process operator is to maintain the control charts. Hence, as described below, the substitution of range for standard deviation, as a measure of dispersion.

'$\overline{X}$ and R' charts

Because of the extra arithmetic of calculating the standard deviation from the sample measurements, instead of the measured mean ($\overline{X}$) and sigma (σ or SD) of the sample, control charts of mean and range ($\overline{X}$ and R) are more usually kept.

The range 'R' is simply the difference between the largest and smallest value in the group of measurement values taken, and $\overline{X}$ their average. Although not statistically as precise a measure as σ, the range does give a useful indication of the dispersion associated with the process. Tables can be used to calculate σ from R, so the latter can be employed for control charting purposes. An example of an $\overline{X}$ and R chart is given in Figure 12.4.

Attributes control charts

Although SPC is usually discussed in terms of measurements information, it can also be applied to the case where attributes information is being collected. In this case the data that are being examined is the number (i.e. the frequency) of faults or faulty items.

The collection of the initial data will establish the process capability in terms of the typical level of errors or defects generated. Observation of cases where counts above or below the control limits will then signify assignable deterioration or improvement in the process.

Attributes charts can warn you of the extent of a problem but give only limited clues as to its cause. Because of this the use of warning limits is particularly useful so that investigations can commence at as early a stage as possible.

CUSUM charting

SPC control charts such as those we have discussed so far are normally plotted by taking a group of measurements periodically and plotting the average

RECORD SHEET

JOB No. 13 DRG. No. 172 DIMn. REF. K MACHINE No. 15 OPERATOR K G
DESCRIPTION SHAFT INSPECTOR J M ZERO OF MEAS.T 2 506
DRG. DIMn. 2 508 DRG. TOL. LIMITS ±0.002 UNIT OF MEAS.T. 0.001 SAMPLING INTERVAL 2 hours

SPEC. No.	No. OF SAMPLE																	
	1	2	3	4	5	6	7	8	9	10	11	12	13	14	15	16	17	18
1	4.0	2.5	2.0	3.5	2.0	2.0	1.5	2.5	1.5	2.0								
2	2.5	2.0	3.0	2.0	3.0	2.5	3.0	2.5	3.5	2.5								
3	1.5	2.5	2.5	2.0	1.0	2.0	2.0	3.0	2.0	3.0								
4	2.0	2.0	2.5	2.0	2.5	3.0	1.5	3.0	2.5	2.0								
5											TOTALS FOR AVERAGES 10 SAMPLES (40 ARTICLES)							
6																		
Totals	10.0	9.0	10.0	9.5	8.5	9.5	8.0	11.0	9.5	9.5		94.5						
Av.	2.5	2.2	2.5	2.4	2.1	2.4	2.0	2.8	2.4	2.4		2.35	=	$\bar{\bar{X}}$				
Range	2.5	0.5	1.0	1.5	2.0	1.0	1.5	0.5	2.0	1.0		13.5	∴ $\bar{R}$	=	1.35			

CONTROL CHARTS

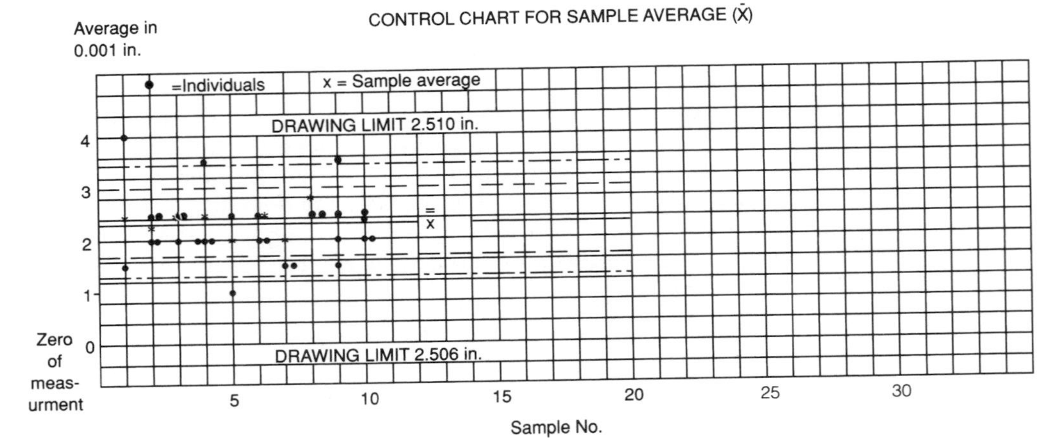

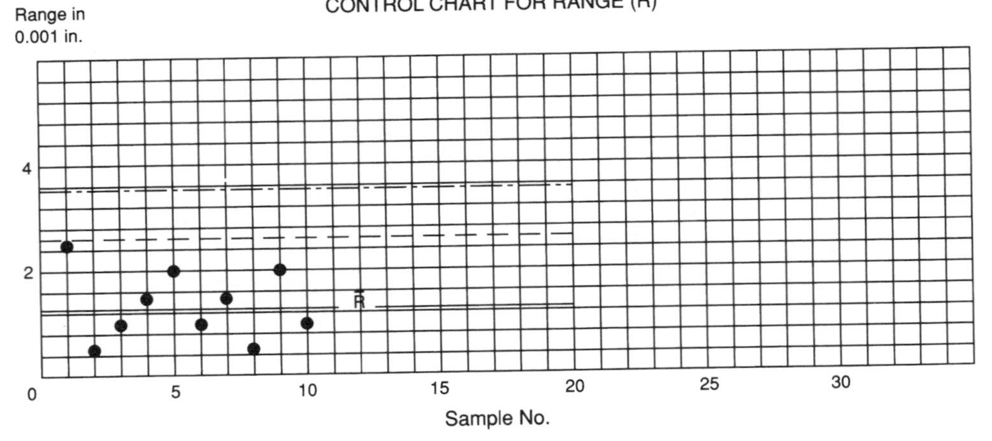

Fig. 12.4 Example of an $\bar{X}$ and R Chart (repoduced from BS 5700, courtesy of BSI.)

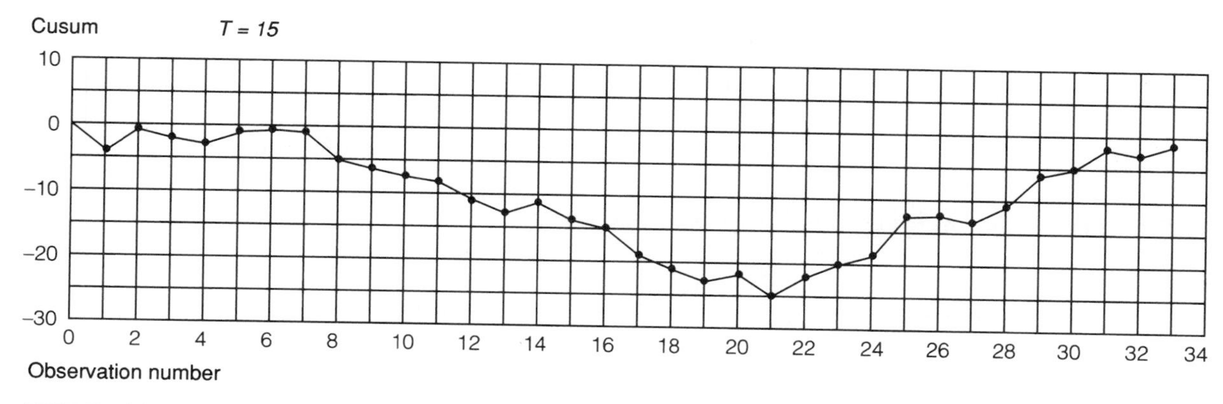

Observation i	Observed value x_i	Deviation from target (15) $(x_i - 15)$	Cumulative sum of deviations $\sum_{r=1}^{i}(x_r - 15)$	Observation i	Observed value x_i	Deviatation from target (15) $(x_i - 15)$	Cumulative sum of deviations $\sum_{r=1}^{i}(x_r - 15)$
1	12	−3	−3	17	11	−4	−19
2	17	+2	−1	18	12	−3	−22
3	14	−1	−2	19	13	−2	−24
4	14	−1	−3	20	16	+1	−23
5	17	+2	−1	21	12	−3	−26
6	16	+1	0	22	18	+3	−23
7	14	−1	−1	23	18	+3	−20
8	11	−4	−5	24	17	+2	−18
9	13	−2	−7	25	20	+5	−13
10	14	−1	−8	26	15	0	−13
11	15	0	−8	27	14	−1	−14
12	11	−4	−12	28	18	+3	−11
13	14	−1	−13	29	20	+5	−6
14	16	+1	−12	30	16	+1	−5
15	13	−2	−14	31	18	+3	−2
16	14	−1	−15	32	14	−1	−3
				33	16	+1	−2

Fig. 12.5 Example of a CUSUM chart (reproduced from BS 5700).

(mean) of each group. An alternative approach is to plot the cumulative sum (hence 'cusum') of the deviations of the measurements from a chosen target value, and compare the slope of the line so created with a reference line. An illustration is given in Figure 12.5, where the raw data have been processed as shown in the table, so that the desired line is the horizontal one 'deviation from desired value = 0'.

CUSUM charts have the following features:

- Greater sensitivity to changes in mean value; and
- They give a clearer indication of the point on the chart, and hence the time, when the change occurs.

These features make them more suitable in some circumstances. For example, they have been much used in monitoring expenses against an annual budget, where the cost of spending relative to budget can be monitored and the onset of the major items of expense can be easily picked out.

Statistical design of experiments

Quality improvement often demands changes to a process. Contemplated changes have to be evaluated by experiment, inevitably in the prescence of unintentional variations in the experimental conditions.

Good statistical design of experiments can help distinguish between the results of the deliberate and the unavoidable variations, so determining whether the results of the experiments are significant, or could have happened by chance.

Tests of significance

The basis of the statistical design of experiments is that the significance of the outcome observed can be tested by various statistical methods. These can give a quantified judgement such as 'there is only an *x* per cent probability of this result having occurred by chance'.

This is obviously a very powerful tool in giving confidence in the outcome of an experiment designed to find out the optimum process conditions. Of equal practical importance is its ability to show that seemingly convincing results can have come about purely by chance.

Taguchi methods

The various statistical techniques for analysing experimental results, especially where more than one independent variable is known to be changing, can be daunting to non-specialists. A large part of the procedures known as 'Taguchi methods' is a set of recipes for experimental design and testing formulated by Genichi Taguchi, to make the use of statistical methods more straightforward for practising engineers.

Designing experiments for statistical analysis

It would serve no purpose to go into details of the statistical methods for analysing the results of experiments here. What is appropriate, is to stress that far more statistical information can be obtained, and much more confidence placed in the conclusions if the experiments are designed with statistical analysis in mind. In other words, seek the advice and assistance of an industrial statistician before planning the experiment, not when you have collected the results.

Designing for yield and process control

It will be evident from the ground we have already covered in this chapter, as well as the relevant chapters in Part One, that it is extremely important for the distribution of values of a parameter to fall easily within the specification limits set. This is dependent on two factors:

1. The specification limits being set realistically in relation to the real need of the design, and not unnecessarily tightly.
2. The process spread being controlled as narrowly as possible.

The parameters μ and σ which describe the central tendency and the dispersion of a normal distribution have nothing to say about its compatibility with the practical requirements of meeting the specification, whatever that may be. Therefore some other figure of merit is needed which combines both factors (specification and capability) and measures their compatibility.

Process capability index C_{pk}

The first index I introduced was C_p, earlier in the chapter. However, using SPC with $\pm 3\sigma$ control limits does not prevent minor shifts in the mean point; indeed fluctuations of the position of the mean point equivalent to $\pm 1.5\sigma$ are possible. To take account of this, and give a more practically realistic figure of merit for a process a modified capability index has been defined.

The modified capability index assumes a mean value shifted by 1.5σ from the centre of the specification, the maximum deviation likely to occur when a process is under statistical process control. This index is C_{pk}:

$$C_{pk} = C_p(1 - k)$$

where

$$k = \frac{target\ (nominal)\ mean\ point - actual\ process\ mean\ point}{{}^1/_2(USL - LSL)}$$

Earlier in the chapter I asked you to check that a capable process must have a C_p value > 1.0. Now confirm for yourself that a 'plus or minus six sigma' process must have a C_{pk} value > 1.0 and a C_p > 2.0.

Complex processes or systems

Controlling a process of demonstrated capability such that it does not stray, through assignable variation, outside its established $\pm 3\sigma$ limits may seen satisfactory. However, let us look at the cases of an assembly made from a large number of components, or a process which comprises a lengthy sequence of operations.

An investigation was done in the USA a few years ago into the price and performance of Japanese consumer electronic products (such as pocket calculators, CB radios, digital watches and TV sets) compared with the products of

US-based manufacturers. The investigation showed that the Japanese had, with remarkable consistency across a number of manufacturers, plants and products, refined their quality level to about one fault per million operations (1 ppm) at each process step. Their US rivals appeared to have accepted with equal consistency an error rate of around 0.5% as 'as good as we can reasonably expect'.

The cumulative effect of this basic yield difference was that for goods with about 60 components (e.g. a digital watch) the Japanese could achieve virtually 100% yield against the US 70%. In the case of a TV set (around a thousand components) Japanese yields would still be in the upper 90%, whereas each set coming off the US production line would have to be repaired before it could be shipped. Think of the cost savings in test and rectification costs alone!

'Plus or minus six sigma'

If we look at what the facts just described mean in terms of process control, an interesting observation emerges. A process capability where the process $\pm 3\sigma$ limits coincide with the specification limits is only going to limit rejects to the level of 0.1%–1%.

To achieve reject levels measured in parts per million means that process control must be refined, to the point where the spread is so narrow that the specification limits are around $\pm 6\sigma$ apart. The Japanese had achieved this through their philosophy of 'never-ending improvement', which we shall meet in Chapter 18, Part Three under the name of 'kaizen'. A leading American electronics multinational has attempted to narrow the gap by training all its engineers in 'Designing for Manufacturability'. This programme was introduced with a goal of 'Plus or minus six sigma capability by 1992'. The integration of design criteria with process capability will be mentioned again in Chapter 25, Part Four, where we shall highlight the commercial and cost-saving aspects of improved quality.

Summary of conclusions

Statistical process control employs statistical sampling for purposes of quality control rather than for sentencing products. Once again it can use either variables or attributes information.

SPC applications distinguish between assignable variation which a process or operator can investigate and regulate, and 'random' variation which demands management action from the process 'owner' to improve, since it is inherent in the limitations of the process method or equipment.

Thus before any process is commissioned it is important to establish its process 'capability'; i.e. what it can and cannot be expected to achieve. Failure to ensure that a process capability is compatible with the specification demanded of its output typically will result in the process operators being blamed for failures which are outside their control. Equally, misguided attempts to adjust process equipment to achieve the optimum results can result in increased variation.

A process is normally considered capable if specification tolerances are wider than 'plus or minus three sigma'. Nevertheless, a realistic goal to aim for, if inspection of the output is to be avoided, is often much tighter control; 'plus or minus six sigma'.

13

The place of the computer

Introduction

The computer increasingly infiltrates everyday life as well as the workplace, and to a greater extent even than most of us are aware.

In our workplace there may be a central mainframe computer, and many engineers and managers have PCs on their desks which may or may not allow access to the central computer. Computerized order entry, inventory tracking, despatch picking and billings may all be in place, as well as computer-aided design and numerically controlled manufacturing tools. In some service environments, such as banks, work practices may be even more strongly dominated by the use of computers.

In everyday life, 'computers' manage various functions of our cars, our washing machines, our central heating and of course our telephone communications.

It is not surprising therefore, that the process of quality assurance also has to direct attention on to computers. We shall focus on two aspects in this chapter of the book:

1. The uses that can be made of computers in aiding the task of quality assurance management;
2. Quality assurance of the generation, procurement and use of computer software within the work environment.

Electronic computer system equipment

The total system can be taken as comprising:

1. The computer hardware itself;
2. The peripheral hardware such as printers, visual display units (VDUs), other input/output and storage units;

3. Software comprising an element of the computer system design (firmware);
4. Operating system software (such as language compilers, assemblers and utilities) designed to control the use of the electronic computer facilities.

However, we shall be concerned mainly with the creation and use of applications software for use on the computer system.

The design and control of computer software

The utility of many products depends very directly on the quality of the computer software programs which control their operation. This is particularly obvious in the defence and aerospace fields. The International Standard which addresses quality systems for software is ISO 9000–3:1991; *Guidelines for the application of ISO 9001 to the development, supply and maintenance of software*.

The place of software in different organizations can be:

1. As the product on which the company's business is based;
2. As the means of controlling the functioning of the product;
3. As the means of controlling part of the design, manufacture and testing of the product;
4. As an aid to administering and managing the company, in such areas as stock control and financial accounting.

In each of these cases the company has to manage the quality assurance of the software, just as it would control and ensure the quality of any other raw material, manufacturing process, documented procedure or shipped product.

There are two elements of the quality system which are especially critical for the control and use of software generated within the company. These are:

1. Design control;
2. Document control.

Indeed, from the point of view of the internal use of computer software packages, you could say that these are critically important operating procedures which happen to be written on magnetic media instead of paper, and as a result are peculiarly susceptible to unauthorized amendments which are difficult to detect.

We shall look at the design control of software packages, and their 'document control' which is often called configuration control.

Design control of applications software

The customers will normally have a clear idea of what they want the software package to do, but not of how that is to be achieved. The development of a solution will take place in several development stages each of which will require a clearly identified input and output.

In designing software solutions to identified needs it is necessary to establish a detailed **development plan** before detailed design and development work is started. The planned programme will include:

1. Development change control procedures;
2. Design reviews to highlight and resolve outstanding technical issues;
3. Progress reviews to ensure that resources are available and time scales being met;
4. Establish a suitable design methodology.

The design and development planning methodology (defined in a design and development 'code of practice') has to include independent design review stages involving program testing and acceptance trials to ensure the software's quality in respect of specification, performance, maintainability, testability and system reliability.

It has to:

1. Be able to distinguish between the levels and activities into which the design has been 'decomposed';
2. Trace software design back to the system specification and statement of requirements;
3. Establish an adequate description of the design which can be kept under configuration control;
4. Permit software modules to be organized into small sections of about 100 lines of code, with a plain language description of its function associated with each module;
5. Permit independent review of each stage (including test stages), before acceptance;
6. Maintain suitable test plans, specifications and records of test;
7. Incorporate a formal software error reporting and correction procedure.

For each stage in the development programme a **design input** must be established; when the development stage is complete the **design output** must be recorded. This can then be examined to verify that it meets the requirements of the design input and also acts as the design input reference of the next stage.

Design verification must be done in such a way that it is:

1. Appropriate to the activity;
2. Repeatable by equally qualified staff;
3. Recorded and followed up;
4. A 'hold point' which prevents unverified work proceeding to the next stage;
5. Able to allow identified design problems to be quickly highlighted.

A software design can be more quickly altered than almost any form of hardware, without the change being detectable, e.g. a few keystrokes at a terminal can produce a change which is quite transparent, compared with a pencilled alteration to a blueprint, which is evidently a later, informal addition in the eyes of anyone who reads it. Moreover it is feasible for different elements or stages of the design to be worked on by different individuals, and a reviewer is in a position to make changes (correct errors?) in the process of checking written coding.

Therefore a rigorous change control procedure has to be enforced. It must ensure that:

1. All design input requirements and output criteria are subject to change control.
2. All material which has been created and verified must be subject to change control.
3. Proposed changes must be subject to critical review before approval and implementation, and recorded on a design change note recording:

(a) authority,
(b) identification of all items to be changed,
(c) identification of all items that will require re-verification,
(d) authorized sign-off when all changes are certified complete.

Document control

The expression commonly used is 'configuration control', 'modification control', or 'baseline control', implying control of design details whether recorded on paper or on magnetic media. Configuration is defined as: 'The complete description of the product and the relationship of its constituent elements'. A particular approved configuration at a specific point in time acts as a 'configuration baseline'.

'Configuration control' is:

> The discipline that ensures that any proposed change and addition, modification or amendment to the configuration baseline shall be prepared, accepted and controlled in accordance with set procedures.

Software development life cycle

The life cycle of software development can be visualized as an activity of decomposition or analysis, starting from the software requirement and breaking this down into smaller and smaller units until small, manageable units of coding are reached. The system is then reconstructed, starting with these individual 'bricks' of code until the full structure has been recreated.

This is illustrated in the associated Figure 13.1. In this 'V' diagram (actually a 'U' as we have represented it) each stage of integration on the right can be verified by comparison with the corresponding stage of decomposition on the left. At each design stage, whether going down or back up the 'V' from writing the code unit, the completed stage is checked until correct (indicated by the feedback loop), reviewed and then becomes a controlled baseline before proceeding to the next stage.

The TickIT scheme

Although ISO 9000–3 provides some guidance concerning the application of ISO 9001 to software development, there was some concern in the UK that this guidance was insufficient, as was the level of software knowledge in many certification bodies and their individual assessors. As a consequence of this, the software industry and Department of Trade and Industry developed the TickIT scheme, which accredits certification bodies and registers auditors as competent to operate in this field. In 1993, day-to-day responsibility for the TickIT scheme was transferred from the DTI to DISC, the part of BSI Standard responsible for information systems.

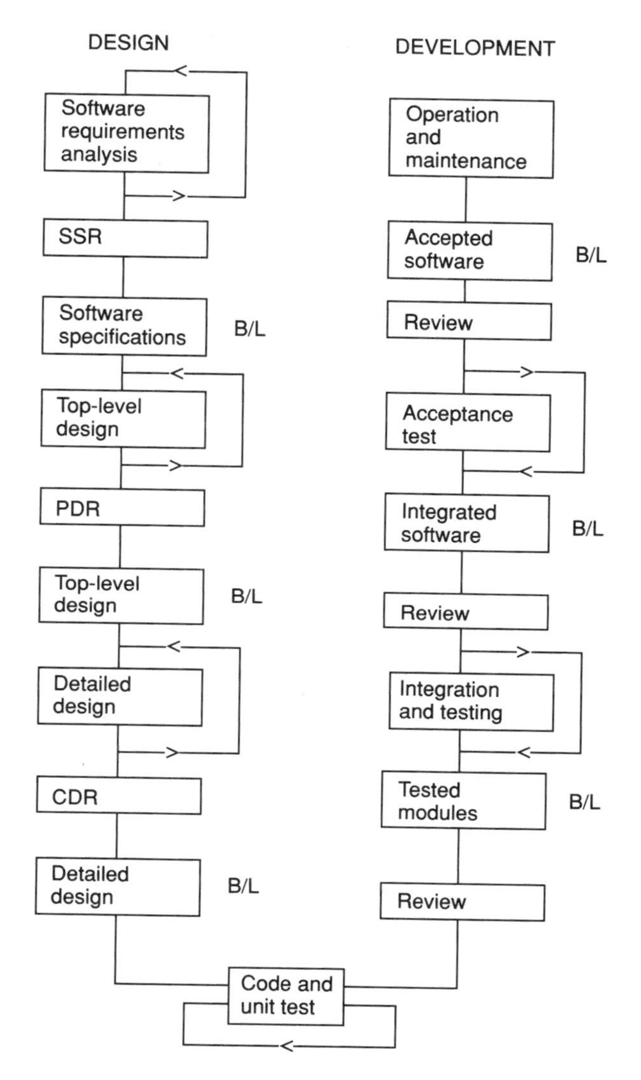

Fig. 13.1 Software development life cycle diagram. SSR = software specification review, B/L = baseline; PDR = preliminary design review; CDR = critical design review. Source: T. Manns and M. Coleman, *Software Quality Assurance*.

A TickIT *Guide to Software Quality Management Systems Construction and Certification* has been issued. Within the UK, TickIT certification to ISO 9001 (or in certain cases ISO 9002) is now required if the business being assessed involves:

1. The development of software;
2. Software embodied in the product;
3. The use of a software-based management information system, if it was developed or modified in-house.

The value of this reassurance has been reflected by the number of overseas companies seeking 'TickIT' certification of their ISO 9001 quality system.

Computer applications within the business

Introduction

Computers are now used so widely that their presence is often taken for granted. This means that it may not be obvious at first that the computer has to be treated as simply another piece of factory machinery and subject to the same quality assurance and quality audit disciplines as any other tool.

Computer-aided design

Computer-aided design (CAD) involves the use of a computer-based system for translating engineering concepts into engineering designs. This is achieved by means of programs holding information on:

1. Design principles; and
2. The key variables which place constraints on how the principles can be employed.

Operating with these data can, for example, allow three-dimensional perspective representations to be constructed in any orientation (computer modelling) or projected into two-dimensional drawings (computer-aided drafting).

The advantages of computer-aided design are:

1. **Speed**, especially in terms of evaluating quickly different alternatives;
2. **Accuracy**, providing the initial ground rules, data and programs for manipulating them are correct.

Computers in manufacturing

Computer-aided manufacturing is another generic term, and relates to any production system which utilizes computer control of manufacturing plant.

Examples are:

1. Computer numerically controlled (CNC) machine tools;
2. Robots.

CNC machine tools can include turning, drilling and other machining and sheet-metal working processes. Choice of workpiece, position, length of operation, sensing and feedback of process, etc. are defined numerically in the computer program.

Robots may be programmed similarly, or may be of the form where the robot is first 'led' through the operation by a human operator and the computer program 'remembers' the sequence it has been taken through. Paint-spraying robots, for example, work on this principle, the object being complex in shape but the positioning of the spray nozzle not exceptionally critical.

The advantages of computer-aided manufacturing are:

- Speed;
- Accuracy;
- Reproducibility;
- Independence of operator skill and fatigue level.

Computers in testing

This may be considered a subdivision of CAM, but since the computer functions through electronic signals, automated test equipment (ATE) is used at an especially direct level in testing computers and their components. Indeed it can be taken to the point where they become self-checking in many respects.

Computers in production control

Computers are widely used to record and make available information on orders received, stores inventory and capacity planning; to determine build schedules, to track work-in-progress, finished stocks and orders waiting to be filled, etc.

The advantages of a computerized system are:

- Improved control;
- Efficient utilization of resources;
- Inventory reduction;
- Reduced labour costs;
- Quicker response.

Implications for quality assurance

Computers can be employed advantageously in all aspects of business, and as computer power becomes progressively cheaper this statement can apply to even the smallest business.

Product or service quality becomes increasingly dependent on the quality (accuracy and fitness for use) of the computer software used. The quality management system has to recognize this.

Software writing and evaluation requires special skills which are not necessarily available within the QA department. However, it must be borne in mind that the same applies to many other skills and technologies over which quality control is exercised. The QA department's function is to ensure that specialists have created and promulgated the rules by which they will work, and that they are adhering to them.

Much of software quality assurance is similar to the quality assurance of written documentation. It features the need for generation by responsible people, vetting and approval, revision control and issue control.

Applications of computers within the QA department

This section will consider some of the applications for computer software within a QA department. The same safeguards on software quality as we have already described will apply in this context also. The purpose of this section is to show how the workload of the department can be lightened, and how its effectiveness can be increased.

There is already a wide choice of software packages for each of the applications to be mentioned, which are:

1. Word processing;
2. Spreadsheet analysis;
3. Data base management.

Word-processing

A word-processing program provides the means to compile, store, edit and retrieve text electronically. While the text is in computer memory it can be recalled and changed as often as desired. Finalized text can be printed on a variety of printers or sent, on disc or via modem to a typesetter for book, magazine or newspaper production.

The point to stress for those unfamiliar with word-processing is that text can be stored on, and retrieved from disc, and edited to an unlimited degree. In normal typing the insertion of a new paragraph would not only require the page to be retyped in full, but also all subsequent pages. The penalties are time, and the possibility of introducing new typographical errors. With word-processing, by contrast, the new paragraph would be inserted at the desired place and all subsequent pagebreaks would be automatically relocated according to the rules desired by the typist (e.g. breaks at paragraph ends only, no 'widows and orphans', etc.) The inserted text could be new, transferred from elsewhere in the document (where the gap would be closed up) or copied from another file.

While word-processing using a PC or computer terminal is useful in any clerical context, it has special advantages over a typewriter in document control applications. Its strengths extend far beyond the ability to correct errors, vary fonts, and produce rapid print-outs. For example:

1. When a set of operating procedures is being maintained, a document has to be updated every time an engineering change note is raised. The change is often minor – it might be a single number or an extra line of text. This amendment can take place without the risk of introducing errors into any other checked and approved text.
2. Quality specifications and operating procedures have the characteristic that they are written in standard house-styles and formats, and that there are often groups of related documents which are very similar in content. With a word-processor, standard 'templates' can be created which establish the required format and can also contain any text which will be common to a family of specifications. New documents can be generated by filling in this skeleton when they are required.
3. It is unnecessary to keep archieved hard copies of all the documents, especially obsolete versions. They can be kept on disc; at more than one location to provide cover in case of fire or other damage, yet still with a great saving in space. New copies can be quickly printed on demand.
4. In business where access to a computer terminal is universal no controlled copies at all need to be issued and recalled or destroyed. Documents can be called up for viewing at a monitor, with the knowledge that the version displayed is the latest issue. If a hard copy is needed for reference in a particular task a print-out can be run by the user on the understanding that it is an uncontrolled copy, dated and only guaranteed to be valid on date of issue.

Database management

A database is a collection of related information. Familiar everyday examples are telephone directories, price lists, library catalogues and mailing lists. Computer database management has been likened to an 'electronic filing cabinet'. The program offers a screen picture analogous to a file card, the format being defined by the user, into which data can be entered, stored and retrieved. This enables the data to be used in numerous ways, for example:

1. Like its predecessor the 'punched card' system it enables searches to be made for entries containing particular information; but with much more sophistication. For example, 'wild card' searches for words of which you know only certain letters.
2. 'Relational' database can compare two or more separate files and create new files for entries containing information drawn from different initial files.
3. Databases are also capable of carrying out calculations arising from the number of entries found, and on numerical data recorded within entry fields.
4. They can produce printed reports according to criteria defined by the user.

It will be clear that database management is the ideal tool for tackling the problem of extracting and making use of the data which can so easily become buried in inspection records. Many companies manage databases in this way, distilling from the masses of data which they generate periodic summaries on a weekly or monthly basis, from which key problem areas, improvement trends or fall-off can be monitored.

Spreadsheet analysis

A spreadsheet is an aid to calculation. It can be represented as an extended grid with rows and columns. If these are labelled 1, 2, 3 . . . and A, B, C . . . then each 'cell' can be defined by its row and column; e.g. cell C4 is the intersection of row 4 and column C (Figure 13.2). There will be thousands of cells in all.

	A	B	C	D	E
1					
2					
3					
4			← e.g. Cell C4		
5					
6					

Fig. 13.2 Spreadsheet matrix.

The user can assign to each cell text, numbers or mathematical formulae, and define the relationship between the cells. For example the figures entered at each of the corresponding cells could be combined in accordance with a mathematical formula to give an answer tabulated in column E. If any numerical entry was changed, or if the formula was changed, the whole matrix could be immediately recalculated.

Spreadsheet analysis is obviously very well adapted to accounting purposes. Its major application in the QA department might be in the preparation and control of the departmental budget. Once the budget has been set out, variables can be changed and the impact of various real or anticipated contingencies quickly assessed.

For example: 'What would be the effect on the QA department's operating costs of a 20% increase in factory production level?' Simply keying in the new figure, in place of the level originally assumed would result in the recalculation of all those departmental costs based on production level (inspectors required, overtime level, consumables used, etc.) according to the determined formulae.

As the capability of spreadsheet and database software has become more powerful, their capabilities have increasingly overlapped, and their usefulness broadened. Packages combining the attributes of both categories are extremely useful aids in, for example, monitoring quality-related costs.

Software specific to QA applications

The availability of personal computers and the activities of software consultants have reached the point where packages can be purchased which are designed specifically around quality assurance needs. Examples are:

Quality manual writing

When writing a quality manual for a quality system conforming to ISO 9001 for example, there are obligatory areas to be covered, an accepted format, and certain necessary statements which are almost stereotyped. Therefore it is not difficult to prepare a text which comprises a selected page format, section headings, an outline of the required contents, and certain passages written in full where they are of general application. It is then left for the purchaser of the package to insert through the word-processor the text which is specific to the particular company and any unique procedures.

Acceptance sampling

It is simple in principle to produce a program which incorporates all the sample size information, tables of acceptance numbers, choices of AQL, and switching rules of BS 6001. A user can work from the computer terminal instead of from the document itself, enter is lot size and AQL, and be told sample size and accept number for single, double or multiple inspection under 'tightened', 'normal' or 'reduced' inspection.

If the program also accepts and stores the inspection data it can indicate whether normal, reduced or tightened inspection should be in force, and whether the sample has passed or failed, on the basis of its data base of historical performance.

Statistical process control

Data can be fed into the program which will then calculate process capability, warning and action limits. It will also indicate to the interrogator when subsequently entered information shows that the process has gone out of control.

Scheduling

'Organizer'-style programs which keep a calendar to indicate when actions or appointments are due provide the ability to keep track of actions due on items from a body of data. These can be employed in scheduling and monitoring status of things like calibration recall and audit scheduling.

Summary of conclusions

In this chapter we have seen that:

1. The quality system demands safeguards on the quality of computer software.
2. ISO 9000–3 provides the basis for quality assurance of computer software.
3. Quality assurance of software is important because of the many uses to which it is put in the areas of engineering and design, in manufacturing and production control, and in record-keeping and stock control.
4. Within a QA department itself, computers and their software are used for word-processing, scheduling, database management and spreadsheet calculations.

14

Reliability assurance

Introduction

This chapter deals with the management of a special aspect of quality; namely reliability which, in essence, is the maintenance of fitness for use over an extended period of time.

Thus reliability is a critical test of the supplier's ability to control quality, since it cannot be fully safeguarded by appraisal activity within the supplier's organization; it demands preventive actions especially careful design, and rapid and effective corrective response to any problems reported from the field.

What is reliability?

The definition of reliability is:

> The ability of an item to perform a required function under stated conditions for a stated period of time. (ISO 8402)

Reliability is thus closely related to quality; it is in fact simply quality maintained for a required period during use under realistic conditions. Many important lessons which this book emphasizes in the context of quality apply equally to ensuring reliability.

There is one major difference between managing quality and managing reliability; whereas quality can, for the most part, be assessed by the maker or provider before the product leaves his/her hands, reliability 'over a stated period of time' can only be assessed directly by the user.

Thus the manufacturer or supplier faces distinct problems in the areas of:

1. Predicting reliability in the user's hands;
2. Simulating the conditions of service in accelerated fashion, in order to provide guarantees of reliability;
3. Obtaining data on the product's actual performance in service, from the users.

We shall review some aspects of reliability in this chapter.

In addition, British Standard BS 5760: *Reliability of Constructed or Manufactured Products, Systems, Equipments or Components* is an extensive (178 pages) introductory guide to the topic. Figure 14.1 is reproduced from BS 5760.

Reliability and the market-place

We can say that reliability is an aspect of quality, but really that it putting the cart before the horse. Quality is a special case of reliability. It is the state of reliability at time zero!

The manufacturer has the product in his/her hands until it is released to the customer. In the meantime all the necessary checks can be carried out to ensure that it fully meets the quality requirements of its grade. After delivery, there is little that can be done except respond to the information the customer cares to send back about it.

This is not the complete story. If a manufacturer also undertakes to service the product (remember that there is provision for this in the ISO 9000 documents) he/she can provide preventive maintenance, do repairs and replace faulty parts. In the case of a serious fault, the product can be recalled and a redesigned sub-assembly can be retrofitted.

The consumer market-place

In our introduction to quality and the market-place, in Chapter 1, Part One, we said that the consumer does not have a clear specification in mind when buying a product, but rather that the supplier's marketing department has developed a specification which they think represents the consumer's unexpressed needs.

Using the illustration of a transistor radio; the consumer may not know or even be able to sense the extent to which the specification on output wattage, signal distortion, tuning discrimination, or whatever aspect of the quality and performance specification has been met. But reliability is a different matter. If the radio develops an irritating intermittent fault, if the owner has to start replacing batteries frequently because of increased power consumption, or especially if the radio ceases to function altogether, the owner is keenly aware of this.

The non-technical consumer can therefore detect deterioration in performance, i.e. unreliability, much more easily than he can compare the initial performance with the manufacturer's specification. Most of the reputations for 'quality' won or lost in the consumer segment of the market-place have been won or lost on perceived reliability.

We will leave the consumer market-place now, and switch our attention to the professional market-place.

Responsibilities of the user

For a product to achieve its planned reliability there are two obligations which the user has to accept. These are:

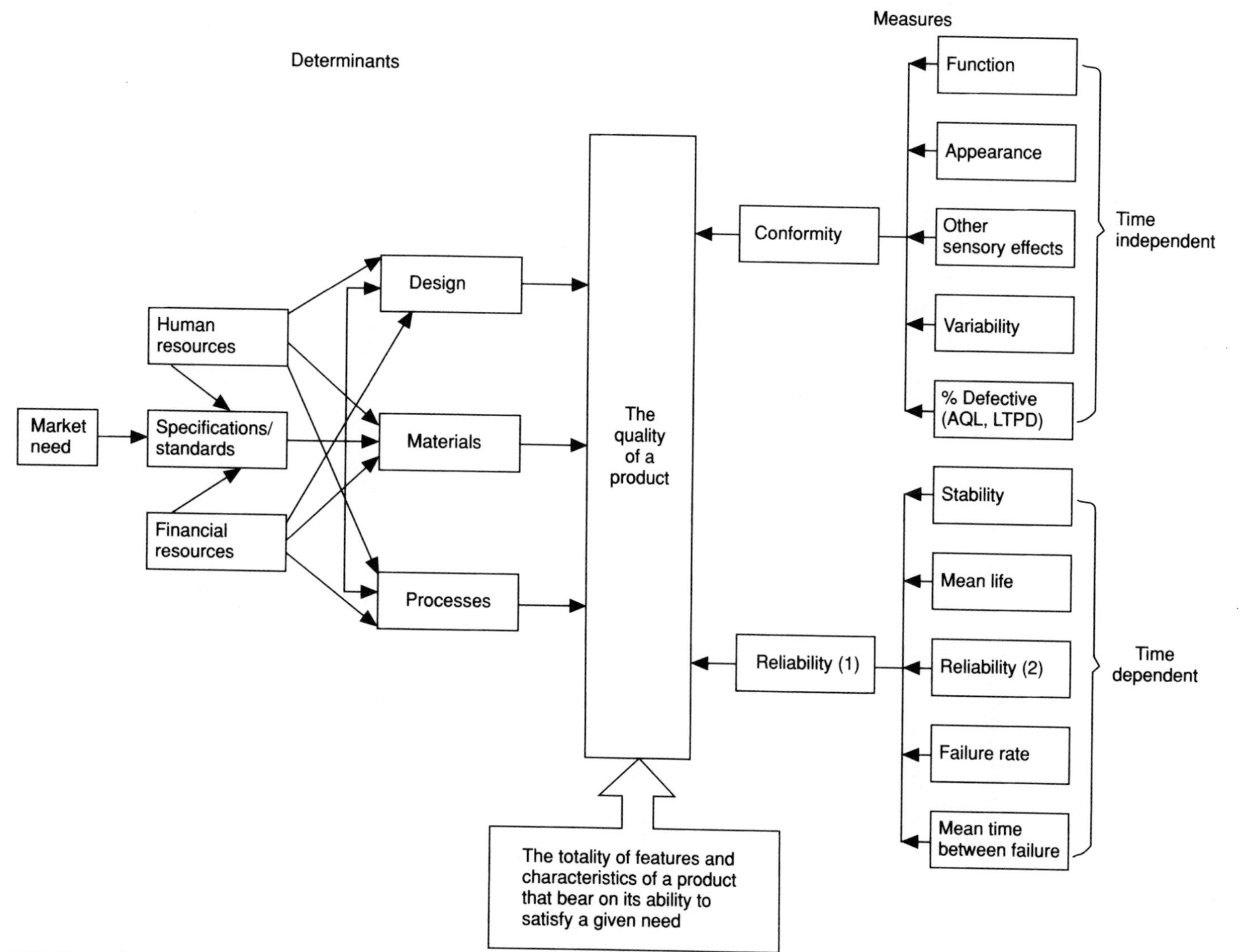

Fig. 14.1 Determinants and measures of quality and reliability. Source: BS 5760.

1. The product should only be used in the manner, and in the environment, for which it was designed.
2. The product should be maintained in accordance with the manufacturer's recommendations.

It also helps the manufacturer to eradicate problems and develop the product if users take the trouble to report back details of any instance of unreliability experienced.

Cost considerations

The comments on regular and appropriate servicing remind us that the need for reliability and the cost of unreliability both involve the customer in expense. Let us consider two cases.

Cost of reliability for the producer

This is represented in Figure 14.2, which is taken from BS 5760. The manufacturer will incur costs if the product is unreliable. By improving reliability the scrap and rework costs incurred can be reduced.

However, the model indicates that there is a point where the cost of building in increased reliability can no longer pay for itself, in terms of saved scrap and rework costs.

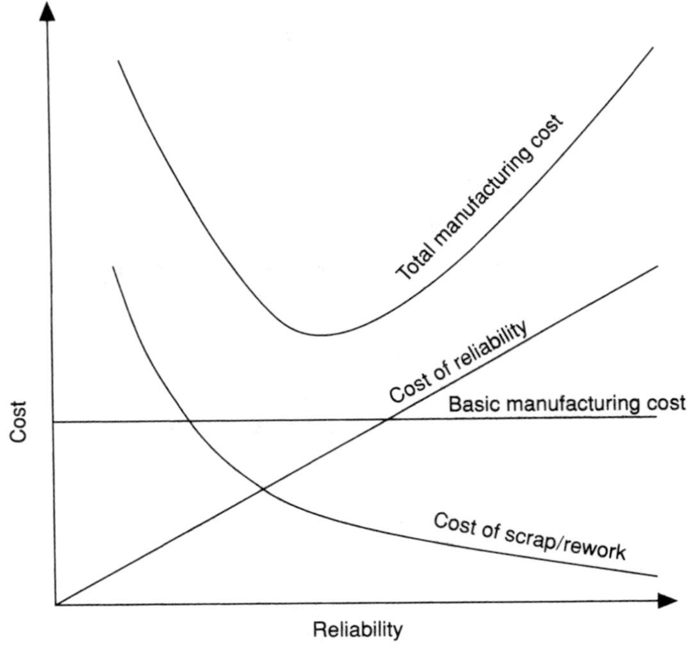

Fig. 14.2 Comparison of cost and reliability for a manufacturer or producer. Source: BS 5760.

Cost of reliability for the purchaser, owner or user

The customer needs to consider, and wants to minimize the complete life-cycle cost. This is the total sum that has to be paid out over the life of the product for:

1. Initial purchase cost;
2. Operating cost;
3. Maintenance cost;
4. Failure and consequential costs.

These elements in combination, are taken as representing the total cost of ownership.

The graphical representation given in Figure 14.3 shows that the maintenance and failure costs plus consequential costs can be reduced by purchasing a product with superior inherent reliability even if, as the graph assumes, the more reliable product will command a higher price.

Once again, the argument assumes a compromise must be established, since it supposes that complete reliability can only be purchased at a price which outweighs the other savings.

Comparison with quality-related costs

We shall deal with quality-related costs in detail in Part Four, Chapters 23 and 24, which focuses more strongly on the cost and profit implications of good and bad

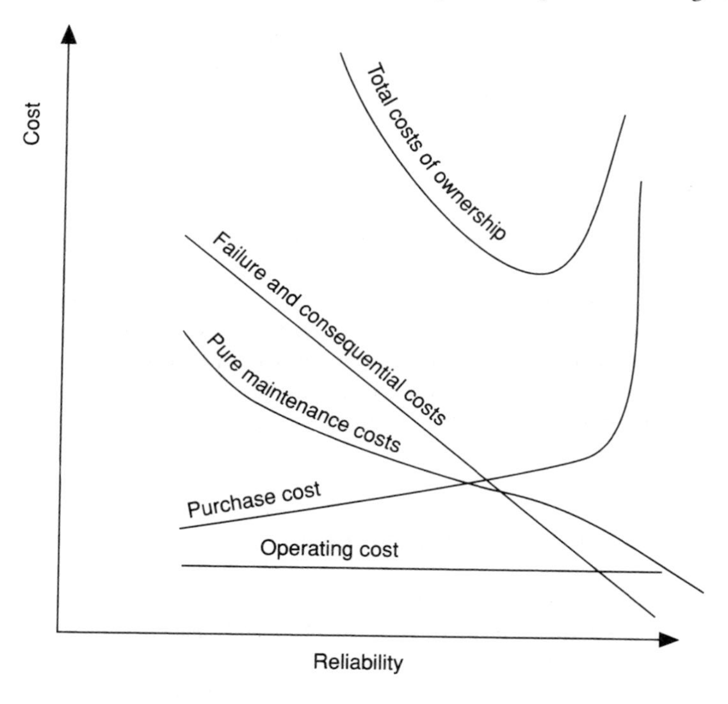

Fig. 14.3 Comparison of cost and reliability for a purchaser, owner or user. Source: BS 5760.

quality. You may wish to refer back to this section when you reach that stage of the book.

Remember that all the actions we are speaking of in this chapter are intended to safeguard reliability, and that all the outcomes of reliability failure cost money. This burden may fall to the user, or be handed back to the supplier or manufacturer.

The cost to the manufacturer/supplier of unforeseen expenses suffered by the customer due to unreliability, and which cannot be recovered from the supplier is – lost future business! The Ford Motor Company in the USA did some market research which showed that Ford buyers who were pleased with their purchase told on average eight other people. Buyers dissatisfied with their cars passed the word on to 22 potential customers!

Organizing to achieve reliability

A prerequisite of achieving reliability, as with achieving quality, is that there is a commitment to this goal from the highest levels of company management. Indeed, the management of reliability is inseparable from the management of quality. Reliability requirements should be an integral part of the quality management system.

Designing for reliability

Quality aspects of new designs can be evaluated at once. This can only be established partially and indirectly for issues affecting reliability: only time will tell for certain. Therefore there is an important distinction to be drawn between 'evolutionary' and 'revolutionary' designs.

Reliability of evolutionary designs

Designs which have evolved from existing, proven designs have a basis of reliability experience on which to build. This will still be valid except to the extent that the new designs differ from the old. If the old design was reliable and the new one will be used in a similar environment, the only aspects of the design which need special scrutiny are the innovative aspects.

Reliability of revolutionary designs

Revolutionary designs involving new principles or techniques do not have this background of prior knowledge and assurance. Moreover, radical new designs are often associated with the need to operate in more demanding environments, e.g. in the development of military supersonic aircraft, where not only was it necessary to use new materials in some areas but also there was no direct evidence that could be referred to concerning the behaviour of the more orthodox parts of the design at sustained supersonic speeds.

In contrast to the revolutionary design of the early military supersonic aircraft, when a civil supersonic plane (Concorde) was produced the design was more evolutionary, because the lessons learned in designing and operating military planes at supersonic speeds could be applied.

In such situations intensive testing is essential in order to discover any weaknesses in radical new designs.

Performance rating and reliability

As implied earlier, there is a risk of reducing reliability in the search for enhanced performance. An increased level of stress, e.g. uprating the power output of an internal combustion engine or increasing the permitted temperature rating of a silicon microchip can cause a significant reduction in reliability. Equally, a minor derating of performance can cause a significant improvement in reliability.

Measurement of reliability

Reliability is specified either as:

1. A mean **time** to failure (MTTF) or between failures (MTBF) (e.g. 1200 hours running time);
2. Or as a failure **rate** per unit time (e.g. 0.01% per 1000 hours).

Of these the latter is a convenient measure for components, and the former for systems.

Behaviour of failure rate with time: the 'bath-tub curve'

We have defined reliability as a frequency or rate of failure: but this frequency or rate is not itself constant, so we need to examine the behaviour of failure rate with time.

The 'bath-tub curve' shown in Figure 14.4 is a familiar term for a widely followed behaviour pattern of reliability with respect to time. It was first observed as the distribution of statistical data collected for Life Insurance purposes on age at death of human populations.

There are three phases, or segments of the 'bath-tub':

1. An initial failure rate which is high but falling;
2. A much lower level which can be sustained for a long time;
3. A stage where the rate of failure starts to rise increasingly rapidly until every unit has failed.

Using the analogy with the human condition, the first stage where defects are due to individual defects in material or assembly is analogous to young children dying

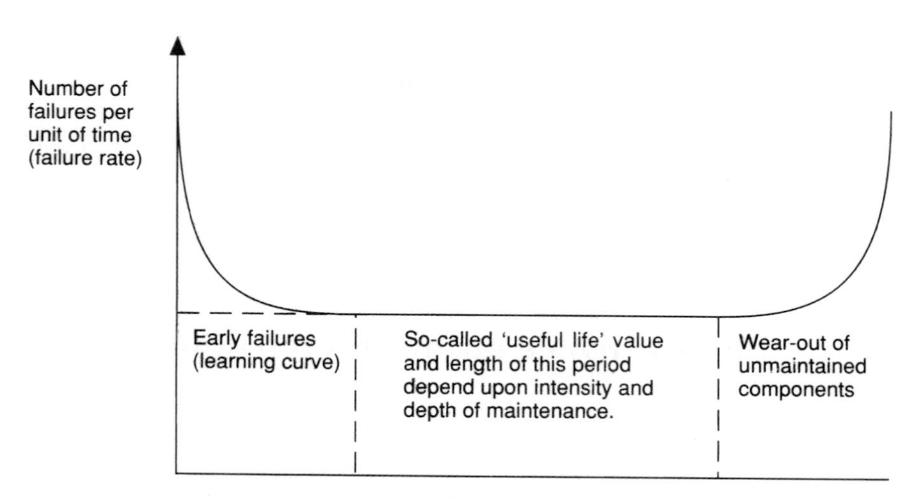

Fig. 14.4 The 'bath tub' curve. More frequent maintenance lowers the bottom of the curve subject to diminishing returns. Taking more components into maintenance prolongs the flat portion. Scheduled preventive maintenance correctly carried out acts on **components** to reduce average failure rate of the **system**.

as a result of defects existing at birth. This portion of the curve is often referred to as 'infant mortality' even in an engineering context.

There then follows a prolonged period when the failure rate is low, and fairly constant. For many products the failure rate is so low in this region that it is difficult to quantify it at all, from available data. This is 'maturity'.

In the final region all surviving units fail over a relatively short period, determined by the limitations inherent in the materials, methods and design used, relative to the environment in which the product has to operate. By analogy you might expect this region to be called 'old age' but in fact it is referred to as 'wear out'.

The considerations I have mentioned above apply to all the components of the overall product, but what can we say about the complete system?

Reliability of a system

Redundancy

Firstly, we can say that the system is only as strong as its weakest link. Therefore, to safeguard the functioning of critical parts of the assembly it may be necessary to introduce redundancy. That is, to duplicate sensitive parts of the system so that if one fails, the other, still functioning, enables the system to continue operating.

The expression 'belt and braces' may come to mind, and it is very apt, describing a system for keeping trousers up which involves redundancy.

Redundancy can involve both sub-systems operating simultaneously (active redundancy) or alternatively one being held in reserve (standby redundancy).

The concept of redundancy can be illustrated as follows. Suppose we have a system composed of a **series** of subsystems as shown in Figure 14.5.

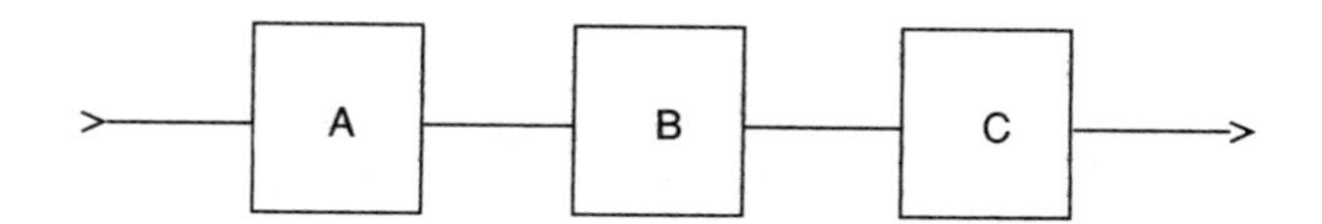

Fig. 14.5

We can assign a probability of failing within a given time P to each subsystem. (The probability of it **not** failing is $1 - P$). The probability of the whole system not failing is the product of the individual probabilities, i.e. in this case P_S is the probability of the system failing. P_A, P_B, P_C the probabilities of the subsystems failing, then:

$$(1 - P_S) = (1 - P_A) \times (1 - P_B) \times (1 - P_C)$$

If P_A and P_C are each 0.02 but P_B is 0.2 then:

$$1 - P_S = 0.98 \times 0.8 \times 0.98 = 0.768$$

So the probability of the system as a whole failing is:

$$P_S = 1 - 0.768 = 0.232\text{, nearly a 25\% chance.}$$

However, for a system of **parallel** alternative subsystems, as in Figure 14.6, the probability of all failing is the product of the individual failure rates:

$$P_S = P_X \times P_Y \times P_Z$$

In Figure 14.7 the most unreliable subsystem, B has been duplicated. The probability of failure of B is now:

$$P_B = P_{B1} \times P_{B2} = 0.2 \times 0.2 = 0.04$$

and for the whole system:

$$1 - P_S = 0.98 \times 0.96 \times 0.98 = 0.922$$

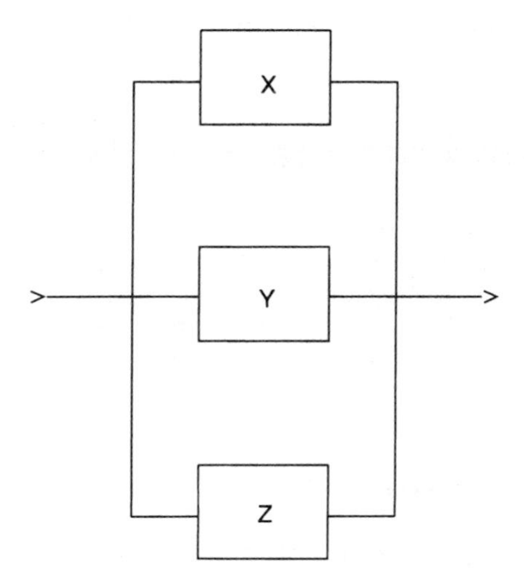

Fig. 14.6

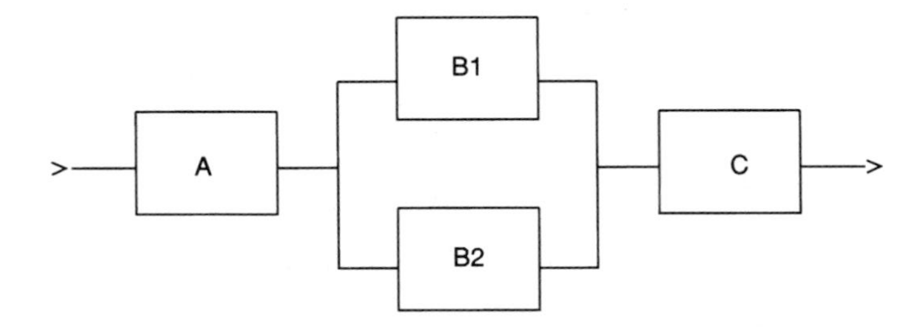

Fig. 14.7

Probability of system failure is now:

$P_S = 0.078$ compared with the previous 0.232.

Maintainability

Because of redundancy, a system does not necessarily fail when a component fails. However, repair is necessary if all alternative redundant paths are not ultimately to fail. Thus instead of mean time to failure we can talk about mean time to repair (MTTR). Routine maintenance is also required, and the frequency of this may also be specified (MTTM).

Availability

What actually counts in practice is how much of the time the system is actually available to earn money. This is called its availability, which is thus a measure of the overall impact on the user of reliability, and maintainability including both repairs and preventive maintenance.

A reliability and maintainability programme

A programme to achieve optimum reliability and maintainability from a product requires methods and disciplines very similar to the quality management system explained in Chapters 7 and 8, of Part Two.

BS 5760 identifies the elements of a programme as follows, taking the phases of the programme in chronological sequence and identifying activities within each phase.

Definition

1. Reliability feasibility studies;
2. Statement of reliability objectives and requirements;
3. Reliability specification and contract formulation.

Design and development

This phase includes initial (prototype or pilot) manufacture:

1. Analysis of parts, materials and processes;
2. Analysis of established and novel features;
3. Failure mode, effect and criticality analysis;
4. Incident sequence analysis (fault tree analysis);
5. Stress and worst-case analysis;
6. Redundancy analysis;
7. Human factors;
8. Design change control;
9. Design review;
10. Design audit;
11. Safety programme;
12. Maintainability programme;
13. Parts and sub-assembly testing;
14. Performance testing;
15. Environmental testing;
16. Accelerated testing;
17. Endurance testing;
18. Reliability growth testing;
19. Data collection, analysis and feedback.

Production

1. Preservation of reliability achievement;
2. Quality conformance demonstration;
3. Screening (also known as run-in, bed-in, burn-in, etc.) of components and assemblies;
4. Reliability demonstration;
5. Additional software check.

Installation and commissioning

1. System acceptance testing;
2. Commissioning tests;
3. Quality assurance;
4. Reliability growth;
5. Reliability and maintainability demonstration;
6. Data collection;
7. Reliability and maintainability assessment.

Function and maintenance

1. Data collection, analysis and feedback;

2. Redesign/modification;
3. Maintenance.

Notes

1. **Failure mode, effect and criticality analysis**
   The FMECA technique, mentioned under 'Design and development' (point 3), is discussed later in the chapter.
2. **Incident sequence (fault tree) analysis**
   This technique, mentioned under 'Design and development' (point 4), is similar to the Ishikawa/fishbone analysis of which an example will be given in Chapter 18, Part Three.
3. **Reliability growth testing**
   This figures in 'Design and development' (point 18) and 'Installation and commissioning' (point 4). It is the name given to a programme of testing, failure investigation, instigation and appraisal of intended corrective actions.

Figure 14.8 is a simplified block diagram, reproduced from BS 5760, illustrating the reliability and maintainability programme.

Failure mode, effects and criticality analysis

Sometimes known merely as Failure Mode and Effects Analysis (FMEA), FMECA and FMEA are in fact the same technique, except that FMECA attempts to assess the seriousness of the effects induced, i.e. their criticality.

Application of FMEA/FMECA

FMEA or FMECA is essentially an analysis of a product carried out at the design stage, in order to predict the effects of any possible source of failure of a component or sub-unit of the complete product. The information obtained is then used to anticipate and avoid or mitigate the results of such failures, especially the ones which have the most serious effects. Actions are then taken on the findings before the design is finalized.

FMECA is a tool of quality assurance for use at the design stage. Its particular aim is to look at the likely impact of delayed failures, after acceptance by the customer and it is, therefore, especially important as part of a reliability programme.

Example

Let's take as an example a heart pacemaker, which contains a particular transistor. In use, the pacemaker is surgically implanted into a patient in order to strengthen and regulate the heart-beat.

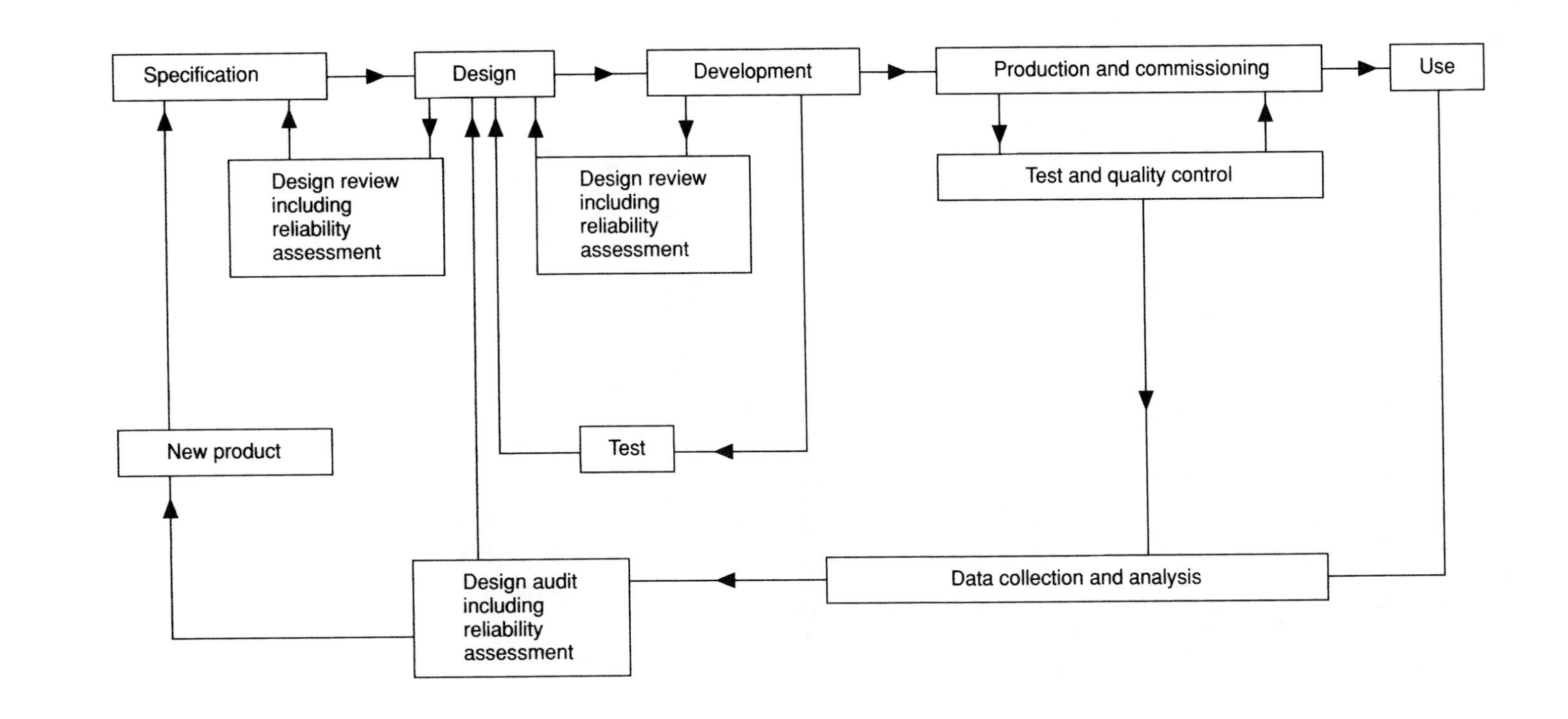

Fig. 14.8 Simplified reliability programme concept. Source: BS 5760.

The analysis is not concerned with the failure **mechanism** of the transistor, only with the mode, i.e. the symptoms of the failure. The mechanisms of transistor failure might include broken bonding wire, ingress of moisture, etc. but FMECA is only concerned with how each of these problems might manifest itself. For a multiplicity of possible failure mechanisms there might be only three possible failure modes:

1. Impaired amplification;
2. Open circuit;
3. Short circuit.

The task of FMEA would be to work out the effect on the functioning of the complete pacemaker circuit of each of these three contingencies. Criticality analysis would make a judgement on the seriousness of each of the various possible effects. Thus a failure mode which had the result of slowing down or speeding up the pacemaker might disable the wearer until medical attention can be administered, but a failure mode which caused the pacemaker to cease functioning altogether would be fatal.

In regard to the likely outcome of a component failure, if the risk of a critical system failure is unacceptable, the offending component must be designed out of the system, or else the system has to be redesigned to be more tolerant, e.g. by introducing redundancy.

Summary of conclusions

In this chapter we have seen that:

1. Reliability is the extension of quality over the useful life of the product;
2. Consumer satisfaction embraces not just reliability, but availability and lifetime cost of ownership;
3. The assurance of reliability demands a programme of safeguards not unlike those built into the quality system;
4. Prevention is essential to achieving reliability;
5. One of the key techniques in prevention is FMECA, which anticipates how things **could** go wrong, how serious the effects would be, and tries to counter these possibilities at the design stage.

Part Three

Using People Effectively

Introduction

In Parts One and Two of this book we have discussed the nature and management of quality, and the specialist 'quality assurance' functions and their techniques. We have stressed that quality is 'providing what the customer wants', and that the provision of quality requires the participation of all employees. This being the case, the effectiveness of any quality system demands that the whole workforce understand the principles involved, and have adopted 'ownership' of its responsibilities.

For an organization to become 'world class' arguably the most difficult step is how to achieve what Crosby calls the 'quality innoculation'. For the quality message to be accepted outside the quality assurance department itself, the missionaries must understand something of the psychology of people at work, both as individuals and of groups. This is not to say that one can, or should ever attempt, to change their personalities or even set out with the explicit objective of modifying their attitudes or behaviour. It is simply a warning that the task can be slow and prone to disappointments, and that there are better and less good ways of getting there.

15

The sociology of people at work

Introduction

This first chapter of Part Three will highlight the work of some theorists who have had a major effect on the ways people have tried to increase the effectiveness of their workforces. You will see that the methods of each of them has been strongly criticized by other theorists, who have questioned their methods, or failed to reproduce their findings. That will not trouble us unduly. Their ideas are a framework around which others have created working environments, and we can consider what has been found to happen in practice.

The team

Researchers have determined that the ideal team size is between 4 and 12 individuals; this has even been likened to the manageable 'hunting group' size of our forefathers before they settled down to agriculture, with the implication that this has a very strong cultural or even genetic grounding.

On this view, many of the problems of an industrial enterprise relate to maintaining communication, motivation and common goals within a much larger group than the optimum size.

Scientific management

The expression 'scientific management' refers to a particular approach to organizing tasks arising from the work of F.W. Taylor in the USA, around the beginning of the present century. These methods have proved very powerful and productive, yet they appear to be totally opposed to the precept of a small, cohesive working group.

A well-documented illustration is Taylor's reorganization of the materials stockyard at the Bethlehem Steel Works.

The 'assembly line'

The principles of Taylor's work underpin the 'assembly line' philosophy, with each job carefully studied and defined by specialists, deskilled and restricted in scope so far as was possible, and integrated into the needs of the production machinery. This was particularly attractive in the USA with its abundant resources, socially mobile population, but shortage of skilled craftsmen.

To this approach are owed such successes and benefits as the Ford 'Model T' automobile, with the whole operation planned to maximize throughput and minimize cost (as an illustration, it is said that Henry Ford specified the exact size and construction of the wooden packing cases in which his bought-in components were to be delivered, so that the planks could be reused as the floor-boards of the car).

The way in which Model T Ford production was organized enabled a motor car to be offered, for the first time, at a 'people's' price. At the same time there was a penalty: inflexibility. This is well illustrated in Henry Ford's dictum that a car could be 'Any colour the customer likes, so long as it's black'.

Work study

Taylor's methods gave rise to the development of project management and work-study methods. His disciples included Gantt and Gilbreth, whose names are remembered in the Gantt (or Bar-)Chart and the Therblig (a work-study unit).

So, Taylor and his followers developed a rational synthesis of the 'one best way' of performing a job, resulting from an analysis of the job into its individual task elements and attempting to optimize each of these. The organization itself comprised managers, specialists (tool-makers, inspectors, expediters, work study engineers, etc.) and included a workforce whose job was defined in detail, and who were paid by throughput.

Examples of Taylor's experiments

Taylor conducted his work in the US around the period 1890–1910. His ideas were essentially practical rather than theoretical, so they are best described by examples.

Large steel works require vast quantities of raw materials: coal, iron ore, pig iron, lime, etc. In Taylor's time, these were brought in by rail, unloaded, stacked and moved to the place of consumption by manpower – no bulldozers or dumper trucks then! The yard where the materials were stored at the Bethlehem Steel Works (where Taylor was employed) was enormous, about 1 to 2 miles long by up to 1 mile wide. He set out to revolutionize the productivity associated with manual handling of these materials, developing what he called a 'science of shovelling'. His experiments took the form of:

1. Firstly, establishing the capability (in terms of tons shovelled per shift) of an experienced labourer using the standard shovel provided, for each of the familiar materials;
2. Then, using shovels of different sizes and weights, establishing the optimum shovel for each material and how much could be moved with it;
3. Also, by studying the styles of different labourers, establishing how the shovelling should be done for maximum efficiency.

The yard was then reorganized to take account of Taylor's findings.

- Stocks of shovels of appropriate sizes for the different materials were kept, instead of each man owning his own shovel.
- A daily work roster was created, so that when each man clocked-in he could draw from store the appropriate shovel for the job in hand.
- Training was given in the correct way of handling the shovel until each man was capable of shovelling the standard amount of material per day, i.e. that which the research had shown was achievable.
- Each man's daily output was measured and if he shifted the standard amount, the next day, when he clocked-in, he received a chit which entitled him to a pay premium of 60% above the standard rate.
- If he failed to shovel the target amount, he would receive a distinctively coloured warning pay chit, and within the next few days he would be taken off the job for retraining under the direction of an experienced labourer.

The method incurred extra expenses in work study, planning, tools and storage (all those extra shovels), clerical work and higher wages; but after three years, the company's statistics showed that the cost of handling materials had been reduced from between 7–8 cents to 3–4 cents per ton. The number of labourers needed, initially between 400–600, had been reduced to 140.

Taylor claimed that after the work had been restructured the morale of the men (the 140 survivors I suppose, though possibly not the 300 or more made redundant) and their regard for management was very high, because management 'had taught them how to earn 60% more wages' and because, if their output slipped, they were coached until they regained the standard rather than being summarily dismissed.

However, we draw attention to the way in which Taylor's approach established a division between the experts who defined in minute detail what the job was and how it should be done, and the operators who were paid a premium for doing exactly as prescribed without thinking or experimenting.

Benefits and defects of 'Taylorism'

Taylorism enabled the 'assembly line' to be achieved, and encouraged all jobs to be visualized in an 'assembly line' fashion, even if there was no tangible conveyor driving the sequence of events. Hence the 'tyranny of the production line', with 'stopping the line' the most unforgivable of crimes. Even faulty product must be passed on to be inspected later, and rectified at the end of the line, rather than disrupt the flow of production.

Hence to Taylorism we owe two of the problems besetting quality specialists: 'Pass it on, if it's really wrong the inspectors will find it' and: 'Sorry, guv'nor, I'm not paid to think'.

Taylor viewed workers as almost entirely economic in their motivation. If the labourers in the stockyard adhered to the method of working prescribed for them by Taylor they could achieve their target quota with the least expenditure of energy, and earn higher wages than in rival labouring jobs which were not scientifically managed and hence less efficient. They were not required to understand the experimental basis on which their jobs had been formulated, and they certainly weren't allowed to look for a better way themselves. The experts had already decided the best way. Thus scientific management has brought penalties as well as rewards, as illustrated in Figure 15.1.

It has been implied that Taylor's 'economic' picture of worker motivation gave rise to the attitude which McGregor (1960) described as 'Theory X', which we shall discuss later. In the meantime we want to highlight a very influential series of experiments carried out during the 1920s and 1930s, which changed management ideas about motivation.

Fig. 15.1 'The successes and failures of Taylorism.' Source: Hutchins, *Quality Circles Handbook*.

The Hawthorne experiments

This well-known and influential series of experiments was carried out over a number of years at the Hawthorne, Illinois plant of the Western Electric Company. Initially planned on strict 'scientific management' lines, the anomalous findings from the early stages resulted in the whole direction of the programme being changed.

The experiments were initially conducted by Western Electric personnel, and later Elton Mayo of the Harvard Business School was called in to advise, and interpret the results.

Level of illumination

Two of Western Electric's own engineers conducted these experiments. They varied the level of lighting in one of the assembly rooms, in order to monitor the effect on output. Initially the tests monitored the effects of increasing the level of illumination, and corresponding increases in output were noted.

As the experiments progressed, the researchers found that although original lighting levels were restored or even reduced, output continued to rise. Eventually two groups of workers were segregated. One was not subjected to any variation at all, in order to serve as experimental controls, but output in both groups continued to improve.

Follow-up tests

In a follow-up to the original tests, six women who worked in the relay assembly test room were rehoused in a separate room and the effects on output of changing times and durations of rest and meal breaks was monitored. Again, in a situation where experiments were conducted to evaluate the effect on productivity of the subjects' well-being, productivity rose steadily as each change was made regardless of whether the newly introduced conditions were more or less congenial than before.

While the relay assembly test room experiments were still in progress, Mayo recognized that the improvements in productivity were the result of the subjects responding to the fact of the experiments, not the result of the experimental conditions. Management had never before demonstrated any concern about the conditions in which the employees worked, nor consulted them about how they could be improved. The improved productivity was the subjects' response to their new perceived status.

Interview programme

Mayo began to interview employees to examine what they felt about their work, conditions and supervisors. The interview programme showed that relations with fellow workers was an important factor in their satisfaction, or otherwise, with their jobs.

Bank-wiring experiment

A bank-wiring experiment segregated the 14 men working on this job from other groups, and monitored their interrelationships and performance. The researchers found that the group, as a team, established what was accepted as a reasonable daily output and forced those members who might have preferred a higher or lower work rate to conform. Neither supervisors' targets nor incentive schemes could influence this or increase output beyond what the group or its informal leaders considered was proper.

Conclusions

We can conclude that:

1. Individual employees must not be seen in isolation but as a group;
2. The need to belong to a group, and have an acknowledged position within it, may be more important than monetary incentives or improved working conditions;
3. Informal groups, without official recognition, can exercise a strong influence on behaviour;
4. The needs of the company and of the informal groups have to be in harmony for an enterprise to be as productive as it is capable of being.

Theories of human motivation

We are now progressing from the practical achievements of Taylor and the pragmatic studies of Mayo to theories of motivation. We briefly mention some of these, chosen more because of their popularity with practising managers than because of their approval by academics.

McGregor – Theory X and Theory Y

McGregor used the tags 'Theory X' and 'Theory Y' to label two contrasting management principles, that he saw exemplified in practice:

1. Theory X managers assumed their workpeople to be lazy, needing coercion and constant supervision, avoiding responsibility, seeking security and motivated only by money;
2. Theory Y managers believed their workpeople were reliable, trustworthy and concerned for the company's welfare, willing to accept responsibility and taking pride in their work.

McGregor found that these assumptions tended to be self-fulfilling; employees' behaviour tended to mirror the employer's assumptions, because those assumptions were the basis upon which company rules and methods were based.

Maslow – the hierarchy of needs

Maslow was a psychologist who studied human motivation. He proposed that human needs form a hierarchy – only when the most pressing level was satisfied would attention be directed to the next level. Maslow originally divided his hierarchy into five levels, though other theorists have modified and re-presented these.

The presentation I shall use distinguishes six levels, grouped in pairs, the most basic at the bottom:

1. Self-realization 2. Self-actualization	'Individual needs' or 'Growth'
3. Esteem 4. Love	'Social needs' or 'Relatedness'
5. Safety 6. Physiological	'Biological needs' or 'Existence'

The most primary needs are for the immediate essentials of survival – air to breathe, then food to eat and drink. After that safety and comfort – warmth, shelter and protection.

When these are filled, we automatically start to take stock of our social situation – family, neighbours, workmates; our position in these groups, the affection and regard in which people hold us.

Only when we are secure in, and satisfied with our position within these groups do we examine whether we could achieve more than our group expects from us. This is to enhance our own ambition and self-regard (self-actualization) and perhaps achieve our full potential and self-knowledge of our ultimate limitations (self-realization).

Herzberg: motivation and hygiene factors

Herzberg's theory arose out of a series of interviews with professional people on what they liked and disliked about their jobs. He found that these topics formed two separate groups – that is to say, the things which caused the greatest dissatisfaction when they were absent were not the same as the ones which caused the greatest satisfaction when they were present, and vice versa.

Herzberg called the issues which could give rise to job satisfaction 'motivators', and those which could give rise to serious job dissatisfaction 'hygiene factors'.

The 'satisfiers' or 'motivation factors' were found in things such as:

1. Achievements;
2. Recognition;
3. Responsibility;

4. Promotion;
5. The nature of the work itself.

The 'dissatisfiers' or 'hygiene factors' were things like:

1. Restrictive regulations;
2. Unsatisfactory supervision;
3. Poor relations with fellow employees;
4. Low salary;
5. Poor working conditions.

The difference between these two sets of factors might be summarized as being that the motivation factors relate to what you have done to enhance the job, whereas the hygiene factors relate to external factors which the individual cannot influence. This gives rise to one of the criticisms of Herzberg's theory, namely that we tend to take credit for our job successes but blame the failures on factors outside our control. On this argument, the interview findings were bound to follow the pattern Herzberg uncovered!

Another problem is the question of money. Salary was one factor which frequently figured as both a motivator and a hygiene factor. However, pay is both an economic need and a symbol of status or worth; thus insufficient pay can be a dissatisfier, but generous pay a satisfier, seen as a recognition of outstanding performance.

The suspicion remains that one attraction of this theory to executives has been their belief that Herzberg had proved that pay level was less important to employees than they had assumed. In a television interview Herzberg admitted that it had been difficult to disabuse people of this misapprehension – 'In the end I had to put my consultancy fees up to get the message across.'

A people-based philosophy

Hutchins, in his *Quality Circles Handbook*, calls for a 'people-based' philosophy that can enable quality to be achieved by means of participation rather than control. By this he doesn't mean a 'Human-relations' based style of working, or the provision of abundant 'benefits' (hygiene factors). Rather the aim is that all employees can use their talents to help achieve the difficult goal of a quality product or service. Hutchins sees this as 'Taylorism versus craftsmanship'. He points out, as we have already claimed, that under Taylorism management creates a 'filter' between the workers and the problem-solvers (Figure 15.2).

Plan, do, check, act

Hutchins says that in any form of organized labour, a task has four distinct phases:

1. **Planning** the task to be done;
2. **Doing** the task;
3. **Checking** that the task has been done correctly;
4. **Acting** on the results of the checks.

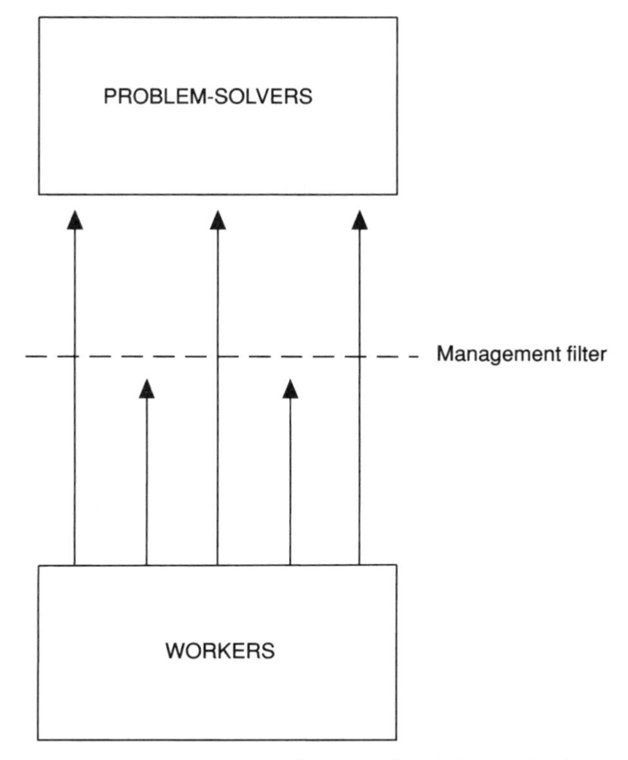

Fig. 15.2 Taylorist management. Source: Hutchins, *Quality Circles Handbook*.

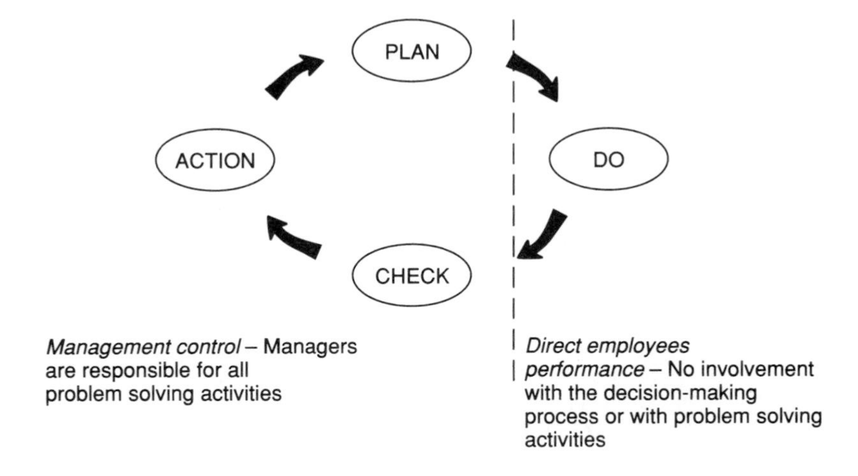

Fig. 15.3 The Taylor regime. Source: Hutchins, *Quality Circles Handbook*.

Figures 15.3, 15.4, 15.5 are Hutchins' representations of these cycles under, respectively 'Simple' Taylorism, where management do not need specialist help; 'Scientific Management' where management are supported by technical experts; and under 'Craftsmanship'. In both simple Taylorism and scientific management,

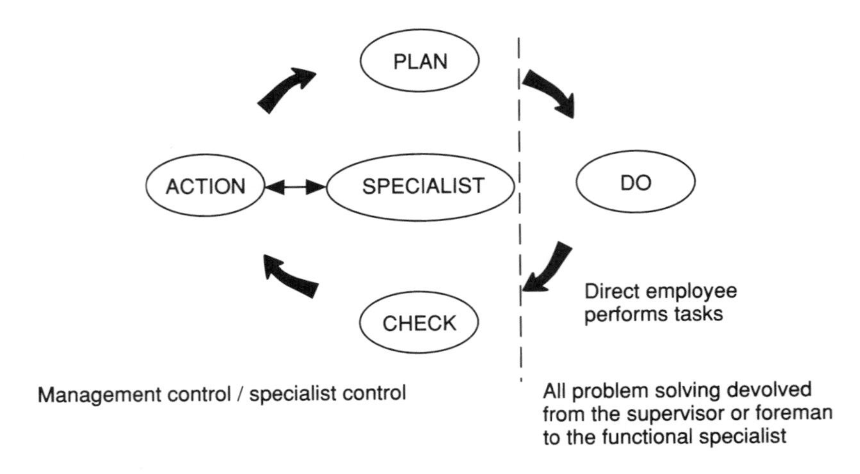

Fig. 15.4 Scientific management. Source: Hutchins, *Quality Circles Handbook*.

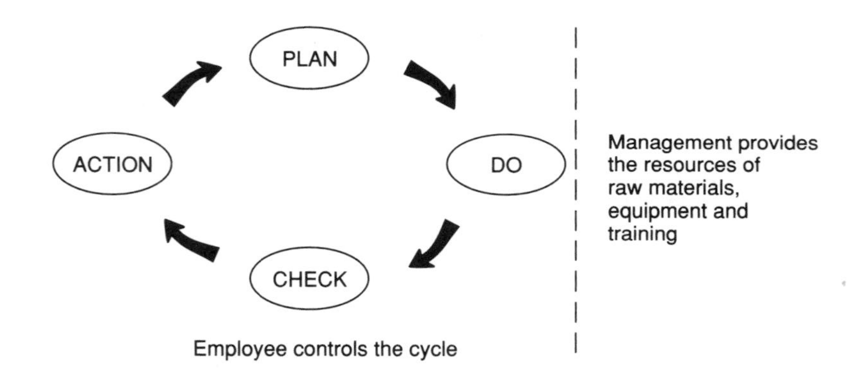

Fig. 15.5 Craftsmanship. Source: Hutchins, *Quality Circles Handbook*.

the workforce are allowed only to 'do', whereas the craftsman controlled all the phases of the task.

Craftsmanship and the assembly line

Twentieth-century industry has been built largely on the strengths of Taylorism. Recently its limitations have become more apparent. This is particularly so as energy becomes scarcer, environmental issues more pressing, international competition fiercer, and as sociological studies show the inevitability of workers being alienated by Taylorism.

Hutchins sees the relative merits and demerits of the two systems as:

Advantages of craftsmanship

1. Self-control;
2. Pride in work;
3. Self-confidence/self-assurance;
4. Loyalty to the work;
5. High level of skill;
6. Sense of responsibility;
7. Motivated and involved;
8. Good quality workmanship;
9. Self-improving;
10. High level of job interest;
11. Unique product.

Disadvantages of craftsmanship

1. High cost of labour;
2. Low output per individual;
3. Low interchangeability of items;
4. Unique product;
5. Scheduling difficulties;
6. Monitoring problems leading to poor control;
7. Low wages.

Advantages of Taylorism

1. Interchangeability of items;
2. High productivity at low cost;
3. Low skill workers;
4. Lower wage costs/unit;
5. Accurate forecasting;
6. Predictable results;
7. Sophisticated, highly trained problem solves;
8. High wages.

Disadvantages of Taylorism

1. Control by others;
2. Low worker morale; caused by boredom, frustration, low self-confidence;
3. Poor quality, absenteeism, lack of job pride, alienation;
4. Poor company image reflected by workers who interface with customer;
5. No opportunity for self-development;
6. Supervisory problems;
7. Loyalty of specialist to specialism rather than to employer.

Especially in order to promote quality (according to our definition of quality as customer satisfaction) a way of working is needed which preserves the advantages of Taylorism while at the same time restoring the benefits of a craftsman system.

The problem to be solved is 'How, in a complex modern environment, can the workperson regain the ability to take part in the planning, measuring and reacting parts of the task in which he is involved?'

Hutchins believes that carefully introduced 'quality circles' are one way. Crosby has formulated his 'zero defects' process as another. We shall look at both of these topics in Chapter 16.

Summary of conclusions

I would like to summarize the conclusions we can draw from this chapter as:

1. Our attitudes and behaviour towards our subordinates and other employees reflect our assumptions about their motivation and their own expected behaviour. When we act on our assumptions we are likely to influence the way they do behave;
2. People's actions and aspirations are governed as much by their social needs as by direct economic motives. Unless they are in economic difficulty the social needs will dominate; in particular, people will behave in a way they feel is expected by the group to which they belong;
3. The manner of organizing work referred to as Scientific Management or 'Taylorism', and the attitudes associated with it are still very prevalent, though the names themselves are no longer in currency. These methods can create frustration and dissociation for the employee.

16

Motivate or involve?

Introduction

The purpose of this chapter is to contrast workforce participation in quality improvement with management attempts to 'motivate' the workforce to produce better quality. 'Quality circles' and Crosby's 'zero defects' process will be examined in some detail.

Motivation or participation?

In Chapter 15 we recapitulated the work of F.W. Taylor, and also of Elton Mayo and others involved in the Hawthorne experiments. The reason for the reminder was that a key element of the maintenance and improvement of quality is; to stimulate people to think about their own job in fresh ways, for them to be to be receptive to changes in working practices, and to be encouraged to contribute their knowledge and ideas in the cause of quality improvement.

I hope Chapter 15 has convinced you that success in attracting people to help your endeavours depends on:

1. Your goals being in harmony with theirs;
2. Their ability to understand the objective;
3. Utilization of the group, not just the individual;
4. Their receiving satisfaction and reward for themselves.

Motivation

What we are talking about is, in a sense, motivation. We should use the expression 'motivation' the way hedgehogs are supposed to mate; namely with reluctance and extreme caution.

In the words of a student on an open distance learning course: 'I think that motivation was designed by management to increase the workload on the workforce.' That was written by an ambitious shop-floor operator studying in order to understand her role better and to seek promotion, so you can appreciate that suspicion and cynicism about motivational techniques are widespread in industry.

Schemes which have motivation as their end result run the risk of being perceived as insultingly patronizing and/or covert productivity drives. In reality, the object of the processes we shall describe later in the chapter, such as zero defects and quality circles is:

1. To remove the barriers to people doing their jobs effectively;
2. To encourage them to 'Work smarter, not harder'.

Recall that methods used in trying to find ways to produce more and/or better work from their employees by 'motivation' have been rooted in the research and the models of workforce behaviour we discussed in Chapter 15, which we can pigeonhole as 'Economic Man' and 'Social Man'.

Economic Man

Economic Man works in order to earn money, the more the better. Therefore, the way to entice him to work harder is to pay him by results. Taylor certainly saw his employees at the Bethlehem Steel Works as 'Economic Men'. There was a contract between management and men in which management and 'scientists' found the best and most efficient way for a man to do a particular job. If the man performed the job that way, unquestioningly, his work became more productive and he was rewarded with higher wages as his share of the extra profit. Taylor said of the men in the Bethlehem Steel yard that they 'looked on them (the employers) as the best friends they ever had, because they taught them how to earn 60% more wages than they had ever earned before'.

It has been said that the picture of Economic Man gives rise to 'Theory X' behaviour, though it is hard to draw a close parallel. Certainly, for example, Taylor didn't think of his workforce as untrustworthy. He said:

> My experience with workmen has been that their word is just as good as the word of any other set of men that I know of, and all you have to do is have a clear, straight, square understanding with them and you will get just as straight and fair a deal from them.

Social Man

Social Man is not just an individual but a member of a group. He cares about his place in the group, and the status of the group as a whole. For example, if paid on piecework, he may actually limit his output and hence take-home pay, to what the group feels is reasonable, rather than 'show up' his less productive colleagues.

However, Social Man is predisposed to work in a Theory Y fashion; he can be reliable, conscientious and concerned for the company's welfare, taking pride in his work and willing to accept responsibility.

Groups and social concerns are powerful. If the employer cannot provide a satisfying group endeavour, then the latent need to contribute to a team may be transferred to the trade union, to a company sports team, or to purely out-of-work activities.

Social Man seems to be the model which is supported by research, and also by most people's personal experience. There is obviously a great deal that managers can achieve with the help of 'Social Man' **provided he is playing on our team.** How do we sign him on?

Process theories of motivation

The theories of McGregor, Maslow, Herzberg and others were **content** theories; they attempted to explain the **needs** by which people are motivated. Other theorists discuss **process** theories, concerned with **how** they become motivated to pursue a particular course of action.

Goal-setting

People's behaviour is directed towards goals:

- Which they have set themselves;
- Or which they have freely accepted from other people;
- And which reflect their own values and desires.

Also:

1. A specific goal results in improved performance.
2. Difficult but achievable goals are the most effective.

And the following can enhance performance:

1. Competition;
2. Feedback of performance.

Harnessing motivation

Motivation is latent, but has to be harnessed. This can prove difficult. Pay, security of employment, a subsidized canteen, paid holidays, health insurance, a sports and social club, a company doctor. Such amenities were provided increasingly in the wake of the 'human relations' school of industrial sociology, but with no identifiable correlation with business success. These amenities were, after all, concerned with 'hygiene factors'.

General motivation programmes, such as those involving poster campaigns. exhortations by senior management, etc. do not succeed either. They fail because of one (or usually all) of:

1. Lack of credibility: the message is not borne out by management's day-to-day preoccupations (e.g. with meeting schedules rather than improving quality);

2. Lack of clear measurable goals (e.g. employees told to 'improve', 'do your best');
3. Lack of guidance, tools and methods (e.g. no training in problem solving, or SPC).

When the workforce fail to respond they are claimed to have (favourite American expression) an 'attitude problem'.

Management's 'attitude problem'

Juran claims that less than 20% of quality problems are due to workers in that case more than 80% are due to management. Deming goes even further; with astonishing precision he claims that management are responsible for 94% of quality problems.

These statistics may seem surprising, but if you reflect on Hutchin's picture of the 'Taylor' or 'Scientifically Managed Organization' (Figures 15.3 and 15.4 of Chapter 15), you will see that they are inevitable. Management have reserved to themselves and their staff specialists nearly all the ways to make things go seriously wrong, and will not delegate them to the workforce at large!

There are far more opportunities for making errors in planning activities (or not), in measuring the wrong parameters, in failing to take appropriate actions on the findings, than in simply **doing** things. Maybe the demotivated employees are on to something. After all, the operator who doesn't want to know about statistics may study historic data and lay out his football pool forecast with great precision; the lead hand who doesn't want to assume extra responsibility may be secretary of the local Pigeon Racing Club: the indifferent machine-minder may be a skilled motor car tuner. They have sown their effort where they believe it will bear fruit.

Participation

Management's task then, is to rid itself of its attitude problem, and to accept that most of the problems in a complex modern business are generated **by management**. The best way to motivate employees is indirect. It is to offer them the opportunity to help management out of the hole it has dug for itself. Then motivation, as it develops, will simply be a by-product of this joint venture.

Two significant methods that have been developed for achieving this need (to use the whole workforce to help solve quality problems) are 'quality circles' and 'zero defects'. The remainder of this chapter will be devoted to describing them.

Zero defects

> The performance standard is Zero Defects. (P.B. Crosby)

Much of Crosby's book *Quality Is Free* is devoted to the concept of zero defects, with his recipe for introducing the zero defects process into a company. It is discussed in detail again in the sequel *Quality Without Tears* where he introduces it as the third of his 'quality absolutes'. (We shall discuss the quality absolutes in Chapter 17.)

The formulation of the zero defect standard

A complex system such as a missile relies on the combination of an enormous number of components and sub-assemblies. It may seem appropriate to define an 'acceptable quality level' or 'lot tolerance percent defective' on each component, (refer back to Chapter 11 to refresh your memory on AQL/LTPD), but there are so many units in each final assembly that some defects in the completed missile are a virtual certainty, even if the AQL or LTPD is very tight. This AQL or LTPD cannot be tightened indefinitely; the sample size increases and sooner or later you are doing 100% or even 200% inspection – you are trying to inspect quality in!

This is the situation that faced Crosby when he was employed by Martin-Marietta as quality manager on the Pershing missiles being built for the US Department of Defense. It was on this project that he first introduced the principle of zero defects. That is not to say that the concept of errors avoidance did not exist previously, but that he was the first executive to insist that it was achievable in principle, and was the essential and natural goal to strive for.

The zero defect philosophy was based on the need to overcome:

1. The unacceptability of even the smallest proportion of defects in one component of a complex system.
2. The confusion introduced at shop-floor level by having a level of defects that was treated as acceptable.

Crosby stresses the importance of the second point. People are not, by and large, prepared to tolerate errors in matters that affect their own lives, e.g. errors in their pay-checks, or in hospital treatment. In contrast, Crosby says, a manufacturer who sets up a new production line and includes a re-work area declares its intention to make some defective product. The fact that employers are prepared to live with a certain number of errors is contrary to their employees' everyday standards, disorientates them, and demotivates them from seeking or offering corrective actions.

Zero defects as the official performance standard realigns the company's attitude with the employees' ethical and common sense. The employees are then reluctant to accept defective material from their 'suppliers' or to pass on their own work to 'customers' if they know it is marred by errors. This creates employee pressure for, and employee involvement in, for example improving methods, materials, training, etc. Thus the wish is self-fulfilling in bringing zero defects nearer.

Validity of the zero defect concept

The zero-defect goal is often criticized on the basis that it is unattainable, or would be prohibitively expensive. We shall discuss the economics of quality in Part Four; Crosby himself insists that 'it is always cheaper to do it right first time, than to do it over [again]'.

The major point, however, is that although complete freedom from defects of any kind may be unattainable except for short periods, it is the **only legitimate standard to aim at**. As Crosby poses it, the performance standard must be zero defects, not 'that's close enough'. If a company's management sets a standard of zero defects, it does not mean they don't expect any defects, only that they will not be satisfied with having any defects, however few. As each class of error is recognized, steps will be taken to eliminate it. It is the workforce's realization of this management attitude that mobilizes their own efforts to assist.

Implementation of the zero defect process

Crosby sees the zero defect process as comprising 14 stages. They will remind you of a number of points we have already stressed.

1. **Management commitment**. We have already written at various points in the book about the need for visible and obsessive concern for quality, and the need for a clear quality policy.
2. **A quality improvement team**. This is needed to guide the process, help it along and clear 'roadblocks'.
3. **Measurement.** Quality must be measured before it can be improved; and before you know if you've improved it.
4. **Cost of quality.** Knowledge of the cost of quality is an impetus to quality improvement (prevention).
5. **Quality awareness.** Management must communicate its intention of quality improvement.
6. **Corrective action.** The root cause must be found and corrected before a problem will go away for ever.
7. **ZD planning.** This phase plans the conversion of the quality improvement programme into a zero defect campaign.
8. **Employee education.** Employees must be taught about quality, and understand what is to be attempted.
9. **Zero defects day.** The management must be prepared to commit itself to a goal of zero defects in front of the assembled workforce and other witnesses.
10. **Goal setting.** Zero defects is the final goal, but for realism and to provide milestones against which to measure progress, intermediate targets and dates must be set.
11. **Error-cause removal.** It is necessary to gain information as to the true source of errors, so that they can be eradicated.
12. **Recognition.** Employees' contributions to error removal must be recognized and celebrated.
13. **Quality councils.** These are needed to bring together quality professionals' contributions to the ZD endeavours.
14. **Do it over again.** When the quality improvement team begins to lose momentum and finds few new problems to tackle, the members should stand down and a new team be appointed. The cycle can then start all over again.

The way in which discussion of the 14 steps is organized within the book *Quality Without Tears* is instructive. The first step, management commitment, appears in Chapter 11 on 'Implementation' (of quality improvement). Steps 2–6 are the subject of the following Chapter 12 on 'Team actions' and steps 7–14 are the subject of Chapter 13 on 'Team executions'.

Thus given the will (step 1, managerial commitment), the following steps relate to the setting up of a team, finding the facts, and preparing the workforce. After the team has been 'run in' and the ground prepared, steps 7 (ZD planning) onwards are 'team executions' implementing the drive towards zero defects.

Zero defects day

Step 9 (Zero Defects Day) with its associated razzamatazz is the step in the strategy which evokes most criticism or reluctance to emulate, especially outside the US culture. It is a 'fun' day to which all employees and their families and bigwigs from outside the company are invited, but it is also a ceremony of dedication performed before witnesses. Management declares its commitment to zero defects, and issues a declaration to this effect to each employee. Employees are invited to reciprocate by 'signing the pledge' themselves, saying that they will do all they can to improve quality in their work. (According to Groocock, within ITT Europe the employee pledge was omitted.)

Critics focus on the demand for employee commitment, Crosby on the manager's public commitment.

Quality circles

Quality circles arose in Japan. In 1962, as an attempt to get supervisors and workers to study the concepts and techniques of statistical quality control, the journal *QC for Supervisors* suggested that supervisors and their workers should get together to read and discuss the articles in the periodical. They also suggested that each 'QC Circle' should register its existence with the JUSE (Japanese Union of Scientists and Engineers). That is how quality circles started.

For optional reading on quality circles, the book *Quality Circles Handbook* by David Hutchins is recommended, and much of the following summary of the nature of quality circles is based on his prescriptions.

Although the concept of 'quality control circles' was developed in Japan, it slowly became adopted throughout other parts of the world, firstly in other parts of the Far East through the influence of Japanese subsidiaries and assembly plants there, then to America, more slowly into Europe.

Definition of a quality circle

A definition of a quality circle is offered by the UK Department of Trade and Industry in its booklet *Quality Circles* prepared as part of the National Quality Campaign. The definition is:

> A group of four to twelve people, coming from the same area, performing similar work, who voluntarily meet on a regular basis to identify, investigate, analyse

and solve their own work-related problems. The circle presents solutions to management and is usually involved in implementing and later monitoring them.

Professor Ikuro Kusaba has described the advantages to the circle members as follows:

> Through the meeting, understanding of their own jobs is deepened, better human relations and the leader's leadership is established, and the members' participation-consciousness and fellowship-consciousness are highly developed. Furthermore, the ways to solve the problems by using quality control techniques are rightly understood by them.

Following on from Chapter 15 you may well feel that this is a sound recipe for motivation of the members. But that is not its purpose. The purpose is not employee satisfaction, but **to solve quality problems.** It is a pragmatic attempt to create an environment in which any member of the organization can be co-opted into the problem-solving activities, and their special knowledge and insights used, especially through:

1. Top management support;
2. Training;
3. Provision of resources;
4. Delegation of responsibility;
5. Credit for achievements.

Needs of quality circles

There have been many quality circle success stories in Western companies, but also very many experiments that have been unsuccessful, and where circles have withered away or been abandoned by management.

To flourish and be successful circles need:

1. Top management support, so that the circle has credibility both to itself and outsiders;
2. Truly voluntary participation, letting people join, abstain or leave without being under any pressure;
3. Specialist support at the start-up phase, e.g. from a quality circles consultant.
4. A quality circle facilitator. The facilitator is a trained in-house specialist who can provide advice, mobilize resources to help the circle, coordinate the activities of existing circles, and help new circles to get started.
5. Operational management support, providing time and facilities, advice, interest and cooperation.
6. Training in problem-solving techniques, in running meetings, in presentation skills, and in any relevant skill area where the circle feels itself to be weak.
7. Recognition of their work, by being allowed to implement their recommendations, and by acknowledgement by management to the workforce at large, of their successes.

Constraints placed on quality circles

The following constraints are usually placed on quality circles:

1. The circle deals only with problems whose solutions lie within its own control.
2. The circle will not address issues of pay, disciplinary matters, etc.

Evolution of a quality circle

Hutchins claims that contrary to Western understanding, quality circles are not simply problem-solving groups, but are the means whereby a work group can themselves control their work. According to Hutchins, a circle, once established, will pass through three distinct stages, finally arriving at a fourth. The stages are:

1. The initial phase. During this phase the circle will have been trained in simple techniques enabling the members to identify and analyse some of the more pressing problems in their own work area. These will generally be the problems which, at the outset, were uppermost in the minds of the members of the circle.
2. Monitoring and problem-solving phase. The more straightforward problems will already have been solved, the circle will monitor their achievements and maintain or enhance improvements already made.
3. Innovation – self-improvement and problem-solving. Rather than simply seeking to solve evident 'problems', the circle will look at the established ways of doing jobs to see if they cannot be improved, using the techniques they have learned and employed in analysing and solving 'problems'.
4. Self-control. Hutchins does not believe that any Western quality circle has reached this stage of maturity. He sees it as a stage where, through the medium of the circle, the members can achieve 'self-control'; by which Hutchins means setting their own goals and controlling their own performance.

Tools of the quality circle

To be successful a quality circle has to be given the necessary tools in order to function, and be trained in the competent and confident use of the tools. These tools are recognized to be:

1. Brainstorming;
2. Data collection;
3. Data analysis;
4. Pareto analysis;
5. Cause and effect analysis;
6. Histograms;
7. Control techniques;
8. Presentation techniques.

Management's role in assisting quality circles

Although (perhaps because) the manager will not take part in the circle's meetings it is important to stress the manager's role and responsibility in making the quality circle successful. This is important because in Western companies many circles have been unsuccessful through failures in this area.

Top management's role

According to Hutchins top management's role comprises:

1. Establishing corporate policy and plan;
2. Setting corporate goals and objectives;
3. Management commitment;
4. Allocation of resources;
5. Monitoring;
6. Auditing;
7. Support.

Middle management's role

Middle management's responsibility as related by Hutchins is to:

1. Let the circle select its own problems;
2. Try to ensure meetings do not have to be postponed;
3. Not prevent a member attending except in exceptional circumstances;
4. Try to enable circles to meet at a regular set day and time.

Supervisors' role

As in many Japanese-initiated approaches to quality, the Supervisor (or 'foreman') plays a central role. In Japanese practice the work-group's supervisor is usually the initial chairman of the circle, though when the circle becomes established and skilled, he/she may step back into an advisory role.

Similarities and distinctions between quality circles and zero defects

The following comparison is an extract from *The Chain of Quality* by Dr J.M. Groocock, describing the conclusions drawn from debates on the merits of the two systems conducted in managerial training sessions at TRW's Quality College. Many similarities were noted, though it was concluded that ZD places more emphasis on quality costs, and quality circles more on training both in human interaction skills and simple quality control techniques. Groocock adds that US circles seem to emphasize the interactive skills and Japanese circles the techniques. We may wonder if this reflects the influence of the US 'Human relations' school of theorists, and the tendency of US managers to see quality circles as an investment in 'motivation' rather than a pragmatic quality improvement activity.

Overall, says Groocock, many of the TRW groups concluded that it was most practical and effective to use elements from both QCs and 14 Steps (to zero defects):

But they also conclude – as the Japanese have always emphasized – that either 14 steps, quality circles, or a combination of the two, is not enough for a rich and comprehensive quality philosophy.

Summary of conclusions

In *Quality Without Tears* Phil Crosby observes 'Getting people turned on has become a major industry . . . why do we need a special program to motivate our people? Didn't we hire motivated employees?'

This chapter has tried to persuade you that motivation is the by-product of a process, not an end in itself. It compared the models of economic man and social man, and has suggested to you that if the goal of the process is quality improvement then social man, equipped and empowered to contribute, is able and willing to do so.

Two examples of a framework in which this synergy can develop have been sketched out: quality circles and zero defects. Characteristic of each is the gradual elimination of instances of known problems; by whoever has the knowledge, desire and capability to do so. Frequently this is not a single individual but has to be a voluntary team effort.

17

American quality 'gurus'

Introduction

This chapter introduces the ideas of the most notable post-war US lecturers on quality.

Introducing the 'gurus'

This chapter will look at the teachings of the leading American quality 'gurus', as they have become known. A letter writer to *Quality News* who deprecated this term pointed out that his dictionary said a guru was a Hindu religious teacher, therefore he would approach one only if and when he desired consultancy on Indian religion. Nevertheless the term may be apt, for in reading their books or watching their videos one can get the impression that each 'guru' is claiming to teach the one true path to enlightenment.

To understand this you should refer back to what you know of marketing, and the necessity for a product to have an associated 'unique selling proposition' which distinguishes it from its competitors. The 'gurus' of whom we shall speak have written books, do lecture tours, offer consultancy, and in some cases run their own training establishments. If some of their insistence on particular recipes for success, or their catchphrases, seem gimmicky remember that they are selling a product and have to establish its distinctiveness in the minds of potential customers.

After reading this chapter, and dipping into the writings of some of these 'gurus' and their disciples (yes, some of them are the subject of gospels written by their followers) you may well come to the conclusion that they are trying to give the same message in different ways; and that for practical implementation of good advice you would prefer to turn to the exponent whose style you feel most comfortable with, and whose message has most conviction for you personally. Certainly, many companies have chosen one adviser from the many on just those grounds.

The 'gurus' whose ideas I shall discuss (and to try to avoid any imputation of bias I will introduce them in alphabetical order) are: Crosby, Deming, Feigenbaum Juran and Shainin. Most of my discussion will concern Crosby, Deming and Juran. Japanese and British writers will be introduced a little later, in the following two chapters.

Crosby

Philip B. Crosby is arguably the clearest, certainly the most entertaining writer and speaker of the three gurus on whom I focus attention. He has been described as having the best intuitive understanding of quality among writers on the topic, and his background is not academic but 'hands-on' quality management. He worked with Martin-Mariatta where he developed his 'zero-defects' approach while working on the Pershing missile programme, and with ITT where he became a vice-president, and director of quality.

After the success of his first book devoted specifically to quality issues (*Quality Is Free*) he left ITT to found Philip Crosby Associates (PCA), a business created to provide training and consultancy based on his precepts. He has since written *Quality Without Tears* and *Let's Talk Quality* which respectively develop and provide a commentary on the topics introduced in the original book; and a TV film/video for the British Broadcasting Corporation: 'The quality man'.

Despite having developed his views about achieving quality in a practical industrial environment, his books do not deal in any depth with specialist quality assurance techniques. His main concern is the meaning of quality and its attributes, and how by coming to a common understanding of what quality is, people can work together to achieve it rather than disrupt each other's efforts through misunderstanding.

Four absolutes of quality

The distillation of Crosby's ideas into what he judged to be the fundamental nature of quality (the 'absolutes') was formulated in *Quality Without Tears*. His absolutes were:

1. The definition of quality is conformance to requirements;
2. The system of quality is prevention;
3. The performance standard of quality is zero defects;
4. The measurement standard of quality is the price of non-conformance.

The definition

Quality has to be defined as conformance to requirements, not as goodness.

Crosby's definition was introduced and examined in the very first chapter of Part One. If quality means goodness, one person's perception of goodness will differ from another person's perception, and rational analysis, or constructive, concerted actions in pursuit of quality are impossible: 'Conformance to requirements enables quality to be assessed in terms of what the (customer's) requirements are; the conclusions can be recorded (specified) and conformance measured.'

The system

The system for causing quality is prevention, not appraisal.

We have stressed in many parts of this book that maintaining quality is not enough. In a free market, with able competitors, quality and value for money have to be constantly improved for business survival. Therefore inspection is wasteful, and must be only a stepping-stone to eradicating the causes of the errors and defects discovered.

The performance standard

The performance standard must be zero defects, not 'That's close enough'.

All the results in a business are created by people, and each task must be done correctly for the end result to be what was required. People must be able to depend on each other.

The measurement

The measurement of quality is the price of non-conformance, not indexes.

Crosby distinguishes between the 'price of conformance', the cost of doing things right, and the 'price of non-conformance', the cost of doing things wrong. He uses the word price where I have used cost to imply that companies decide how much non-conformance to tolerate – thus the cost is their price-tag, not something external that they cannot control. We shall develop this idea in Chapter 23, Part Four.

What Crosby says, and its relation to indexes, is that although the achievement of quality could be measured by many indices or figures-of-merit the most direct and fundamental measure of (lack of) quality is 'What is it costing the company?'

Other absolutes

It may be of interest to cite two other rules, which Crosby felt sufficiently strongly about to have included in his list of absolutes at some point, but later removed.

One was 'There is no economics of quality.' The significance of this statement will be explained in Chapter 24, Part Four, in relation to quality costs.

The other was 'There is no such thing as a quality problem.' That is to say, problems may affect quality but they are not 'quality problems' in the sense of the quality manager being required to correct them. 'Quality' problems are the result of errors in design, procurement, manufacture or whatever, and their solution must be addressed by the management of the responsible department.

Crosby later replaced these two absolutes by 'The system of causing quality is prevention, not appraisal.'

Zero defects

The concept of zero defects as a rational goal, and Crosby's prescription (the '14 steps') for quality improvement towards zero defects, have been discussed as a major part of Chapter 16. Refer back to this point unless they are still fresh in your memory.

Quality vaccine

To provide antibodies which prevent non-conformances occurring, Crosby's analogy says that an innoculation must take place. To administer the vaccine you require:

- Determination;
- Education;
- Implementation.

The formulation of the vaccine serum comprises:

Integrity

1. A chief executive officer dedicated to having the customer receive what was promised.
2. A chief operating officer believing that quality is first among equals'.
3. Senior executives who take requirements seriously.
4. Managers who know they have to get things done right first time.
5. Professionals who know that the accuracy and completeness of their work determines the effectiveness of the workforce.
6. Employees who know that their commitment to meeting requirements is what makes the company sound.

Systems

1. Quality management dedicated to measuring and reporting conformance.
2. A uniform quality education system.

3. Cost of non-conformance used to measure how well processes are conforming.
4. Feedback of customer's experience of the product or service.
5. Company wide emphasis on defect prevention.

Communications

1. Quality improvement information available to all employees.
2. Recognition programmes acknowledging achievements at all levels.
3. Quality concerns expressed to top management receive immediate response.
4. Management meetings begin by reviewing quality.

Operations

1. Suppliers educated and supported.
2. Procedures, products and systems qualified prior to use.
3. Training as a routine activity.

Policies

1. Unambiguous quality policies.
2. Quality function at same level as the functions being reported on.
3. Advertising does not claim more than the specification ensures.

Deming

Dr W. Edwards Deming's fame derives from his having been one of the experts who visited and taught in Japan from 1950. His topic then was statistical process control, which became a major vehicle for quality improvement in Japanese industry. In acknowledgement of the power of this technique and the prestige accorded to Dr Deming for having made Japanese industry aware of it, an annual 'Deming Prize' was instituted for major contributions to improving quality in Japanese industry; and in 1960 he was decorated by the Emperor of Japan.

Deming's prestige in his own country had to some extent to wait for American industry to experience the impact of Japanese quality and investigate its sources. His own works have been supplemented and interpreted by other writers, e.g. *The Deming Route to Quality and Productivity* by W.W. Scherkenbach and *Right* Every *Time* by Frank Price.

The technique of statistical process control (SPC) has been examined in Part Two of this book.

Having taught SPC to the Japanese, Dr Deming turned to advising American businesses how to resist Japanese competitors more effectively. To a degree this also stressed the effective use of SPC techniques, but he was also critical of US managerial practice. He said:

> The basic cause of sickness in American industry is failure of top management to manage. Loss of market, and resulting unemployment, are not foreordained. They are not inevitable. They are not acceptable. The day is past when people in management need not know anything about management – by which I mean to include problems of production, supervision and training.

Deming's prescription for improved quality, productivity and competitive position was based on 14 points. These should not be confused with Crosby's 14 steps. I should also warn you that, in looking at different accounts of the 14 steps taught by Deming, e.g. in Deming (1982) and Scherkenbach (1986), it is clear that they had evolved and been re-stated over the course of time. Deming had long taught that there were 14 points to obey in order to achieve competitiveness through quality, but they haven't always been quite the same 14 points, nor were they always presented in the same order.

The 14 points

1. Create constancy of purpose;
2. Adopt the new philosophy;
3. Cease dependence on mass inspection;
4. Constantly improve the system;
5. Remove barriers;
6. Drive out fear;
7. Break down barriers between departments;
8. Eliminate numerical goals;
9. Eliminate work standards;
10. Institute modern methods of supervision;
11. Institute modern methods of training;
12. Institute a programme of education and retraining;
13. Stop awarding business on price-tag;
14. Put everyone to work on accomplishing this transformation.

In a little more detail:

1. **Create constancy of purpose** towards improvement of product and services, with a coherent plan to become competitive and stay in business.
2. **Adopt the new philosophy**, that historically accepted levels of delays, mistakes, workmanship defects, etc. are no longer tolerable. Top management must acknowledge their responsibilities, and accept the role of leaders in instituting change.
3. **Cease dependence on mass inspection**. Instead of reliance on large scale end-of-line inspection, use SPC to obtain evidence of built-in quality.
   In this context Deming places particular emphasis on choice and control of suppliers, hence the next point.
4. **End the practice of awarding business on the basis of price**. Instead, include considerations of quality, and eliminate suppliers who cannot provide statistical evidence of their quality.

5. **Find the problems**. Management's job is to continually improve the system (in the design, incoming materials, composition of materials, maintenance, machine improvement, training, supervision); and in so doing continually decrease costs.
6. **Institute modern methods of training**. This includes adequate training of hourly workers on how to do their jobs, including use of statistical techniques.
7. **Institute modern methods of supervision**, at all levels in the organization. Provide leadership, to help people and machines to do a better job.
8. **Drive out fear**, so that everyone can work effectively for the company.
9. **Break down barriers between departments** (research, design, sales, production) since they must communicate and act as a team in order to anticipate production and application problems that may be encountered.
10. **Eliminate numerical goals**; and posters, slogans for the workforce, which ask for new levels of productivity but without providing the methods. Substitute leadership.
11. **Eliminate work standards** that prescribe numerical quotas.
12. **Remove barriers** set up between the hourly worker and the right to pride in workmanship. Supervisor priorities must be changed from stressing quantity to quality. Similarly for managers and engineers: restore their pride in workmanship.
13. **Institute a vigorous programme of education and retraining**. To increase flexibility and allow the redeployment of people (such as traditional inspectors) who must learn new skills.
14. **Create a management structure** that will push all the above points on a day-to-day basis.

You will notice that this is simply a list; it is not a chronological sequence like Crosby's '14 steps to zero defects', nor is it grouped in related tasks, as in the case of the elements of Crosby's vaccine. If you read Scherkenbach, be warned that he alters Deming's normal sequence, presumably to give what he thinks is a more logical order.

The Deming cycle

The 'Deming Cycle' is illustrated in Figure 17.1. You will recognize its affinity with the cycle Plan-Do-Measure-Act, and it is also closely related to the 'corrective action loop'.

Feigenbaum

Armand V. Feigenbaum's book *Total Quality Control* is now in its fourth edition, having been first published in 1951. It is the only serious rival to Juran's *Quality Control Handbook* as a universal reference source on quality assurance, and it is on this book that his reputation largely rests. The book divides its subject into six sections:

- business quality management
- the total quality system

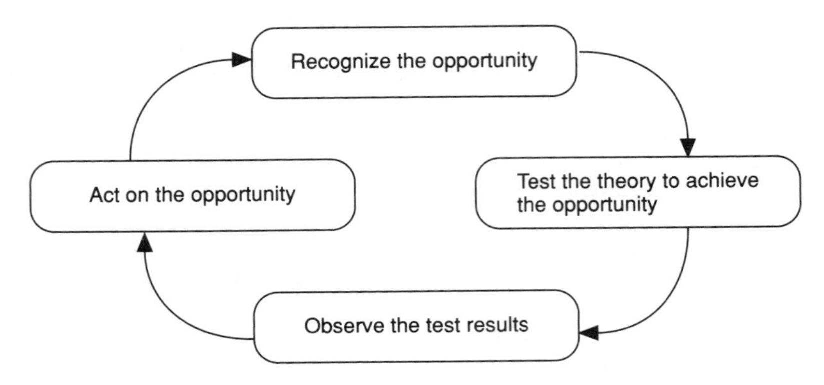

Fig. 17.1 The Deming Cycle. Based on Fig. 4.2 of Scherkenbach, *The Deming Route*.

- management strategies for quality
- engineering technology of quality
- statistical technology of quality
- applying total quality control in the company.

Total quality control, the philosophy underlying the book, clearly has much in common with total quality management (TQM) which will be discussed in Chapter 20, especially in a manufacturing context. Feigenbaum himself evidently sees this as being the case, maintaining that 'Quality is in its essence a way of managing the organisation'. He defines total quality control as:

> An effective system for integrating the quality-development, quality-maintenance, and quality-improvement efforts of the various groups in an organisation so as to enable marketing, engineering, production, and service at the most economical levels which allow for full customer satisfaction.

In the 40th Anniversary edition of the book Feigenbaum summarizes four 'Management Fundamentals' of total quality. They are:

- 'There is no such thing as a permanent quality level'. Competition and customer expectations demand constant efforts to improve, even on success.
- 'Personal leadership is the hallmark of good management'; in mobilizing the quality knowledge, skill and positive attitudes of everyone in the organization.
- 'Quality is essential for successful innovation'. This is increasingly true as a result of the increasing speed of new product development, and because a product design is likely to be manufactured multinationally.
- 'Quality and cost are complementary, not conflicting business objectives'. Good quality fundamentals lead to good resource utilization; and hence to good productivity and low quality costs.

Juran

Joseph M. Juran is probably the most celebrated US expert on quality assurance. He is well known for:

1. His monumental, nearly 1900 page textbook, *Quality Control Handbook*, and other writings;
2. His contribution to the quality achievements of Japanese industry;
3. His lectures and training courses conducted by the 'Juran Institute' throughout the world.

Like Deming he was involved with Japanese industry shortly after the war, and he is now over 80 years old.

'Breakthrough' in quality improvement

Juran defines quality, briefly and succinctly, as 'Fitness for use', and acknowledges that the fitness is as judged by the customer, for the specific use intended. Juran acknowledges the waste of resources that occurs in producing quality in any particular company: excessive inspection, redesign, customer returns, price concessions, service costs, inventory waste, delays, loss of market share resulting from dissatisfied customers.

He sees these as failures of the traditional concepts of 'controlling' quality. He preaches 'breaking the chain of the past' to achieve continually improving quality levels. He describes breakthrough as 'Organized creation of beneficial change'. The quality improvement process perceived by Juran is an organized, step-by-step approach to breakthrough and its benefits.

The ten steps to quality improvement

The ten steps are:

1. Build awareness of the need and opportunity for improvement;
2. Set goals for improvement;
3. Organize so as to reach the goals;
4. Provide training;
5. Carry out projects to solve problems;
6. Report progress;
7. Give recognition;
8. Communicate results;
9. Keep score;
10. Maintain momentum by making annual improvement part of the regular systems and procedures of the company.

Juran sees the ten steps as demanding a new orientation from top management (hence 'breakthrough') and demands that each member of the management team personally engages in the quality improvement process. The prime objective of the quality improvement 'process' is to help the management team develop the habit of annual improvement in quality and reduction in quality-related costs. In parallel with this he sees the 'process' as a means of training managers in the special quality-related concepts, skills, tools, and techniques essential to continuous quality improvement.

The project team

Juran sees project teams as the vehicle with which the goals can be reached. First he advocates the creation of a steering committee or quality council, comprising key senior managers with operational responsibilities. This committee decides the broad policy and strategy for the quality improvement programmes, the training programmes to be implemented, and creates the management project teams.

Once a team is formed it decides what project(s) to address, and typically meets weekly for about three months, until the quality improvement problem is solved. The team is then dissolved, to be replaced by another group who will choose a different problem to address. Juran claims that the typical team achieves a saving of around $100 000 as a result of the quality improvement it initiates.

Shainin

Dorian Shainin is a respected quality consultant in the USA, whose advice is sought by many major companies. However, since he has not publicized his ideas in a major book, he is less generally well known than Crosby, Deming or Juran. The direct quotations in this section of the chapter are taken from an interview published in *Quality and Reliability*, September 1982 issue.

Shainin began his career as an Aeronautical engineer. Having received his degree from MIT in 1936 he worked throughout the war for a division of United Aircraft Corporation. One critical assignment early in his career was a 'materials review board' function, deciding whether certain parts were usable (either 'as-is' or after rework) when they had first been rejected by inspectors on the shop floor.

It seems that this was a formative experience, and as in the case of Crosby the day-to-day hassle, antagonism, and evident waste of energies this entailed caused him to question the received ideas on quality and how it could be achieved.

In Shainin's case this was directed in particular to the use of SPC, but Shainin also emphasizes, in a memorable phrase, 'Talking to the part'.

'Talking to the part'

Shainin sees the essence of eliminating defects and reducing variability as identifying the real primary causes of the difficulty. In other words to characterize as fully as possible the variables which affect the 'part' (component, material or sub-unit) and how they influence its performance. In this way, Shainin claims, the sources of unexpected variation can often be discovered and controlled, whereas the traditional approach would be the more costly one of redesigning the part.

Achieving improved quality

Shainin asserts that the following three technical disciplines are essential for success:

1. Establish realistic tolerances, after statistical study of the interaction of material, process and part tolerances.
2. Having established the tolerances, study the process to reduce the variability within the process capability.
3. Monitor the process continuously to eliminate each new cause of assignable variation as soon as it appears.

You will see in the following chapter that Shainin's message shares much in common with Ishikawa, Taguchi and Shingo.

Management's role in achieving quality

Shainin stresses that the three techniques he advocates require training and patience to achieve. They involve looking much closer into the 'fine detail' of what is causing variation than most American companies have been prepared to do. When they have been defined, the problems can then be tackled in a systematic fashion.

Senior management has to support and nurture these activities through:

1. Support, involvement, commitment;
2. Acknowledgement that accomplishment can only be achieved by the operating staff;
3. An intensive education programme.

The payoff, Shainin claims, is simultaneous better quality and lower cost: 'When tolerance limits are regularly met; scrap, rework and sorting all move toward zero.'

Summary of conclusions

This chapter has introduced some of the best known and most celebrated American writers/speakers on quality topics. Bear them and their teaching in mind while you read about other approaches during Chapters 18, 19 and 20 of Part Three. Although each of these 'gurus' speaks with an individual voice, you may already be forming the conclusion that much of the underlying message is common to all of them.

18

Japanese quality management style

Introduction

In this chapter we shall examine characteristic features of the Japanese approach to quality.

The Japanese business

> Japanese and American management is 95% the same and differs in all important respects. (Takao Fujisawa, co-founder of Honda)

The Japanese reputation for quality in all market sectors is now so high that it is easy to forget that not long ago 'Japanese' was a derogatory term, indicating a cheap and shoddy imitation of a European or American product. Now that Japanese quality has become a mystique, associated with ideas of Japan's unique culture, it is salutary to remind ourselves how relatively quickly our perception has changed. While it may well be that aspects of the Japanese culture enabled the ideas to take root and flower more rapidly in Japan than elsewhere, we must remember that:

1. Much of the methodology was learned from Western advisers.
2. Japanese subsidiaries overseas using locally recruited workforces, and Western companies wholeheartedly adopting methods used in Japan, have achieved comparable results.
3. Successful Western companies are found to have independently developed cultures much closer to those of successful Japanese, than to those of unsuccessful Western companies.

So success does not depend on national identity, nor on national culture, but on company culture. The company has to be patient and thorough, and looking towards long-term benefits as Japanese companies typically are. There is no 'instant pudding'.

The historical background of japanese business

Firstly, the briefest possible sketched history of Japan. For many centuries Japan had been an essentially 'closed' society, governed by a military aristocracy. Craftsmanship was highly prized, as was perfectionism in all the arts; but with few natural resources or overseas contacts, Japan took no part in the industrial revolution. Western colonial and economic expansion in the 19th century brought contacts with, especially, the USA and awareness of the technological strength and expansionist policies of the Western powers. Under the Meiji dynasty, restored to power in 1867, there was a desire to emulate the Western powers and fill the power vacuum in South East Asia; to become an Eastern imperialist state rather than risk Western colonization. You could say it was a policy of 'if you can't beat them join them'. Its motto was 'Western technology, Japanese spirit'.

Japanese expansion during the century following the Meiji Restoration was marked by a successful war with Russia, occupation of Korea and parts of China, culminating with entry into World War II on the side of Germany and Italy, through the surprise attack on the US fleet at Pearl Harbor followed by occupation of British possessions in the Far East. The Japanese surrender after the dropping of the atomic bombs on the cities of Hiroshima and Nagasaki left the country ruined and demoralized.

As Japan, under American occupation, began to attempt to reconstruct its industry it faced a formidable task. The problems can be summarized as:

1. Lack of indigenous raw materials;
2. Run-down or destroyed plant;
3. Disillusioned labour force, poor industrial relations;
4. Shortage of raw materials;
5. Shortage of foreign currency;
6. Poor quality reputation in Western markets.

The USA was anxious for Japan's economic recovery, believing that a disarmed but revitalized Japan under a constitutional Emperor was the best safeguard against communism or anarchy in Japan. Consequently much American aid in help and consultancy in setting up new plants was supplied.

However, Japanese businessmen were very conscious of the waste and inflexibility associated with American-style mass production. We have talked about the contrast between the attitudes of Taylorism and craftsmanship in Chapter 15, Part Three. The Japanese culture was one which highly valued the craftsman and, especially at that juncture, anything which could waste precious raw materials or restrict product variety and adaptability to the tastes of different overseas markets was a serious objection. Hence, dissatisfaction with American methods of managing industry. Hutchins has put it this way, that the Japanese experienced the defects of Taylorism before they had enjoyed its benefits. They were therefore less ready to tolerate its disadvantages than Western businesses.

Thus there was a predisposition to adopt Western methods, but to adapt them in a way which fitted a Japanese strategy of:

1. Extreme thrift in the use of raw material;
2. Recognition that Japanese people were their only significant natural resource;
3. Purchasing the resources Japan did not possess (e.g. oil, minerals, food surpluses) from overseas using the revenue from selling Japanese finished products into the wealthy and sophisticated Western markets.

Quality as a Japanese business strategy

In the 1950s most European industry was still recovering from the devastation of the Second World War; American industry had switched back from war work to consumer products, but demand was unprecedented since the war effort had finally brought the Depression and subsequent stagnation of the 1930s to an end. Thus the whole industrialized world was a seller's market, and quality standards were low.

At this point, it happens that two of the industrial experts sent out to Japan by the US government were Deming and Juran. In both cases their teachings had an enormous impact; quality was the missing ingredient in Japan's policy. Quality would enable Japanese products to gain acceptance, would enable them to make the best use of their scarce resources and compete on price, and the dedication and respect for craftsmanship of the Japanese people would enable quality to be delivered.

How are Japanese companies managed?

White and Trevor quote a Japanese authority as saying that the most distinctive policies of large Japanese companies in their own country are:

- Lifetime employment;
- Pay increases based on length of service;
- Company-based trade unions.

White and Trevor add, as additional well-documented characteristics:

1. Heavy emphasis on training throughout a worker's career;
2. Extremely wide range of welfare benefits;
3. Consensus style of decision-making;
4. A specially important and prestigious role given to the first-line supervisor;
5. Finely differentiated status structure;
6. Heavy reliance on qualifications both in recruitment and in promotion to higher posts;
7. A degree of regimentation (uniform, keep-fit, songs, slogans);

They add:

> There is much evidence that Japanese management was influenced by American business ideas after the war. As the 'human relations' movement was at its zenith in the USA at that time, it seems possible that Japanese personnel practices were not entirely the fruits of their own social soil, but in part a transplant from across the Pacific.

Large and small companies

It must be emphasized that there is actually much diversity in Japanese companies, and that the remarks quoted above apply particularly to the large corporations, as they operate in Japan itself. The statements do not necessarily apply to smaller companies, nor to the way in which overseas Japanese subsidiaries are operated.

The large companies hold an especially influential place in the Japanese economy, even more so than similarly sized companies in the USA or Europe. When the Meiji emperors began to develop Japanese industry the first enterprises created were state owned, but in time the nationalized industries were sold off to the wealthy noble families. Banking in Japan was not equipped to finance large operations, and so the new conglomerate corporations, the 'Zaibatsu' such as Mitsui and Sumitomo created their own banks.

Although General McArthur went some way towards having the zaibatsu dismantled after the war, there is still a big gap between the major corporations which still have preferential access to bank loans, etc. and the average company. The prestige of the big companies also aids them in recruitment; both business and education are very 'rank-conscious', and the 'best' companies inevitably hire the brightest students from the 'best' universities.

One example of popular misconceptions is that all Japanese companies guarantee life-time employment. This is not so; life-time employment is limited to permanent male employees of the large companies. It is assumed (and if necessary enforced) that any female staff will resign when they marry; fluctuations in workload which are seasonal or due to changing business activity are cushioned by the hiring of seasonal and temporary staff, and by the wide use of subcontractors to cope with excess demand. It is the temporary staff and the staff of small companies acting as subcontractors to the large ones who bear the burden of layoffs and redundancy.

Obligations between employer and employee

The concept of mutual obligation and harmony of interest between employer and employee is often seen as deriving from the pre-industrial feudal society of Japan. The peasants owed respect and obedience to the landlord, but the landowner in turn was expected to acknowledge their work for him, treat them fairly and promote their welfare. When the Japanese textile industry was mechanized and country people came to work in the mills, the same relationship was assumed.

Also, the original small Japanese trading companies were run by the merchant and his extended family. Legally the business and its assets belonged to the 'house', not to an individual. Valued subordinates such as a confidential clerk, or a branch manager might be metaphorically or even literally 'adopted' into the family.

How far the stereotypes are valid is disputed, but they do give a basis for accepting that Japanese companies expect:

1. Complete identification of employees with the firm and its objectives;
2. Social relationships and out-of-work activities involving fellow-workers;
3. Respect for age and seniority;

4. Seeing the firm as having at least equal claim to loyalty as the employee's family;
5. A corresponding obligation on the firm to promote the employee's and his family's career opportunities, education, welfare and social life;
6. An obligation on the firm to consult workpeople who could be affected by any proposed change in practices.

Management by consensus

Another common understanding (or perhaps a misunderstanding?) in the West concerning the way Japanese companies operate is that decisions are made by a consensus of managers, who then take full joint responsibility and commitment for the proposal.

According to Trevor, the reality is more complex. A proposal may be initiated typically by a 'section chief' (a position higher in status than a supervisor but lower than a deputy departmental manager). The members of the section will have helped develop the proposal, which the supervisor records and passes up the line. This 'Ringi' document is stamped by the various levels of management up to the final authority, the President. Trevor says that 'whether the formal stamping of the document means "seen", "consulted", or "approved" only those with an inside knowledge of a given organization can tell'. According to Trevor, it is equally difficult for the outsider to say how responsibility would be apportioned if the agreed change proved a failure.

Preceding the 'ringi' process there is the activity of 'nemawashi', literally 'binding up the roots', a metaphor indicating that an idea, like a tree, cannot be successfully transplanted unless it is carefully prepared. This 'preparing the ground' involves, according to Trevor, 'Bargaining, persuasion, seeking support, the long-term trading of favours' in addition to simply arguing the technical merits of the proposal. Since this is normally conducted before the 'ringi' is prepared, the 'ringi' may be no more than a rubber-stamping exercise.

Trevor adds:

> The use of personal relationships and the long-term trading of favours, resembling the establishment of credit that can be called in even much later, is specific to Japanese society and assumes long-term membership in the same organization.

The role of the supervisor

In Japanese industry the role of the supervisor is seen as crucial. We have already met this in our discussion of quality circles where, especially in Japanese practice, the supervisor of an area is generally the chairman, leader and father-figure of the quality circle. As the lowest 'management' level the supervisor is seen as the crucial link and information pathway between the workforce and the higher levels of management, with a special responsibility for acting as a communicator, and for training, motivating and developing the subordinates.

In the context of quality, the supervisor must ensure that the quality message permeates the workforce. A striking indication of this is the existence of a flourishing monthly technical journal titled *Gemba to QC* ('QC for the Supervisor'). Hutchins again contrasts this approach, which he sees as a direct inheritance from the 'craftsman' system of a group working under the direction of an acknowledged 'master craftsman', with the organization under Taylorism. In the latter case, he claims, the supervisor's job is diminished and the workforce may be expected to work autonomously with the support of the production/quality engineers. Thus he says a direct link in the chain is replaced by an indirect support function, and hence a communication gap in the direct reporting line.

Kaizen

This Japanese expression represents 'continuous improvement'. It is critical to much of what we have to say about quality improvement, especially as it is practised in Japanese companies, and discussed by teachers such as Ishikawa and Taguchi.

'Kaizen' improvements may be very modest in scope, the anthithesis of Juran's 'managerial breakthrough'. Many 'kaizen' improvements are initiated and implemented at the level of the supervisor and the workgroup. Indeed Nissan Motor UK are reported to have chosen to call their quality circles 'Kaizen Teams'. We shall refer to the concept of Kaizen again from time to time.

A quotation from Pascale and Athos sums up 'kaizen' very well, even though the word is not mentioned:

> The most significant outcome of the way Japanese organizations manage themselves is that to a far greater extent than in the US they get everyone in the organization to strive by virtue of each small contribution to make the company succeed.

Kanban/'just in time'

This very important topic is a production control method where an operation only works on demand, never simply to build up stock. Work-in-progress is 'pulled' through when it is needed by the next stage, rather than being 'pushed' by the speed at which the preceding stage can work.

Just in time, or JIT, has enormous implications for quality, and for quality-related costs. Because of its importance we are deferring JIT for special attention in Chapter 25, Part Four, when we talk about it in the context of quality and profitability.

Japanese quality authorities

We have seen how Deming and Juran visited and lectured in Japan at a crucial point in Japanese post-war reconstruction, and how their ideas were adopted increasingly as their effectiveness became evident. Crosby's ideas were also explored and followed by Japanese businessmen and doubtless the other experts were also monitored and their advice implemented if it appeared useful in a Japanese context.

As time went on Japanese industry began to make its own contribution, until the tide turned and American businessmen began to pay attention to the tenets of Japanese thought leaders.

We are going to look briefly at the work of three people – Ishikawa, Shingo and Taguchi. Each of these is now well-known in the West.

Ishikawa

Kaoru Ishikawa was a very influential person in the development of Japanese attitudes and practices in relation to quality. He is acknowledged as the father of quality circles, since he was editor of the journal *QC For Supervisors* when, in an editorial article, it proposed the introduction of 'quality control-circles' in which supervisors and their workers could study and try out the statistical techniques being taught in the journal.

He remained very much involved in the QC movement. His name is probably best known here in relation to the 'Ishikawa Diagram', more descriptively known as the 'Fishbone Diagram'. This was a visual problem-solving tool, designed by Ishikawa to help quality circles to formulate and examine all possible causes of the problems they were addressing.

Ishikawa was also seminal in developing the idea that bad quality, meaning wastage, had an effect not only on the business creating the waste but also on its customers and on society as a whole. This is the concept that Taguchi attempted to quantify in his 'Loss function'.

Shingo

Shigeo Shingo's background was in industrial engineering, and it is in the field of manufacturing quality that his interests and prestige lie. Several books written by him have been published in English translation, and the material I present here is based on one of them – *Zero Quality Control: Source Inspection and the Poka-Yoke System*. Shingo first came to prominence in Japan through his achievement of 'SMED' (Single-Minute Exchange of Die) in the Japanese motor industry.

At the time the US auto industry would take approximately two hours, if there were no hitches, to change the die in a press for pressing out steel panels. This was also accepted as the norm in Japan. Because of the delay, inflexibility, and disincentive to do short runs of particular models that entailed, it was seen as a serious barrier to efficiency. Encouraged on learning that the similar operation only took 30 minutes in Germany, Shingo set himself to study the process and refine it as much as he could with the intent of decreasing change-over time.

SMED, his final solution, was almost instantaneous. It involved a modular design of press where the die and the associated parts of the press whose alignment was critical were adjusted off-line and then replaced as a unit.

Japanese technical approaches to quality in engineering, following Deming, are heavily committed to statistical process control (SPC) techniques. Unusually, Shingo is scathing of overdependence on SPC. His concept of zero quality control is a zero defect philosophy based on:

1. Finding where in the process, and what kind of defects are liable to be generated;
2. Utilizing 100% in-process inspection aimed at the particular anticipated fault;
3. Devising methods which prevent operator errors occurring.

It is this last aspect which Shingo refers to as Poka-Yoke ('mistake-proofing'). For example, if it is possible for an operator to position an item in a jig the wrong-way round, Shingo would alter the shape of the jig (and, if necessary the item itself) so that misorientation became impossible. The cited book illustrates numerous examples of Poka-Yoke devices with which Shingo has been involved. In the great majority of cases they are concerned with manually loaded or fully manual assembly operations, and make use of sensors such as load-cells, photoelectric cells and microswitches to detect if an operation has been prepared incorrectly.

Taguchi

In the West Genichi Taguchi is perhaps the best known, and currently the most fashionable of these personalities. The 'Taguchi Methods' of applying statistical theory to manufacturing problems have raised controversy; their attraction to many people as an engineering tool being counterbalanced by statisticians critical of their validity. I don't propose to go into the technical aspects in this book, which is not their place. The controversy relates to Taguchi's procedures for the statistical design of experiments in furtherance of process optimization; the objectors claim lack of rigour, lack of originality, lack of accuracy and lack of efficiency. Against this should be put the fact that many more engineers have been trained in, and use, Taguchi methods then ever exploited the techniques when taught in the traditional way.

More relevant in the context of this book is what Taguchi says about the wider implications of quality assurance.

The Asian Productivity Organization have published an English translation of Taguchi's book *Introduction to Quality Engineering* which is subtitled *Designing Quality into Products and Processes*. I shall quote from this book, but for a bird's-eye view of Taguchi's ideas I shall also refer to an article 'Taguchi's quality philosophy: analysis and commentary' by R.N. Kackar; originally published in the USA in *Quality Progress* in December 1986 and reprinted in the UK in *Quality Assurance* in July 1987.

1. **Total loss to society**. An important dimension of the quality of a manufactured product is the total loss to society generated by that product. Taguchi's definition of quality is 'The loss imparted to society from the time a product is shipped.'

   Loss can be one of two things: either loss caused by variability of function (of the product), or loss caused by harmful side-effects.

This is an unusual definition first, in that it defines quality in a negative sense – lack of quality. Second, that it only measures the impact of the finished product: avoidable costs, wastage, pollution occurring within the manufacturing plant and passed on to the consumer are not included by Taguchi.

How does Taguchi visualize the loss to society? An example from his book provides a simple illustration. It relates to a dirt-resistant shirt, and it is quoted in Chapter 23, Part Four of this book, which examines different models for estimating quality-related costs.

2. **Staying in business**. In a competitive economy, continuous quality improvement and cost reduction are necessary for staying in business.

   A quotation from a Japanese businessman forms an apt commentary:

   > A business should quickly stand on its own feet, based on the service it provides to society. Profits should not be a reflection of corporate greed but a vote of confidence from society that what is offered by the firm is valued. When a business fails to make profits it should die – it is a waste of resources to society. (A. Takahashi, Matsushita)

3. **Incessant reduction in variation**. A continuous quality improvement programme includes incessant reduction in the variation of product performance characteristics about their target value.

   A high quality product performs near the target value consistently throughout the product's life and under all different operating conditions.
4. **The customer's loss**. The customer's loss due to a product's performance variation is often approximately proportional to the square of the deviation of the performance characteristic from its target value.

   Traditional inspection; acceptance of product within the specification tolerance, rejection of product outside the tolerance, encourages us to think of quality as an all-or-nothing thing. You have it, or you don't have it; pass or fail. Taguchi's approach follows on from his point 3; the closer an item is to its design value the better it will work regardless of where we put the specification limits; and the more closely we control the variation the more we can improve the product.
5. **Design and manufacture**. The final quality and cost of a manufactured article are determined to a large extent by the engineering design of the product and its manufacturing process.

   Product design and process design play crucial roles in the success of the product. The dominance of just a few companies in high-technology fields is due to their successful combination of product engineering and process engineering: designing for manufacturability.
6. **Reduction of performance variation**. A product's (or process's) performance variation can be reduced by exploiting the non-linear effects of the product (or process) parameters on the performance characteristics.

   An illustration of this is the non-linear performance characteristic of a transistor. Designing a circuit to operate at one part of the characteristic will produce a much more stable output than operating at another (Figure 18.1).
7. **Statistically planned experiments**. Statistically planned experiments can be used to identify the settings of product (and process) parameters that reduce performance variation.

   Taguchi has developed an original approach to setting up statistically planned practical experiments to optimize product or process design.

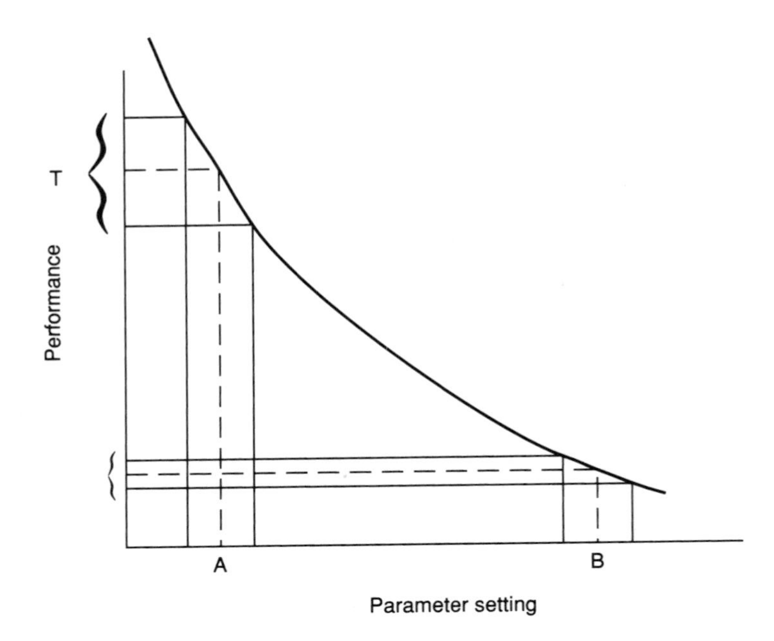

Fig. 18.1 The effect of non-linearity on design robustness. This is a generalized non-linear system. The same degree of fluctuation in the input value at operating point B has a much smaller effect on the output stability than at operating point A. Thus performance can be enhanced by careful choice of the operating conditions. Source: based on Fig. 4 in B. Gunter, *A Perspective on the Taguchi Methods*.

Summary of conclusions

I can leave the last word to two writers on the impact of Japanese industry. Firstly Wickens:

> We have concluded that much that is good about Japanese management practice *is* transferable, with modification, to a western environment. Indeed, those elements which are transferable can almost be regarded as international rather than Japanese.

And then Wilson:

> What could profitably be applied by western managers is not Japan's imaginary magic formula for industrial success, but such time-worn universals as the practice of a little more care and thoroughness, a lifting of sights to the slightly longer term, and a greater consideration for the self-esteem of employees.

19

Quality in the United Kingdom

Introduction

This chapter looks at the extent to which United States and Japanese ideas have been accepted in the United Kingdom, particularly in relation to 'getting the best from people'. It also explains how the BS 5750 system came about, this being a pioneer National Standard for Quality Management Systems.

Quality in the United Kingdom

Chapters 15–17 of Part Three have reviewed the theoretical and practical methods of achieving quality in business which have been used this century. The facts as we have presented them indicate that through first Taylor and his followers, and then the Humans Relations school, it was the US which introduced most of the new ideas and practices.

The success of American business validated the methods on which it was based, until the penetration of Western markets by Japanese industrial products starting in the 1960s. Japanese ideas on using people to build quality into products and processes were studied in Chapter 18.

Europe has not produced its own 'gurus' to the same extent as the USA and Japan, and has been heavily influenced by American theory and practice. US multinational companies have long manufactured in Europe, and Japanese manufacturers are now setting up within the European Community, particularly within the UK. Thus Britain provides a convenient focus for further comparison between recent US thinking, Japanese and 'traditional' British approaches to quality.

We shall also point out that in the BS 5750 scheme Britain has the longest-established and the most demonstrably successful **national** quality system standard and accreditation scheme, which has served as the prototype for the ISO 9000 international scheme.

Leadership and quality management in industry in the United Kingdom

The United Kingdom has always had a high reputation for craftsmanship, and the ability to produce dependable goods of high quality. This is still the case; Savile Row suits, Rolls-Royce cars, Purdy & Purdy shotguns, and many other examples spring to mind. Selection of the highest quality materials, excellent craftsmanship, and attentive personal service are the characteristics of the traditional (and usually small) UK world-leading company.

It is in the larger company, and in providing high quality levels in lower-grade articles, that UK industry is seen at its weakest. It is as though American mass-production methods had been adopted, but never fully assimilated. Yet, with the establishment in the UK of off-shore operations of leading American and Japanese firms with strong quality reputations, their UK plants have shown themselves able to match their counterparts worldwide.

What are the reasons for this paradox? Apart from references by UK authors such as Frank Price, there is little written examination of British industry from a specifically quality-oriented standpoint, except in a recent booklet issued by the Department of Trade and Industry as part of its 'Managing into the 90s' programme. It is entitled: *Leadership & Quality Management, a Guide for Chief Executives*.

The DTI booklet, written by management consultant Ron Mortiboys, identifies the 1950s as the years when UK industry started to decline, by uncritically adopting the methods of the USA, which had not changed much since the 1920s.

(Notice the juxtaposition of these dates: as we have seen in earlier chapters, in the USA in the 1920s 'Scientific Management' was at the peak of its acceptance, before the doubts engendered by the Hawthorne experiments. During the 1950s the new American theories of Human Resources management were still being developed, and the techniques of process control developed in the 1930s and during World War II were being exported to Japan.)

So on this viewpoint the UK only seriously adopted American methods and styles when they were obsolete, and between 1960 and 1980 British and American shares of world trade both fell by nearly a half (USA: 18% to 11%, UK: 13% to 7%).

Deficiencies of traditional management

The DTI booklet alleges that the UK's traditional management style is characterized by:

1. Emphasizing short-term profitability;
2. Clamping down on costs, but tolerating high levels of waste;
3. A 'take it or leave it' attitude to customers;
4. Treating employees as productive robots;
5. Competing on price alone;
6. Buying at the lowest price;

7. Discouraging change – but changing arbitrarily when forced to change;
8. Macho management – acclaim for the 'troubleshooter'.

(American critics of US industry make a virtually identical diagnosis.)
Such attitudes were only viable with:

1. A docile workforce;
2. Demand exceeding supply;
3. Slowly increasing customer expectations;
4. A static worldwide situation.

After World War II the international business situation was far from static. When European, notably German industry began to recover, and Japanese industry to develop, competition became increasingly severe:

1. Unprecedented competition;
2. Fiercely competitive strategies;
3. Fluid and unpredictable financial systems;
4. Increased customer expectations;
5. Increased employee expectations;
6. Political changes;
7. Businesses and service organizations fighting for survival.

The reaction of much UK and US management was to intensify many of the traits which had caused them to lose market share. For example to try to buy components even cheaper, to regulate expense (but still not wastage) even more closely. Only the successful companies eliminated their weaknesses. Many companies, surviving but not prospering, still suffer from these defects such as:

1. **Compartmentalization**. Each department or function works for itself. Each individual function is usually efficiently managed internally, but does not understand the needs of other departments which are its suppliers or customers.
2. **Controlling people through systems**. Treating people as robots and trying to direct them by means of bureaucratic systems does not work. Systems are to help people do their job, not to control them.
3. **Acceptable quality levels**. The only absolutely acceptable quality level is zero defects, but having agreed a shipping AQL with a customer, no attempt is made to improve it.
4. **Firefighting is macho!** The manager who gains attention and credit is the one who is resolving problems. Less credit is given for managing in a way which plans, trains, consults, and thereby prevents problems arising in the first place.

The DTI booklet continues by warning against:

1. Doing the same things over again but expecting to get different results from last time;
2. Trying to learn from experience when past experience is no longer relevant;
3. Adopting 1950s American theories to solve problems of the 1990s.

Leadership and quality needs

The solution that the DTI proposes to British chief executives is discussed under twelve headings; three leadership, and nine quality imperatives.

Leadership

1. Develop and publish clear documented corporate policies and objectives;
2. Develop clear and effective strategies and supporting plans for achieving the objectives;
3. Encourage effective employee participation.

Quality

1. Commit yourselves to satisfying your customers' needs and expectations;
2. Get very close to your customers;
3. Plan to do all jobs right first time;
4. Agree expected performance standards;
5. Implement a company-wide quality improvement programme;
6. Measure performance;
7. Measure the cost of quality mismanagement, and the level of firefighting;
8. Demand and lead continuous improvement;
9. Recognize achievements.

The DTI supports these recommendations by actively encouraging companies to seek BS EN ISO 9000 registration, and supporting the practice of total quality management.

Working for Japanese companies

The books we shall quote from in this section are Trevor, *Japan's Reluctant Multinationals*, White and Trevor, *Under Japanese Management* and Wickens, *The Road to Nissan*. The sections of these books which are significant to us are those dealing with the UK workforce's perception of what quality means to them, and to the company; and also how their perception of the style of Japanese and UK managers differs. Although highlighting the situation in the UK, the purpose of the extensive quotes is to examine whether the methods contributing to Japanese success are transferable to other cultures.

Misconceptions

Before presenting this information, there are two misconceptions we have to address:

1. The first misconception is that the human relations practices of Japanese multinationals resemble those applied by the major Japanese companies within Japan. On the contrary, they tend to follow those of the local economy. Thus Japanese companies in the UK:
   (a) do not guarantee lifetime employment;
   (b) do not offer substantial fringe benefits;

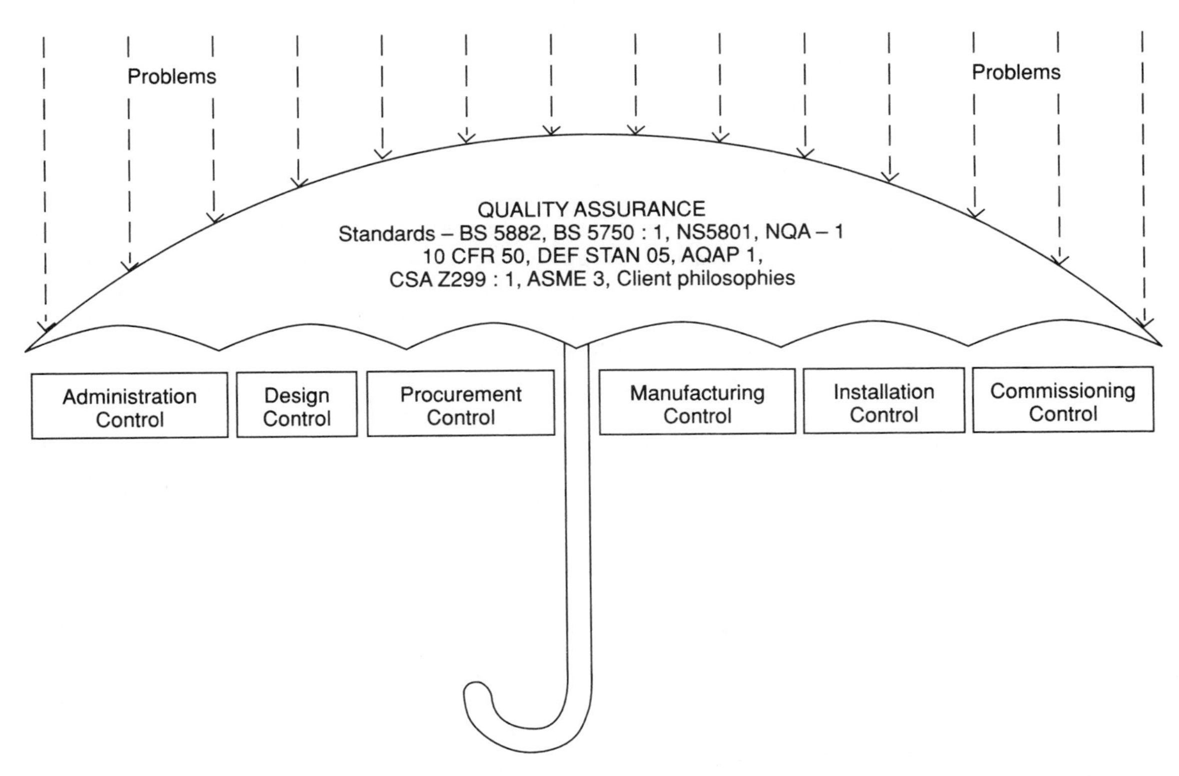

Fig. 19.1 The quality assurance umbrella. A quality assurance management system, such as one based on BS 5750, and embracing the control of administration, design, procurement, manufacturing, installation and commissioning protects you from the stream of 'quality' problems that would otherwise shower you. Source: L. Stebbing, *Quality Assurance*.

   (c) do not seek to set up 'company' unions;
   (d) pay wages which are comparable with the local labour market.
2. The second misconception is one held by Japanese companies before they set up in the UK, namely that labour relations with the workforce will be difficult, and critical to their success. In fact, UK workforces have responded well to Japanese operating styles; if anything it is their UK managers who have found it difficult to adjust to Japanese expectations.

Workforce attitudes and perceptions

Is working for a Japanese company less 'foreign' to a UK workforce than working for an American subsidiary would be? Your first reaction would probably be 'no, they are much **more** alien'; but it is worth examining the studies which have been published on Japanese companies operating within the UK. Some conclusions which are surprising (in the context of the conventional wisdoms) emerge. They will help us to understand the typical UK company as well as the Japanese implant. I justify these assertions with quotations from one of the cited books, describing workforce perceptions of Japanese management methods, choosing passages which make direct reference to 'quality'.

> Another aspect in which the employees of JEL [note: a fictitious name] perceived the company as being exceptionally proficient was in relation to quality. Virtually the same result was obtained, whether quality was seen in terms of the individual's work or of service given to customers. At the same time, employees at JEL did not perceive such a strong emphasis by the company on the quantity of production achieved, or on producing work by a deadline. There appeared to be a consistent priority given to quality standards. Once again, spontaneous remarks by employees confirmed this; a feeling was expressed that the company would never compromise over quality. (White and Trevor)

> What then were the main elements in the profile of the Japanese-style working practices? A short list of four can be discerned: An organised or orderly approach, an emphasis on detail, an over-riding priority attached to quality and a punctilious sense of discipline. . . . The workers concerned saw these practices as highly unusual; and in manufacturing at least, as giving their firms a great competitive advantage. Well organised, smooth-running production and low levels of product defects were their guarantees of output. (White and Trevor)

> Management behaved in a way which was highly consistent with the working practices demanded of workers. For instance, just as the latter were required to be highly quality-conscious, so the managers took a most active interest in quality problems, and never allowed substandard product to go out to customers in the interests of expendiency. In short, the Japanese firms were characterised by a high degree of leadership by example. (White and Trevor)

Acceptance of Japanese style

In short, we see that the Japanese companies manufacturing within the UK have not achieved their success by importing the practices popularly supposed to be characteristic of Japanese firms, but by their commitment to quality and attention to detail at the work-station; and that UK workforces have responded to these priorities of managers and engineers. They see Japanese management as demanding a full and concentrated day's work, but also as 'egalitarian'. White and Trevor say:

> It must be stressed that values relating to the 'work ethic' or to 'egalitarianism' are unlikely to be acquired in a short period. There can be no suggestion that the Japanese company 'taught' these values to its employees. It is more reasonable to assume that they are deeply rooted in the British working class and can be readily evoked by a system which both requires and rewards them.

The impression of 'egalitarianism' the workers received may be surprising, since Japanese culture is very rank conscious: from the studies examined it appears to arise from:

1. The tendency for all ranks to wear company overalls, eat in the same canteen, etc.;
2. A continuous spectrum of grades, and small promotions based on seniority or gaining extra skills, with the supervisor as the fulcrum of the organization

instead of being uncomfortably poised at the boundary between 'us' and 'them';
3. Visibility of managers: 'management by walking around', eager to ask questions, listen to problems, explain company policy;
4. Personal touches like the company sending flowers for a bereavement, managers lining up to wish 'happy holiday' as workers leave for annual shutdown, etc.

In contrast, in many companies where expatriate Japanese had been phased out and replaced by UK managers the workforce detected barriers being set up again: managers ceasing to wear company uniform, less often seen outside their offices, etc.

No magic formula

Thus there is no 'magic' in Japanese culture which enables only Japanese-staffed companies to achieve the highest standards of quality and competitiveness. The ingredients of the quality recipe are those which are being highlighted in this book; many were developed in the West, and they are as accessible to Western as to Oriental companies. Several observers have commented that the most successful Western companies have independently evolved styles which appear to have more in common with successful Japanese companies than with less successful Western ones. The difference is that whereas so far as we can see, the well-run Japanese company is the rule, in the West it is the exception.

United Kingdom 'gurus'

UK quality experts have not received the same overseas acclaim as their American and Japanese counterparts, and the same is true of continental European experts. Certain UK exponents have reached an international audience through their books, these being mainly people who have worked overseas for a significant part of their careers. I will introduce some of them, people who have written books useful for further study. Once again I shall introduce them in alphabetical order.

Groocock

John Groocock's international experience (he was responsible for quality for ITT Europe and for TRW in the USA) is reflected in his books. He worked for ITT during the Crosby era.

The concept which is the title of his book *The Chain of Quality* is, in essence, one cycle of the 'quality loop' or 'quality spiral' but without the ends closed up.

Hutchins

David Hutchins played a key role in the introduction of quality circles to the UK. His book *Quality Circles Handbook* is quoted extensively in Chapters 15 and 16.

Oakland

John Oakland is Professor of Total Quality Management at Bradford University Management Centre. The brevity of our note here is simply because I recommend you to read his book *Total Quality Management* if you have the opportunity, and because it is discussed further in the next chapter. He has also written a book on SPC.

Price

Frank Price is well known for his lectures and in particular for his book *Right First Time*. This unusual and entertaining book is divided into two sections: the first is a step-by-step introduction to statistical sampling and process control; the second, a discussion of 'Quality and People'. Thus it covers two of the main lines of development of quality ideas of the last few decades.

The book also caricatures, in an entertaining and sometimes scurrilous fashion, the attitudes to quality and its control, to be found in the least admirable kind of UK company.

Price has expressed the role of quality control ('quality assurance' in the terms we are using) as seeking to answer four questions. These are.

1. **Can we make it OK?** (Process capability analysis).
2. **Are we making it OK?** (Process control monitoring).
3. **Have we made it OK?** (Quality assurance).
4. **Could we make it better?** (Product research and development, and process evolution).

This a useful vision, but be warned that (as pointed out in Chapter 1) Price uses 'quality assurance' in a much more restricted sense than do most other writers, including ourselves.

Another handy maxim of Price is his 'Three Rules' of quality control:

1. No inspection or measurement without proper recording;
2. No recording without analysis;
3. No analysis without action;

Obeying these three rules will save you from the activity trap of collecting data but not using it to any useful purpose. It also underlines Price's concern that items should not simply be **examined;** they should be **measured** and logged, and the measurements used for **process control**.

Stebbing

Lionel Stebbing's book *Quality Assurance*, from which Figure 19.1 is taken, is written from a standpoint which reflects the author's experience of the quality management of large projects such as those in the power and offshore oil/gas industries, rather than the quality management of repetitive processes. He has worked outside the UK, for example in Canada and Australia.

Possibly the most interesting and original part of the book is Chapter 5, 'The organization for quality'. Taking as its premise the belief that 'Quality is the responsibility of everyone' he argues for a quality assurance department which performs a strictly limited range of activities:

1. Quality system auditing;
2. Auditing inspections and process control;
3. Ensuring that all procedural non-conformances are resolved;
4. Ensuring that working methods are documented, and the instructions available where needed;
5. Verifying that procedures are reviewed and updated as needed;
6. Collecting and reporting the causes of quality losses;
7. Determining when improvements are required, and where appropriate recommending the corrective action.

Figure 19.1 represents Stebbing's portrayal of a quality system. Thus in his ideal organization, the QA department does not involve itself in any inspection or measurement activities. The functions which do reside within the QA department are only:

1. Internal audit;
2. External assessments;
3. QA training;
4. QA personnel assigned specific projects.

Stebbing's views will be discussed further in Chapter 21, where we shall compare them with other, more traditional approaches.

A national standard for quality management systems

We have seen that the quality reputation of UK companies is very variable one to another, and that UK quality experts do not have the standing in the rest of the world, even the English-speaking countries, that the best-known American and Japanese exponents enjoy.

Even so, there is aspect of quality where the UK has established a lead during the last decade or so, and that is in creating a national Quality System Standard together with the infrastructure to support it.

The purpose and requirements of a quality system were dealt with earlier. An effective National Standard for quality systems requires.

1. A National Standards Organization which can formulate and issue the standard, accepting inputs from potential users of the scheme.
2. One or more independent Certification Bodies, accredited to award or withdraw certification of organizations as meeting the requirements of the scheme.
3. An organization which can accredit the competence of certification bodies to assess specified types of organization.

Let us see how the infrastructure for operating such a scheme was established.

Origins of the standard BS 5750 (now BS EN ISO 9000)

BS 5750 was first issued in 1979. It was by no means a wholly original document being strongly influenced by NATO's corresponding 'AQAPs' (Allied

Quality Assurance Publications). The inspiration for the AQAPs, in their turn, had been US Department of Defense documents governing their requirements for contractors' quality and inspection systems, such as MIL-Q-9858 and MIL-I-45208.

The aim of the BSI committee which produced BS 5750 was to create from the defence standards a document which could be adapted to any industry and scale of production or service provision. It was also necessary for compliance with the standard to be testable by a skilled 'second-party' (purchaser's) or 'third-party' (independent) auditor.

Contractual function of the national standard

Although it was not original in concept, BS 5750 was one of the first attempts toapply such a requirement on a general and national, rather than industry-specific basis. Parts 1, 2 and 3 of the standard were respectively addressed at the situations (i) design plus manufacture, (ii) manufacture without design involvement, and (iii) inspection plus test only. They were written in a way which could be contractually enforced, and subject to second or third party audit.

The national standards organization

This is the British Standards Institution, the earliest-founded national standards body. BSI are concerned with the establishment of **written** standards for a wide variety of products and services, and should not be confused with the National Physical Laboratory (NPL) which is the repository of national **material** standards.

BS 5750:1979 was a purely national scheme, administered by BSI. Its text was agreed and amended through BSI committees on which interested parties 7 were represented. Now that BS 5750 has become harmonized with ISO 9000, BSI submit UK national proposals to an ISO committee, vote on other proposals put forward to or by the committee, and implement agreed changes.

Certification under the national standard

In order to establish a national register of approved firms, such as is maintained by the UK government's Department of Trade and Industry, organizations deemed competent to assess compliance of the firms' quality systems with BS 5750 had to be identified. These are known in the UK as Certification Bodies and the organization which approves them is the National Accreditation Council for Certification Bodies (NACCB).

There are over 30 Certification Bodies active in the UK, the largest number of certifications having been awarded by BSI's own certification branch, BSIQA. Most of the other certification bodies have their origins in either:

1. Insurance/inspection companies e.g. Lloyd's Register QA.
2. Consultancy organizations e.g. SGS Yarsley.
3. Industry sectors e.g. Ceramic Industry Certification Scheme.

Guidance documents

Many sectors of industry have developed guidance documents to respond to regulatory requirements or desired standard practices in specific industries. They accept that guidance documents should not contain requirements other than those which are specified in the standards and that proliferation of documents is confusing to the user. The UK chemical industry, for example, was one of the first to use guidance developed by its trade association to aid interpretation of ISO 9001. At the other end of the spectrum, the recent demand for attention to problems experienced in small businesses led to the very firm conclusion that new guidance documents should be produced as a matter of urgency.

BS 5750 in use

Approval schemes are normally generated by or for contractual purposes, e.g. product liability insurance. BS 5750 is no exception. The objective of the scheme is to **ensure sources of supply whose quality can be relied on.** This is a goal which is equally essential for both the consumer and the supplier. Just as important is another aspect which is less explicitly recognized – the supplier's success is in any case dependent on having a quality system, so as to achieve quality targets as efficiently and predictably as possible.

Thus BS 5750 is customer-driven. Customers with strong purchasing power make registration under the scheme a mandatory requirement for vendor approval. Suppliers therefore seek registration – some welcoming the chance of scrutiny and improvement; others reluctantly and with misgivings, only because they see no other way of gaining the business they desire.

It is important to distinguish these positive and negative motives for seeking approval. The positive motive is probably the healthier, since if you achieve good quality, competitiveness and customer satisfaction will follow naturally. In contrast, organizations which reluctantly install a quality system as defined in BS 5750 because it is a condition of doing business, and feel it is an expensive investment in gaining market share will find their expectation is a self-fulfilling prophesy. It costs them money because they install the formal disciplines only in response to external pressures. Because they do not 'understand' the disciplines, they do not 'exploit' their power to improve quality and reduce quality costs.

Approved suppliers are also responsible for establishing the capability of their own material suppliers. This in turn brings pressure onto those subcontractors to quality, and so involvement snowballs.

Multiple assessment

Before the introduction of BS 5750, businesses often found that many customers independently required to assess their quality system, before according them approved vendor status. You will appreciate the customers' motives but will also understand that it was frustrating to the supplier to have frequent visits from assessors representing different customers. The establishment of an independently administered national scheme has greatly reduced the need for 'multiple assessment' within the United Kingdom.

Preparing to be assessed

Many would-be applicants see their major task as 'writing a quality manual'. This is not the case – if there is already a working quality system and a committed workforce, you can prepare a quality manual quite easily. The quality manual is just the tip of an iceberg. More significant is the substructure of operating procedures which underpin it. These define how all the operations and control systems work. Do these procedures exist? Are they comprehensive? Are they up-to-date? Are they available to the people who need them? Are they understood? Are they followed?

If the documentation system is not healthy then you may have to do a vast amount of preparatory work – not a cost and fault of the BS 5750 system, but the cost of your past neglect of quality.

Conversion to a new creed has to start at the top and writing the quality policy acts as the prompt, since the first section calls for the statement of a **quality policy**. This should be a simple but memorable statement of your company's quality objectives. For example, Nissan Motors UK policy, quoted by Wickens: 'We aim to build profitably the highest quality car sold in Europe.'

The quality policy should be spelled out by the chief executive. With the chief executive's commitment and understanding and adoption by the rest of the workforce, company-wide quality is a realistic possibility.

Summary of conclusions

The earlier part of the chapter compared UK attitudes and practices with those of the USA and Japan. Criticism from a booklet issued by the UK's own Department of Trade and Industry was quoted, which identified as a weakness the half-hearted and only partly understood application of US norms. It seems to me (though this insight has only come since I wrote the original text) that the UK has stronger parallels with the Japanese situation.

Whereas the Industrial Revolution of the 18th century was led by the UK, and natural resources of iron, coal, limestone, foodstuffs and physical space were abundant in relation to contemporary needs for materials and manpower, these are now relatively scarce. Yet craftsmanship and innovative design are still respected and flourish. Thus, for example there is arguably no longer an indigenous motor industry; yet British built and designed chassis dominate motor racing.

Although it is now subsumed within ISO 9000, the impact of BS 5750 has so far been relatively far more pervasive than any other national scheme. As we shall see in Part Four, the same principles are now being applied to health, safety and environmental management.

20

Total quality management

Introduction

We have reviewed the messages of a number of quality experts; American, UK and Japanese whose ideas have been translated into English.

Each of these messages is coloured by the background and experience of the author, and the national audience he is trying to reach. Some of the speakers seem to have enjoyed creating controversy: Crosby in promoting 'zero defects' and 'It's always cheaper to do it right first time'. Juran in deprecating these ideas as simplistic. Price in castigating the lack of understanding of quality and statistical methods he sees as a characteristic of bad British management. Shingo in his rare (for a Japanese) lack of confidence in the power of statistical process control. Are there really so many different messages, or are the different voices really giving the same message in different ways?

Since so much of what the messages say is convincing, and since so much of it appears to be common to all the creeds, is there some amalgam which retains all the points of value but which avoids any one individual's particular idiosyncracies? This section of the book has said much about quality improvement, and the role the whole workforce can (and should) play in this task. This concept is often referred to as 'TQM', the acronym standing for Total Quality Management. The problem is that, as G. Vorley has pointed out in *Quality Assurance Management (Principles and Practice)*, TQM means as many things as there are quality management consultants.

The present book in relation to TQM

As author, I confess that for a long time I felt that I was missing the point of TQM. Then one day I was conducting an Assessor training course and had to field a question from one of the delegates which evidently related to the concept of TQM. I was sufficiently diffident to preface my answer with the statement that I wasn't a TQM expert, so he should take my reply as being no more than an 'off the top of my head' response. 'But', he exclaimed 'you have been preaching TQM to us the whole week!'

This comment surprised me, and set me thinking about where I stood in relation to the topic of TQM. The route by which I had entered quality assurance had been through being invited to establish a QA system for a new plant being set up on a greenfield site, in a high-technology industry, drawing on my prior R&D experience in order to set up procedures to monitor, identify and cure problems in collaboration with the other members of the start-up team. I began belatedly to realize that this was indeed a TQM endeavour, even though I don't think the expression had been coined at that time.

My subsequent experiences too, I now recognize, had been in fashioning a Total Quality environment and sometimes suffering the frustrations of which Crosby and others warn, when dealing with other groups which did not 'buy in to' continuous quality improvement. Thus my difficulty in recognizing TQM was not unfamiliarity with it, rather it was over-familiarity; for me there was no other valid way to manage – no 'partial' quality. This book is perhaps a belated attempt to expiate for my slowness of perception.

This puts me in danger of inventing and describing to you yet another version of TQM, the very tendency I have deprecated. So instead, what I shall do in this chapter is to base my review on two relatively definitive views of TQM, these being embodied in documents issued by the British Standards Institution, and (UK) Department of Trade and Industry. At the same time I shall try to establish links between these descriptions and the teachings of the various 'gurus' cited in the immediately preceding chapters.

Definition of TQM

The following formal definition of TQM has been proposed for incorporation into ISO 8402 *Quality Vocabulary*:

> A management approach of an organization, centred on quality, based on the participation of all its members and aiming at long-term success through customer satisfaction, and benefits to the members of the organization and to the society.

That definition seems to me preferable to the one offered in BS 7850:part 1:1992 *Total Quality Management Part 1. Guide to management principles* which is:

> Management philosophy and company practices that aim to harness the human and material resources of an organization in the most effective way to achieve the objectives of the organization.

The first definition highlights various characteristics claimed for TQM: that it makes quality a central management concern, requires all to participate, anticipates future needs and expectations, sees customer satisfaction as the route to success, benefits employees and enriches society. The second definition is no more than a 'motherhood statement'; it expresses sentiments to which no-one could object, but which could apply to any desirable process, without saying anything distinctive about TQM.

In fact, different presentations of TQM contain different degrees of emphasis in what are sometimes distinguished as the 'soft' and 'hard' aspects of TQM. The 'soft' elements relate to the team-building and personnel development aspects of TQM, the kind of concern we have been examining in Part Three. The 'hard' elements relate to specific techniques applicable to quality improvement, to which Part Two provides a starting point.

Common elements in TQM

It seems to me that the following features apply to many of the TQM models which have been expounded by various people:

1. Total Quality Management is outward-looking; it is customer-oriented rather than manufacturing-oriented.
2. It analyses the organization serving the customer into a sequence or network of interlocking processes.
3. Each process must have clearly identified suppliers, process-owners and customers; suppliers and customers may well be internal to the organization.
4. TQM emphasizes continuous quality improvement in regard to both quality level and grade; the goal is to delight rather than merely to satisfy the customer.
5. Improvement is typically achieved through an accumulation of (often) relatively modest individual steps.
6. To sustain steady improvement requires long-term commitment and involvement from top management, providing a visible and credible role-model to the workforce.
7. It also requires the contribution of all the members of the organization, typically operating within teams. Especially, teams drawn from a variety of disciplines and bridging the divisions which are assumed in the bureaucratic picture of the organization.
8. Teams can only function effectively if they are supported with appropriate training, and if the techniques taught are exploited to the full.
9. TQM will only continue to advance if progress can be measured objectively, achievements being identified and honoured.
10. TQM demands, and builds on, the foundation of a suitable and effective quality system.

Characteristics of a 'TQM-led' company

For an organization to achieve quality consistently, all parts must work properly together. The way we have looked at quality in this book, which is the way it is treated in TQM, gives every member of the organization a common point of reference, and a common language for improvement.

Some of the most exciting and unforeseen rewards of TQM have come in departments to which quality initially appeared only weakly relevant, e.g. sales and accounting departments whose error-rates have been dramatically improved, saving costs and enhancing their standing with customers and suppliers.

Management cannot rely on exhortation to the workforce: most problems in achieving quality are management problems, and many relate to the boundaries and interfaces between departments. The nature of the problems, therefore, has to be recognized by management, and solved by cooperation between departments. Truly, total quality management.

The contrast between a company which has total quality and the one that hasn't is detectable to the visitor. The TQM company is calm. Material problems are corrected with suppliers. Equipment down-time is addressed by improved planned maintenance programmes. People are trained, change takes place smoothly. In the other company everyone is busy 'firefighting', attacking their latest crisis. Of this situation, Crosby says 'Quality Assurance is ballet, not hockey', comparing the two kinds of approach with witnessing a ballet (choreographed, meticulously rehearsed, seemingly effortless) and an ice-hockey match (maximum effort, but totally improvised and of uncertain outcome).

The terminology of TQM

Different 'brands' of TQM emphasize particular concepts and the use of particular techniques. Some conceptual terms which have come to acquire special connotations associated with TQM will be mentioned in this section.

Process

A process is defined in BS 7850 as:

> Any activity that accepts inputs, adds value to these inputs for customers, and produces outputs for these customers. The customers may be either internal or external to the organization.
> NOTE 1. Every activity within an organization comprises one or more processes.
> NOTE 2. Inputs, controls and resources are all supplied to the process.

Figure 20.1 illustrates a simple process diagram.

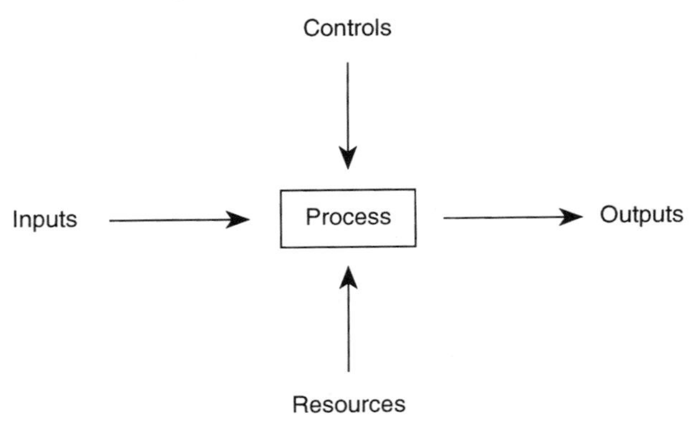

Fig. 20.1 Simple process diagram. Source: BS 7850:part 1:1992.

Process ownership

A process owner is defined in BS 7850 as:

> The person responsible for performing and/or controlling the activity.

Figures 20.1 and 20.2 illustrate the role of the process owners, as also customer or supplier to other process owners.

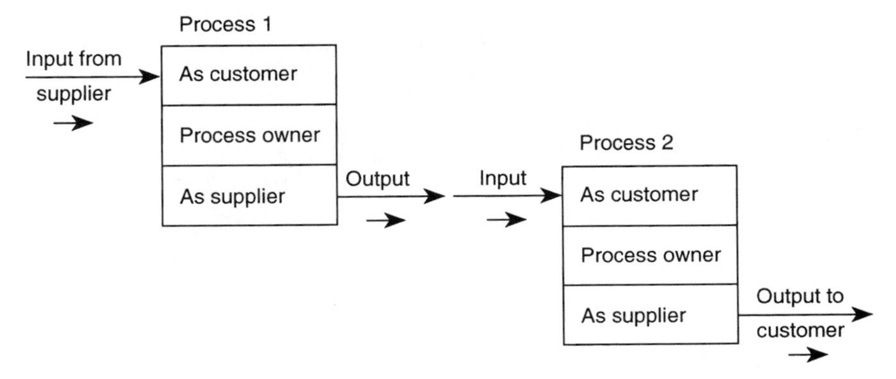

Fig. 20.2 Changing role of the process owner. Source: BS 7850:part 2:1992.

(Non-)value adding processes

All activities represent processes, but these can have harmful side-effects as well as beneficial effects. Moreover, not all processes add value to the product (e.g. cross-checking someone else's work, or rectifying an error). Non-value adding processes should be avoided or eliminated wherever possible.

Internal and external customers

These are sometimes distinguished as the 'big C' (the external or final customer) and the 'little C' (my immediate internal customer), especially in the USA.

Quality chains

The networks of customer/supplier relationships are sometimes referred to, or represented as, the quality chains, as in Figure 20.3.

Empowerment

Empowerment implies delegating the authority to make changes to as low and direct a level in the organization as possible. You could argue that the operator

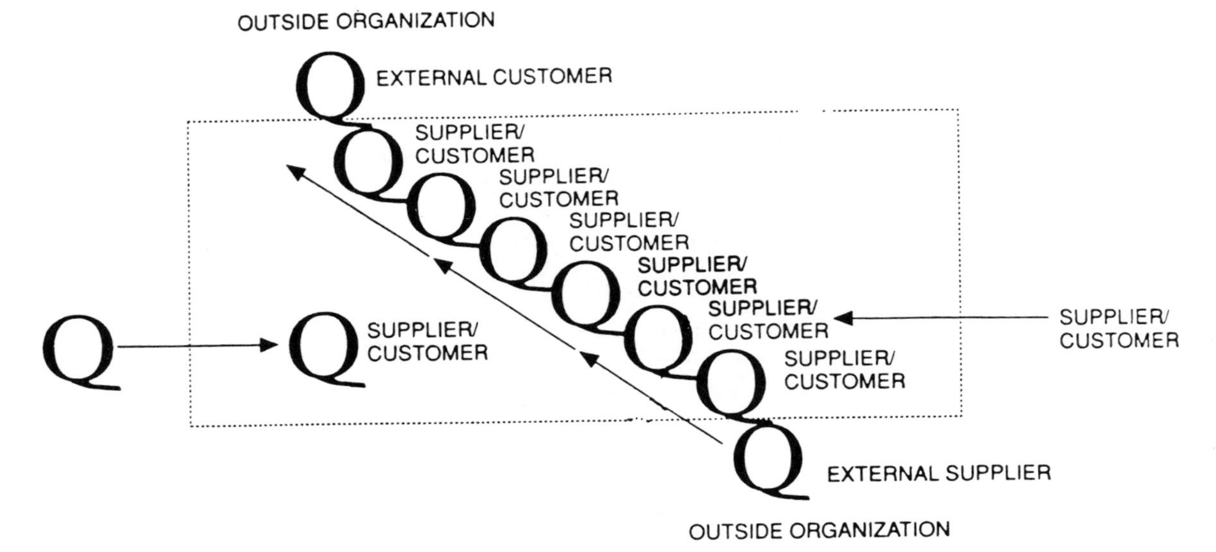

Fig. 20.3 The quality chains. Source: *Total Quality Management*, DTI booklet.

of a process is only truly the process 'owner' (see definition above) if they have the authority (i.e. 'are empowered') to control the process (change settings, method or documented procedure) without appeal to a higher level of management.

Personal quality performance indicators

To help people take responsibility for controlling and monitoring the quality of their own work, 'personal quality indicators' are advocated. They are set by the person concerned, in consultation with their immediate customer(s). They should thus be:

- Relevant and important to the customer, not minor aspects of the measured person's job;
- Unambiguous and measurable in objective terms (such as quantities, ratios, costs and times);
- Challenging but achievable targets;
- Within the individual's personal control.

Metrics

In order to provide a 'hard' emphasis to TQM, measures of progress are needed, such as the personal measurable goals mentioned under the preceding heading. In TQM-speak the word 'metric' is often used instead of measure.

Excellence and delight

Concerted quality improvement gives the opportunity to exceed customers' immediate expectations. This is a measure of excellence and does not merely satisfy but delights the customer.

Commitment

Commitment to quality has to begin at the very top of the organization. If quality is the priority of the chief executive, and is, visible as such, it will be the priority right down the organization. Lip service is not enough, it must be obsessional. This is difficult to achieve.

In the course of a speech made in 1986 Pier-Carlo Falotti, President of Digital Equipment Corporation's operations in Europe, said:

> We as managers are responsible for too many compromises. We compromise every day and we think it is not important but people see that management is compromising and, therefore, it is all right to compromise. Extend the chain all along and the result is a poor quality company.

Notice how this echoes what White and Trevor, quoted in Chapter 19, in the section, 'Working for Japanese companies', said about the power of leadership by example in Japanese firms, and what Chapters 15 and 16 had to say about people adopting the norms of the group with which they identify.

Mind-set or paradigm

A person's mind-set is an expression used to indicate the way the person views quality. It is used in contexts implying that the correct mind-set is one which is disposed to look at quality from the viewpoint of prevention rather than inspection. An alternative expression is paradigm, and a change in mind-set is then referred to as a paradigm-shift.

A practical approach

The heading of this section is taken from the title of a UK Department of Trade and Industry booklet on TQM issued as part of an initiative 'Managing into the '90s'. The booklet's full title is *Total Quality Management – a practical approach,* and it is essentially a condensation of the first edition of a book *Total Quality Management* by Professor John S. Oakland, though this has subsequently been extensively revised in its second edition.

Introducing TQM

Total Quality Management is introduced as:

> A way of managing to improve the effectiveness, flexibility and competitiveness of a business as a whole. It applies just as much to service industries as it does to manufacturing.
>
> TQM involves whole companies getting organized, in every department, every activity, every single person, at every level. For an organization to be truly effective, every part must work properly together, because every person and every activity affects and in turn is affected by others.

The benefits of TQM are summarized as helping companies to:

- Focus clearly on the needs of their markets;
- Achieve a top-quality performance in all areas, not just in product or service quality;
- Operate the simple procedures necessary for the achievement of a quality performance;
- Critically and continually examine all processes to remove non-productive activities and waste;
- See the improvements required and develop measures of performance;
- Understand fully and in detail its competition, and develop an effective competitive strategy;
- Develop the team approach to problem solving;
- Develop good procedures for communication and acknowledgement of good work;
- Continually review the processes to develop the strategy of never-ending improvement.

Meeting the requirements

Taking 'meeting the requirements' as the meaning of quality, the booklet points out that the implications of this statement are very wide. Everyone in the quality chain has to ask themselves who their suppliers and customers are, and what they actually require. Otherwise there will be constant confusion and argument, and little hope of systematic quality improvement:

> Every day in organizations all over Britain, people scrutinize together the results of the examination of the day's work, and commence the ritual battle over whether the output is suitable for transfer to the 'customer'. They argue and debate the evidence before them, the rights and wrongs of the specification, and each tries to convince the other of the validity of their argument. Sometimes they nearly break into fighting.

A model for total quality management

This section of the booklet reiterates that the organization should be seen as a chain of supplier/consumer relationships, starting with the external suppliers and ending with the external customers. This is perhaps the most distinctive and characteristic element of the TQM message, though it is implicit in many other recipes for quality improvement.

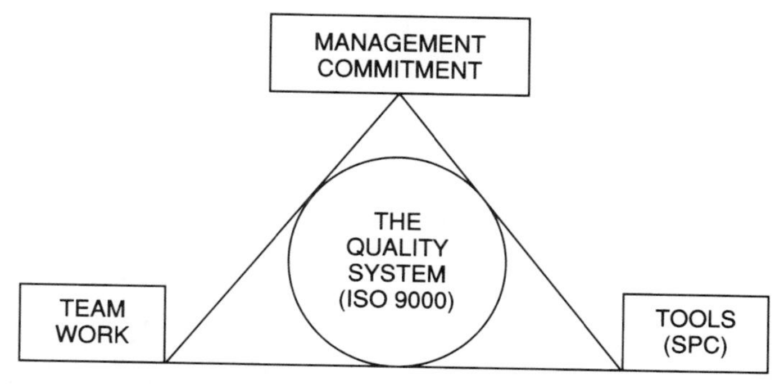

Fig. 20.4 The TQM model. Source: *Total Quality Management,* DTI booklet.

Diagrammatically, (Figure 20.4), the TQM model shows the three prerequisites:

1. Management commitment;
2. Teamwork;
3. The necessary tools (especially SPC);

bound together by:

4. A quality system based on ISO 9000.

In the second edition of the book on which the DTI booklet was based, Oakland puts the process at the heart of the diagram, with systems, teams and tools at the corners of the triangle, bound together by culture, communication and commitment. The figure 'Launching quality improvement' (Figure 20.5) presents a similar message.

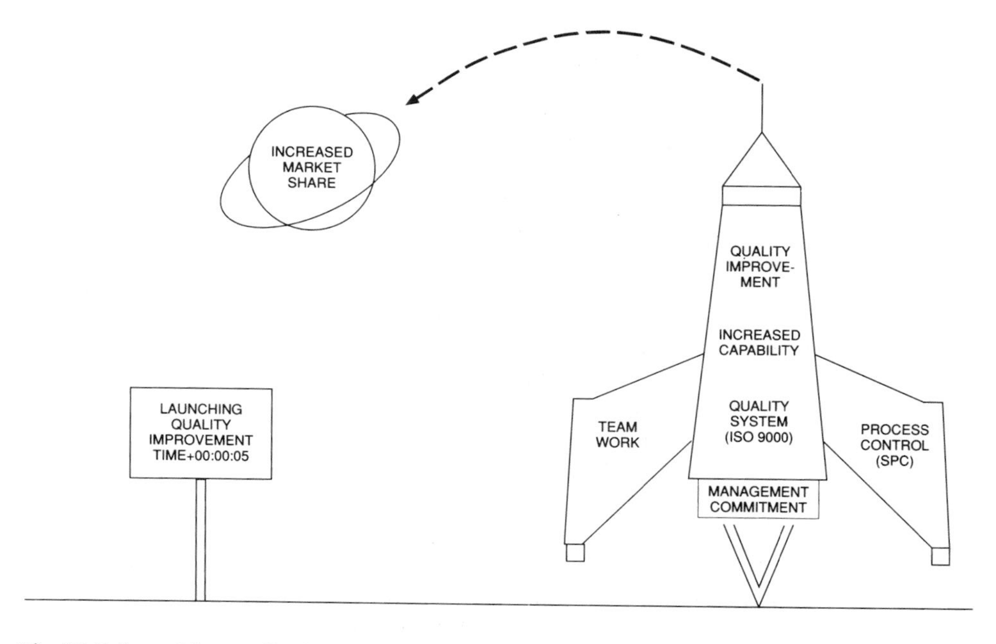

Fig. 20.5 Launching quality improvement. Source: *Total Quality Management,* DTI booklet.

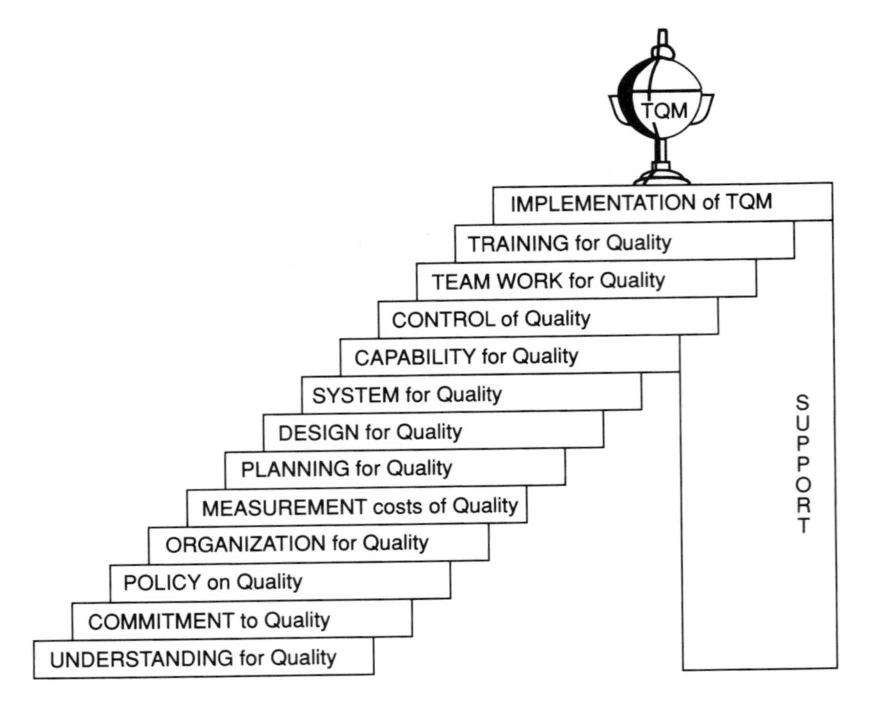

Fig. 20.6 The steps to TQM. Source: *Total Quality Management*, DTI booklet.

Planning the implementation of TQM

The DTI booklet presents planning the implementation of TQM as a series of steps (Figure 20.6) which are:

- Understanding quality;
- Commitment to quality;
- Policy on quality;
- Organization for quality;
- Measurement of costs of quality;
- Planning for quality;
- Design for quality;
- System for quality;
- Capability for quality;
- Control of quality;
- Team work for quality;
- Training for quality;
- Implementation of TQM.

I am not sure that it wise to represent them as a precise sequence of steps to the goal of TQM, and would caution you against supposing that they should be established in a precise order. If you were charged with establishing TQM in an existing organization you would have to act on several fronts at once. If you were starting on a 'greenfield' site with a new group of people, it is still unlikely that you would address the topics in precisely that order of priority.

British Standards guidance on TQM

The British Standard guidelines to Total Quality Management were issued in 1992, and their official status provides them with a degree of authority which could be reassuring to you, while considering what you might otherwise feel is a rather nebulous topic, subject to different experts' interpretations.

The standard is issued in two parts:

BS 7850:part 1:1992 *Total quality management Part 1. Guide to management principles*
and
BS 7850:part 2:1992 *Total quality management Part 2. Guide to quality improvement methods*

Thus part 1 concentrates on the 'soft' and part 2 on the 'hard' topics.

BS 7850 Definition of TQM

I have already quoted and criticized the BS 7850 definition of TQM. I should add that there are also footnotes to the definition which can be quoted here:

> NOTE 1: The objectives of an organization may include customer satisfaction, business objectives such as growth, profit or market position, or the provision of services to the community etc. but they should always be compatible with the requirements of society whether legislated or perceived by the organization.
> NOTE 2: An organization operates within the community, and may directly serve it; this may require a broad conception of the term customer.

You may recognize here parallels with the Japanese views on the obligations of the company, and the bad-quality costs borne by society which I quoted in Chapter 18.

> NOTE 3: The use of this approach goes under many other names some of which are as follows:
> Continuous quality improvement;
> Total quality;
> Total business management;
> Company-wide quality management;
> Cost-effective quality management.

The list given in BS 7850 is still not exhaustive; for example:

Total quality control (Feigenbaum);
Market-driven quality (IBM);
Company-wide quality improvement (Bristol Quality Centre).

Management Principles of TQM

BS 7850 part 1 claims that the fundamental concepts underlying TQM are:

1. **Commitment** to TQM by the chief executive is essential, and must be transmitted to all levels.

2. **Customer satisfaction** must be a key objective.
3. **Quality losses** are losses caused by failure to utilize resources (whether material, financial or human) as effectively as possible in some process. Examples:
   - Loss of customer satisfaction;
   - Loss of opportunity to add value;
   - Loss due to waste or misuse of resources.
4. **Participation by all** members of the organization.
5. **Process measurements** should be applied to all activities.
6. **Continuous improvements** to the performance of people and processes.
7. **Problem identification**: resolving both existing and potential problems.
8. **Alignment of corporate objectives and individual attitudes**: removal of prejudices and restrictions.
9. **Personal accountability**: the recognition of individual responsibility and authority.
10. **Personal development**: all individuals continuously appraised, trained and developed.

Implementing TQM

BS 7850 part 1 implies that executives attempting to introduce TQM into a company will meet resistance, since it indicates that overcoming resistance to change requires:

- An appropriate organizational structure;
- Implementing process management concepts;
- Systematically measuring performance and recognizing achievement;
- Introducing improvement planning techniques;
- Training.

Supporting techniques

BS 7850 sees a role for what it calls three 'supporting techniques' aiding the introduction of TQM. They are:

- A **quality management system**: primarily in order to ensure that customer needs are accurately translated into an acceptable product or service. BS 7850 reminds readers that such a system exists as BS 5750 (now BS EN ISO 9000).
- **Quality improvement action**: wherever a situation is seen as needing improvement.
- **Tools for analysis and diagnosis**: to identify, prioritize, assign causes and find solutions to problems.

Recognize the correspondence with the Oakland/DTI trinity of system, (improvement) teams and tools.

Supporting tools and techniques

After recapitulating the themes of part 1, part 2 of BS 7850:1992 concentrates on the quality improvement aspects of TQM, emphasizing the methods listed in Table 20.1. Much of this part of the standard consists of brief descriptions of these methods.

Table 20.1 Selecting an appropriate tool or technique

Tool or technique	*When to select*
Data collection form	Gather a variety of data in a systematic fashion for a clear and objective picture of the facts.
Tools for non-numerical data	
Affinity diagram	Organize into groupings a large number of ideas, opinions, issues, or other concerns.
Benchmarking	Measure your process against those of recognized leaders.
Brainstorming	Generate, clarify, and evaluate a sizeable list of ideas, problems or issues.
Cause and effect diagram	Systematically analyse cause and effect relationships and identify potensial root causes of a problem.
Flow chart	Describe an existing process, develop modifications, or design an entirely new process.
Tree diagram	Break down a subject into its basic elements.
Tools for numerical data	
Control chart	Monitor the performance of a process with frequent outputs to determine if its performance reveals normal variations or out-of-control conditions.
Histogram	Display the dispersion or spread of data.
Pareto diagram	Identify major factors and distinguish the most important causes of quality losses from the less significant ones.
Scatter diagrams	Discover, confirm or display relationships between two sets of data.

The tools and techniques identified in BS 7850 and presented in the table include both numerical and non-numerical methods. Wherever possible quality improvement decisions should be made on the basis of numerical data, though this will not always be the case. Most of these tools and techniques for quality improvement have been introduced earlier in the book, and the remainder will be reviewed in Chapter 28, Part Four, under the heading of 'Seven new tools of quality'.

ISO 10014

There is as yet no ISO international equivalent of BS 7850 dealing with TQM in general terms. There is however ISO 10014 *Economic effects of Total Quality Management*. It measures the success of an organization in terms of a 'gain/cost' ratio:

$$\text{Gain/cost ratio} = \frac{\text{gains in terms of profit}}{\text{costs}}$$

or for a non-profit organization:

$$\text{Gain/cost ratio} = \frac{\text{gains in terms of perceived value of service}}{\text{costs}}$$

ISO 10014 continues by developing the ideas of improving gain/cost ratio through good management, paying attention to planning, process ownership, self-imposed performance measurements, relevance, priority, simplicity, flexibility etc. It distinguishes three levels of meeting customer expectations: dissatisfaction, satisfaction and delight; corresponding to failing, meeting and exceeding customer expectations.

Summary of conclusions

In this section of the book we have concentrated on the human aspects of improving quality, and focused on key contributions to the ideas in this field. By now, your heads may be buzzing with all the different prescriptions which have been offered by various experts, in their different ways.

Chief executive officers of companies throughout Europe and the USA are realizing the importance of quality to profitability and survival, and looking for some ready-made package of solutions to put in place. In their books (though not in their publicity material), none of the gurus is offering 'instant pudding', though in popular demand one or another may seem to be 'flavour of the month'.

As time goes on and the messages become more refined (remember that Deming and Juran were lecturing and advising for over 40 years) it seems to me that these messages converge. In *Let's Talk Quality* Crosby says:

> I never missed a chance to explain the concept of zero defects properly. . . . Once one of the 'grand old men' of the quality profession came up on stage out of the audience to ask me how I could consistently defend such a stupid thought. 'You seem bright enough', he said, 'you probably have never listened to what you are saying.'
>
> So I asked him to explain to me what he thought the concept of zero defects meant. He talked about worker motivation, how the idea of exhorting the worker to do better was useless, and how it was causing people to not use the real tools of quality control. I walked across the stage, put my arm around his shoulder and said 'If that is what zero defects is, I don't want any part of it either'. This had no effect on him. To this day he still doesn't understand that we are on the same side.

So perhaps by now, the gurus themselves are the only ones who don't realize that they are in reality singing from the same hymn sheet.

21

Quality managers and their departments

Introduction

Earlier in this book, you learned about the importance of having an effective, documented quality system in order to have confidence that you are meeting your customer's requirements and expectations. Stress was placed on the importance of the organization having a 'management representative with responsibility for quality'. For companies of any size this is a full-time responsibility, bearing with it a title such as Director of Quality, Quality and Reliability Assurance Manager, or the like. Here we will refer to the individual as the Quality Manager.

In this chapter we will concentrate on the duties of the Quality Manager. Although it is not automatically essential that he/she is the head of a department, in all but the smallest companies the responsibilities that devolve to Quality Managers make it necessary for them to deploy supporting staff. Consequently we shall also look at the duties and organization of a typical quality assurance department, distinguishing between the activities which the QA department must control and those which they often do but which could be done outside the QA department provided certain safeguards are met.

It is true to say that the present trend is to rethink the role of the quality department, and to reduce its size. This recognizes that the creation of quality is not the responsibility of the QA department personnel; it is everyone's responsibility. The role of QA is to create a framework within which people can operate to promote quality (the quality system), to help and encourage, and to monitor and report.

The management representative for quality

We have already introduced you to the 'management representative for quality', alias the Quality Manager, who plays a key role in the creation, maintenance and review of the quality system. We also explained that the phrase 'management representative' came from the ISO 9000 series (BS 5750) Quality System Standard.

Apart from ISO 9004–1, references to the management representative appear in ISO 9001, 9002 and 9003 under 4.1.2.3, this section being headed 'Management representative'. What it says is:

> The supplier's management with executive responsibility shall appoint a member of the supplier's own management who, irrespective of other responsibilities, shall have defined authority for
> (a) ensuring that a quality system is established, implemented and maintained in accordance with this International Standard; and
> (b) reporting on the performance of the quality system to the supplier's management for review and as a basis for improvement of the quality system.

By now, if you have read the ISO 9000 series of standards you will have realized that they contain much detail on the requirements for quality **management,** but very little about a quality **manager**. This is contrary to many people's impression that the ISO 9000 standards demand that you employ a full-time quality manager, and that this could place a severe burden on a small company. On the contrary:

1. The Quality Manager post comes into being partly because of the need for customers to have a known management contact on quality matters;
2. The Quality Manager is responsible for ensuring that the quality management system is implemented and maintained, and holds the necessary authority to accomplish this;
3. It is not necessary to exclude all other responsibilities provided they do not conflict with the Quality Manager's quality responsibilities, either in terms of divided loyalty or drain on time.

The position can be filled in various ways, depending on the size of the company:

'Management representative' for a small company

In the smallest companies it would be quite proper for the Proprietor or Managing Director to act as management representative himself. Who better to have the quality of product or service at heart?

In fact one often finds, in the small company in which the owner plays an active day-to-day role, that despite employee titles **all** the decisions are made by the owner. In such a case he/she is the **only** person who can exercise genuine authority for quality.

In a somewhat larger company it often happens that quality, together with design and engineering, is delegated to a 'technical manager'. This too is acceptable, provided that as well as close contacts with manufacturing, he/she maintains sufficiently close links with marketing and other non-manufacturing areas affecting quality.

Management representative for a medium-sized company

Here there will be a full-time Quality Manager, normally reporting to the General Manager, MD or CEO as the case may be. He will be independent of manufacturing, sales, or any group with objectives relating to short-term quantity goals.

Some companies have their Quality Manager reporting directly to the MD but, because this is seen as an artificial status demanded by outside agencies, without according him/her the status of the MD's other 'Direct Reports'. Thus: Sales Director, Production Director, but Quality **Manager**. In so doing the MD weakens the position of the quality manager and is, in effect, saying that he has not unreservedly been delegated responsibility for quality.

Management representative for a large company

A large company, especially one spread over many sites, or indeed multinational, will usually have a Quality Director on the central board, and a Quality Manager at each site. The Quality Managers may report directly to the Quality Director, or jointly to him and to the General Manager of their own site. The Quality Director will be responsible for setting quality policy, and ensuring uniformity in quality standards at the various sites. The site Quality Manager will be responsible for local implementation of the policies.

The quality manager's job

Under the overall direction of the CEO/Managing Director, each critical aspect of the enterprise; production, research, finance, sales, personnel, etc. has its own manager to make use of the resources allocated to it. One of these key areas is the overall quality of the product or service. Quality pervades all areas of the business; as does the deployment of people and the use of money.

So just as there are specialist Personnel and Finance Managers, so it is natural to have a specialist Quality Manager. All the cash in the factory is not held by the finance department, nor do all the people report to personnel. It would be equally unrealistic for all the quality-related matters to be handled by quality assurance.

One of the Quality Manager's duties (if also 'Management Representative') is creation and maintenance of the quality system, as was made explicit in section 4.1.2.3 of ISO 9001 quoted earlier. Thus the responsibilities of the Quality Manager include devising, defining, documenting and administering a quality system that will cover the needs of all departments contributing to the quality of the product or service being generated. Using whatever personnel, material, and other resources over which he needs to exercise **direct** control (those of the QA department), he has to promulgate, coordinate and monitor the system, compare the actual with the targeted quality performance, report and indicate where improvement is needed.

The quality manager and the customer

Under the previous heading we examined the role of the Quality Manager in achieving the overall quality goals of the company. He/she also has responsibilities in respect of the customers, who look to the Quality Manager as the protector of their received quality.

It is easy to become familiar with a company's aim of maintaining, shall we say, an acceptable quality level of 0.15%, and that many customers will never experience any problem as a result. But on the occasion when a customer does receive a defective product, it is no consolation to know of the thousands of good products which have been received from the same supplier by other customers.

To underline the point, the Ford Motor Company established from a survey that someone pleased with his car will on average inform 8 other people; but if dissatisfied with his purchase he will tell 22!

Therefore a key part of any company's strategy must be to do all it can to ensure that **nothing defective is shipped to a customer**. Unless your company has a secure monopoly, this also makes unchallengeable commercial sense. Failures in your customer's hands, or even further along the chain can be very expensive, as we shall see during Part Four of this book.

This is another reason why the Quality Manager must have a direct reporting line to the most senior level of management, independent of other managers with short-term commercial and production goals. The Quality Manager stands, as it were, at the gate between the supplier and the customer. If the supplier has been formally approved, e.g. to ISO 9001 or a comparable standard, he has been made effectively the customer's representative within the supplier's organization.

Profile of the quality manager

I have indicated the Quality Manager's job description and the position he/she should hold in the organization. What sort of person should a company be looking for to do this job? If they find that Superman or Superwoman is not available, which candidate should they choose?

I suggest that the qualities needed are:

1. A strong personal commitment to quality and to the good name of the company;
2. A clear vision of how quality can be achieved, recognizing the interaction of all parts of the company;
3. Integrity and independence, firmness and unwillingness to compromise the quality standard which the company has set;
4. Ability to communicate clearly and persuasively, with diverse individuals and groups from a board-meeting to the shopfloor;
5. The confidence and respect of customers and regulatory agencies;
6. An understanding of the use of standard QA techniques – knowing both when each is valid, and also when it is inappropriate;
7. A capable manager of departmental resources, in particular the training and development of subordinates.

The quality assurance function

Having examined the role of the Quality Manager or 'Management representative responsible for quality' we will now look at the activities which can, often are, and (maybe) should be performed by subordinates in the 'quality department'.

Why have a QA department?

Granting that quality is a major concern for any company, but also acknowledging that the achievement and maintenance of quality are concerns which know no departmental boundaries, why have a distinct quality assurance department? What can it hope to achieve?

It can:

1. Monitor the quality system;
2. Report the level of quality measured;
3. Safeugard outgoing quality, by direct action if necessary;
4. Ensure that corrective actions are carried through.

It can also:

1. Represent the interests of the customer;
2. Represent the broader interests of the company as a whole;
3. Arbitrate between different departments (especially with regard to the acceptability of material transferred from one to another).

It also fulfils a service role to other departments, in the interests of quality, by:

1. Giving them information about their quality;
2. Coordinating their quality efforts, and paying special attention to where quality may be 'leaking away' at the joints between departments;
3. Providing training in quality concepts;
4. Provide 'consultancy' through specialized QA techniques and resources.

The QA department does not have to operate all the procedures for defining and safeguarding quality, but its manager does have to express, in the quality system, what is required. The department also has to exercise the technique known as quality system audit, which evaluates whether the quality system is, in fact, working correctly.

In short, the QA department exists because in the pursuit of quality it can exercise:

- Independence;
- Impartiality;
- Concentration;
- Coordination;
- Special skills.

What activities should the QA department perform?

Since no QA department can, by its own unaided efforts, create a product or service fully satisfying the customer's needs, the range of activities the QA department is required to perform varies from company to company.

A department dedicated to quality still cannot produce quality. It can only plan, and assist, and cajole, and investigate, and report; and as a last resort impound defective material to prevent it being submitted to a customer.

Of the activities which a QA department **might** undertake with its own resources, the ones which are central to its existence as an independent department are, I believe, the following:

1. Verification, through audit, that the quality system is working;
2. Verification, with objective evidence, that tests and inspections are being properly carried out by the groups assigned test/inspection responsibility.
3. Ensuring that no defective product is knowingly shipped to customers without their agreement;
4. Ensuring that all processes are documented and that personnel have access to correct procedures, and are trained in their use;
5. Determining and reporting the principal causes of failures and non-conformances, and ascertaining that corrective action is taken.

BS 4891 view of the scope of the QA department

A quite explicit statement of the function of a QA department is given in the British Standard BS 4891: *A Guide to Quality Assurance*. This document was a predecessor of BS 5750.

I will quote what it has to say about the QA department:

> When a centralised quality assurance or quality control department has been established, it usually has authority for taking action or giving advice on quality matters in the following areas:
>
> 1. Marketing/servicing;
> 2. Design/specification/standardization;
> 3. Value analysis and reliability assessment;
> 4. Research and development engineering;
> 5. Documentation;
> 6. Purchasing and vendor assessment;
> 7. Incoming materials;
> 8. Pre-production and production controls;
> 9. Metrology and test equipment calibration;
> 10. Quality control, inspection and production test procedures;
> 11. Feedback of service difficulties and complaints;
> 12. Defect analysis and remedial action;
> 13. Quality cost measurement;
> 14. Education and training for quality assurance;
> 15. Review and evaluation of management system for quality assurance.
>
> The execution of the majority of the quality assurance functions listed above should remain the direct responsibility of the individual departmental managers concerned. . . . Rather than attempt to usurp these responsibilities, a quality department manager should advise and assist on all quality tasks, and monitor and coordinate them throughout an organisation wherever possible. The following activities are likely to be the direct responsibility of a quality departmental manager:

1. Developing the quality assurance programme and quality manual;
2. Incoming and defective material control;
3. Metrology and test equipment calibration;
4. Quality control system operation;
5. Defect/failure analysis;
6. Quality cost measurement.

In addition, a quality department manager might be required to carry out the training of staff throughout the company in matters relating to quality assurance, or to coordinate such training schemes.

The minimal QA department

Lionel Stebbing (*Quality Assurance*) goes further than do most people in the limits to the direct activities of the QA department which he advocates. He advances the proposition that only specialists can check the accuracy and quality of specialist work, that the checkers themselves must be subject to independent scrutiny, and that the QA Department cannot possibly contain all the necessary specialisms.

Stebbing sees the QA Department's function as limited to auditing, coordinating and following up, with all inspection and much of the detailed auditing being done by qualified personnel co-opted by the QA manager. This picture of the scope of the QA department is illustrated in Figure 21.1.

Under such an arrangement, the functions of the QA department would be:

1. Verifying the operation of the quality management system, by means of quality audit;
2. Verifying the existence of objective evidence that control and inspection of activities have been performed effectively;
3. Ensuring that all procedural non-conformances are resolved;
4. Ensuring that fundamental working methods are established, documented, and accessible to personnel;
5. Verifying that procedures are reviewed and updated as needed;
6. Determining and reporting principal causes of quality loss;
7. Determining, in conjunction with senior management, necessary improvements and corrective actions.

Organizational options charted

Figures 21.2, 21.3 and 21.4, which are again taken from Stebbing's book, show the impact of his philosophy on a company's organization. Figure 21.2 illustrates a quite common situation, in which the quality manager reports to the technical director, and is responsible for all personnel doing QC inspection doing calibration and maintenance of test equipment.

In Stebbing's preferred organization (Figure 21.3), calibration and maintenance are transferred to production, and quality control to engineering and design. QA becomes a purely 'staff' function reporting directly to the MD, with the internal organization shown in Figure 21.4.

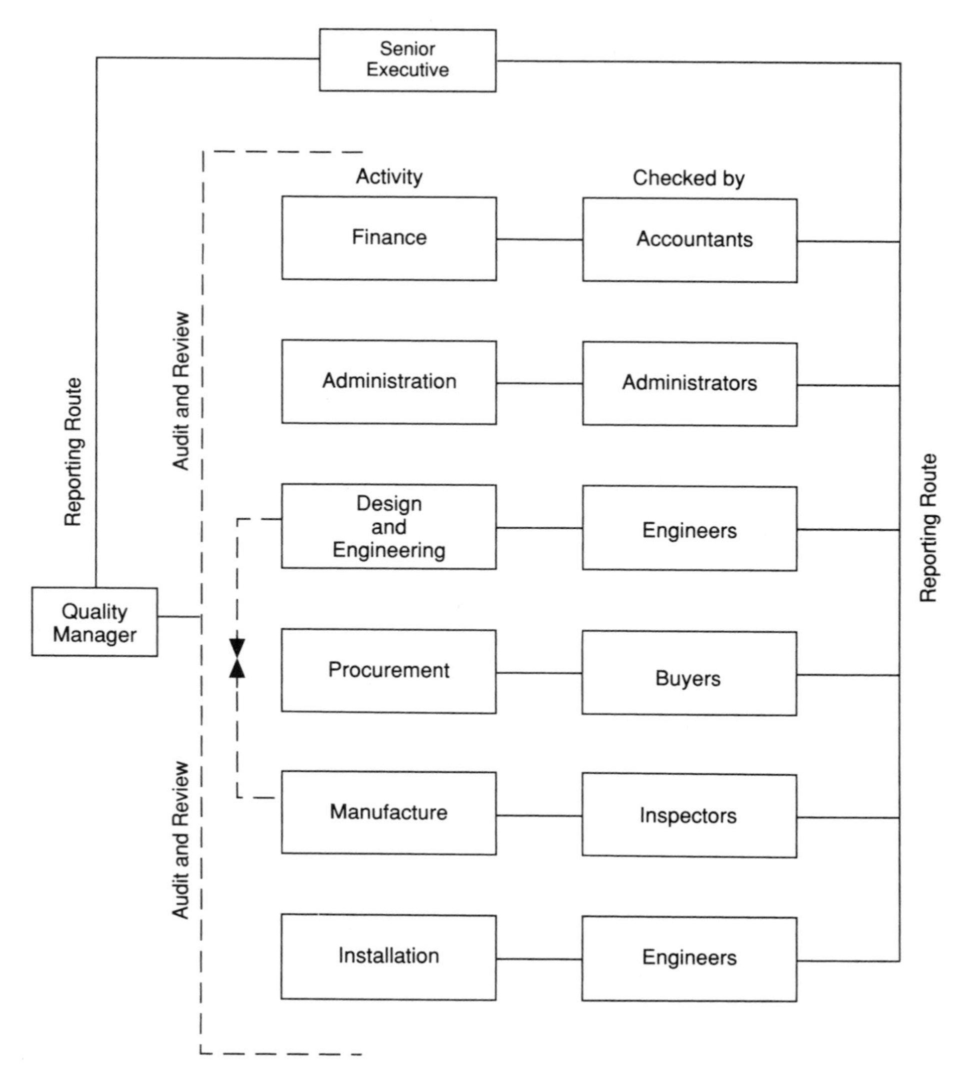

Fig. 21.1 The checking functions. Source: L. Stebbing, *Quality Assurance*.

Would you sign a release note?

You will realize by now that there are quite wide differences of opinion on how broad a range of duties should be undertaken by the QA department. This is an issue on which newly appointed Quality Managers will have to establish their own position. Should they attempt to build up their own departments and costs, or rely on people outside their own subordinates?

A lot depends on the complexity of the company and its products, the top-level commitment to quality, and the enthusiasm and quality-education of all levels of the workforce. I have found that the **release note** helps the manager to focus his judgement on this point. Goods must often be dispatched with a

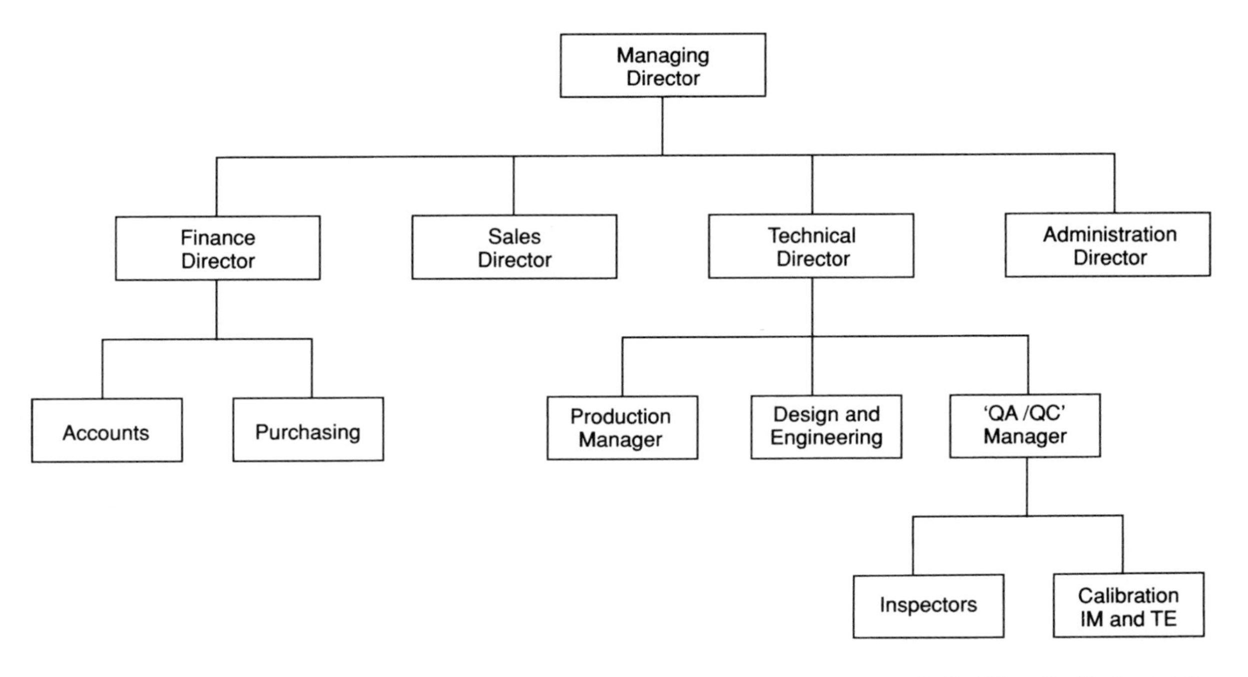

Fig. 21.2 'Typical though not recommended' organization. Source: L. Stebbing, *Quality Assurance*.

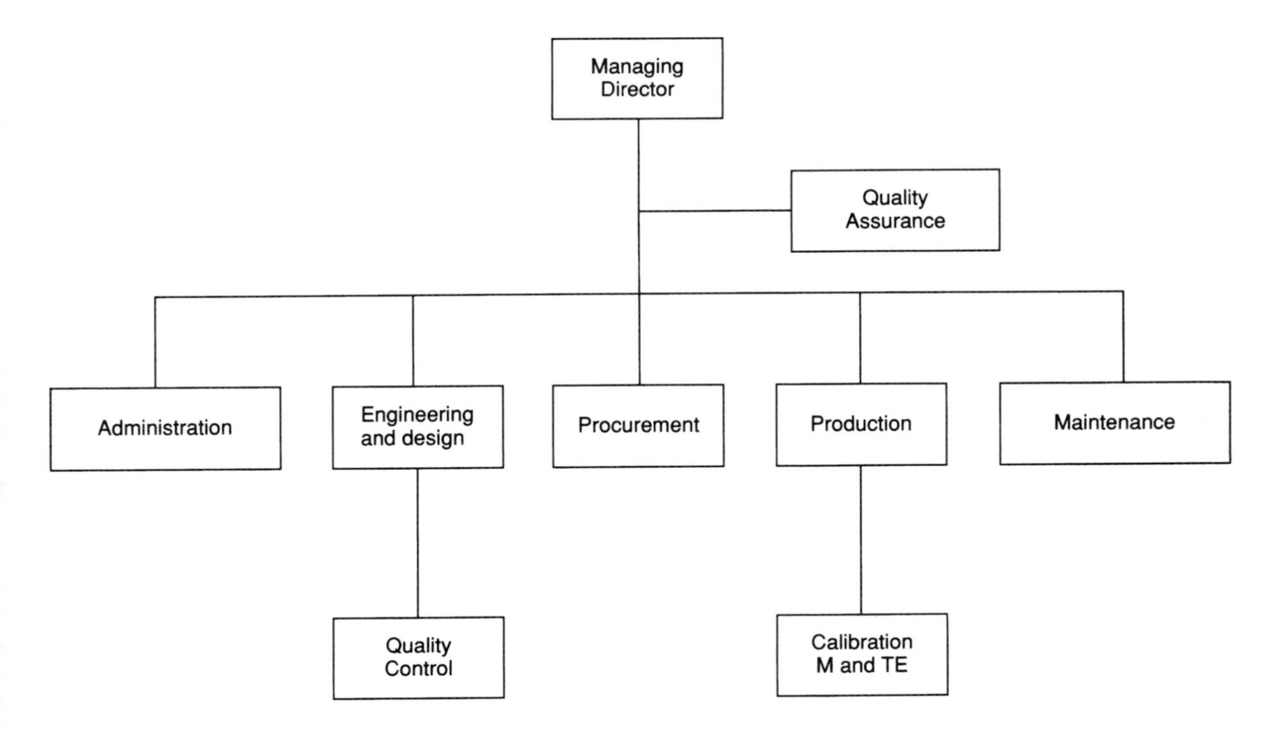

Fig. 21.3 Organization as recommended by Stebbing. Source: L. Stebbing, *Quality Assurance*.

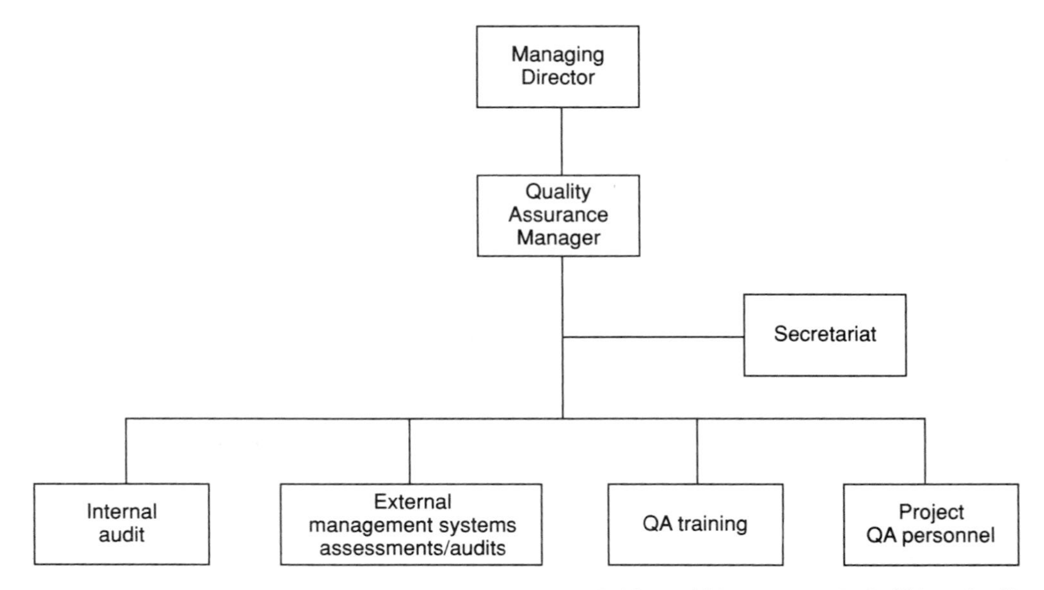

Fig. 21.4 QA department organization as recommended by Stebbing. Source: L. Stebbing, *Quality Assurance*.

release note, sometimes called a certificate of compliance, or of conformity. The wording goes something like this (actually taken from the certificate of compliance issued under the BS 9000 Scheme for Electronic Components of Assessed Quality):

> The components detailed hereon have been manufactured, inspected and tested in conformity with the specification quoted, and are released with my authority.

Thus the Quality Manager is placed in the position of guarantor to the customer. The issue boils down to 'Knowing the capabilities and commitment of the workforce, what am I prepared to take on trust without testing or direct verification done by my own departmental employees?'

The degree to which the Quality Manager is willing to trust the integrity, training and dependability of the rest of the factory personnel will determine the span of activities which the manager feels necessary to reserve to the QA personnel. The extent to which he/she can justify placing trust in non-QA personnel is a measure of how well the company, including the Quality Manager, have equipped the workforce to safeguard the quality of what they produce.

Classifying QA duties

Since there are so many ways in which the boundary between the QA department's duties and responsibilities related to quality outside the QA department can be drawn, it is difficult to make generalizations.

However, I will try to generalize by saying that QA duties can be grouped into what I will call:

1. Quality planning and administration;
2. Quality engineering;
3. Quality inspection.

We will consider each of these in turn.

Quality planning and administration

This includes the 'core' QA activities which are undisputed as being inseparable from the QA Department, such as:

1. The Quality Manager's personal responsibilities, examined earlier in this and the preceding chapter;
2. Drawing up the quality manual;
3. Quality planning;
4. Quality system auditing;
5. Quality information reporting;
6. Quality education;

And another discretionary administrative activity:

7. Document control, **if** part of QA.

All of Stebbing's QA activities would fall into this category. The key titles here would be quality manager, quality auditor, QA clerk.

Quality engineering

1. Drafting QA procedures;
2. Clarifying procedural requirements and inspection standards;
3. Setting up specialized equipment for inspectors;
4. Evaluating customer complaints;
5. Developing new quality test methods.

The key position here is the quality engineer. He/she is a member of the QA department and is experienced in relation to the philosophy of QA, statistical techniques, etc. but has some other specialism be it as a software, mechanical, electrical engineer, which is relevant to the company's product or service.

The engineer decides what the QA inspectors should be inspecting and how, and works with them and their supervisors to train, help set up their tests, validate or arbitrate on marginal inspection decisions. The engineer may have one or more technical assistants under his/her direction.

In the Stebbing model, the person I have just described would not exist as a member of the QA department. Though he/she would report elsewhere, the QA department might still borrow his/her skills. The only QA engineers would be the auditors exercising purely 'QA' skills.

Quality inspection

This is the bulk of the 'appraisal' work, and is done by inspectors under a supervisor. If the work is fragmented geographically, or by shift, there will be some intermediate position such as 'lead hand' who deputizes for the supervisor when the latter is unavailable.

The inspectors may be 'home grown' by QA, or they may be selected ex-operators, often inspecting the operation they used to perform. The supervisor is responsible for work planning and throughput, liaising with production to anticipate bottlenecks and ensure smooth work flow, dealing with inspectors' technical or personal problems, arranging training and evaluating performance.

QA department personnel

Having examined what functions can be expected to be performed within the QA department I will review personnel selection and development to fill those functions.

Job descriptions

Once the Quality Manager has decided what functions must be performed within the QA department, it is necessary to raise job descriptions in order that the personnel who will do the jobs, or who might apply, know what is expected of them.

The format of a standard job description might contain the following sections:

1. Title of the position;
2. Employee grade;
3. Hours of work and overtime requirements;
4. To whom the post-holder reports;
5. Whom the post-holder supervises;
6. Main job responsibilities;
7. Qualifications and experience required.

Internal or external recruitment?

Various decisions and choices have to be made in recruiting a new member for the QA department. Where recruitment is for a relatively senior or specialized position, one question is whether to recruit internally or externally?

Inevitably, each course has both advantages and disadvantages.

Advantages of external candidates

1. External candidates may have exactly the qualifications and experience you require.

2. They may have valuable knowledge of how similar organizations (e.g. competitors?) operate and how they have solved problems similar to your own.
3. They bring a fresh eye to your own organization.

Disadvantages of external candidates

1. You have not been able to see them at work first-hand – interviews can mislead about people's capabilities.
2. You cannot fully verify the experience and qualifications they claim.
3. You are not sure how they will fit into your environment.
4. They may have to move home – involving expense, adaptation for them and their family, which will be unsettling and will distract from assimilating their new job.
5. They will have to build up their own network of friends and contacts at work.
6. Internal candidates will be de-motivated if they feel you always promote from outside.

Advantages of internal candidates

1. You have the opportunity to observe them at first hand, and have your own and colleagues impressions of their personality and capabilities.
2. They already have their own 'grapevine' (network of personal contacts).
3. They know the QA department and its people well enough to believe they can contribute, fit in and enjoy their work there.
4. They are familiar with the company's technology, products and procedures.
5. If well-regarded, their inexperience will be counterbalanced by their colleagues' goodwill for them to succeed.

Disadvantages of internal candidates

1. They may be inclined to take the existing practices for granted.
2. They may have inadequate QA experience/training.
3. They may carry insufficient authority, because people are aware of their inexperience, or of previous mistakes.

QA personnel training and development

This section deals with the training, development and promotion of personnel within the QA department. However, the preceding and following sections will remind you that the personnel themselves may see their career progression as bringing their skills into the QA department, or taking what they have learned out of the QA department (or indeed out of the company) at the appropriate time. This is a valid attitude, of which Quality Managers should not lose sight. It can work to their advantage.

Training

Appropriate training is one of the tools with which the Quality Manager has to equip personnel if they are to do an effective job. Quality system standards such as ISO 9000 demand that you provide training, record the training given, and do not use untrained or inadequately skilled personnel to perform QA functions. So you have to:

1. Train;
2. Measure;
3. Record;
4. Certify;
5. Control the use of personnel.

At the level of operator, clerk or inspector, the majority of the training is in understanding and using the set procedures and work instructions. Engineers need technical training, e.g. in problem-solving and statistical techniques. At supervisor or senior engineer level, training in planning, leadership, employee performance review become relevant. The manager must:

1. Always seek objective evidence that the training was successful;
2. Always follow up the training with opportunities for the students to practise what they have learned.

Development

Good managers invest in the development of their staff. Training is a prerequisite of development but it is not the same thing. To develop, people need the opportunity to try out what they have learned. They also need their own quality standard. They must know:

1. What you want them to do;
2. Your standard for performance – e.g. without error, by what date;
3. That you will let them know how good a job you consider they did.

Goals should be mutually agreed if possible, and performance appraisal made a two-way communication, so that employees can express their problems and their ambitions. In *Up the Organisation* Robert Townsend suggests that one of the functions of a manager is 'To take away his people's excuses for failure'. If you can remove the external impediments to the job you want them to do, the success or failure is the subordinate's own.

'Flying the coop'

The time may come when one of your staff is approached and offered an attractive transfer to another department. The person is a particularly able member of the department and would leave a gap you would find difficult to fill quickly. Do you try to persuade the individual to stay, or do you encourage them to accept the position offered?

There is a natural impulse to try to dissuade the person, but is that **really** in the best interests of the department? An effective QA Department is always, in a way, striving towards its own destruction. As a company becomes imbued with a desire for quality, and accumulates a greater store of wisdom on how to achieve it, it becomes possible to transfer more and more responsibility for controlling quality to the people directly concerned with the work, and the QA department then necessarily shrinks in size.

The most successful QA managers are not necessarily the ones with the biggest department, they are the ones with the most and strongest allies outside their department. An able ex-QA person in a responsible post outside the QA department, concerned with quality and truly understanding the role of the QA department, is not only a valuable asset but also is a convincing salesperson for quality and an advertisement for the QA department.

QA departments should not become inbred. The larger the QA department and the lower the personnel turnover there is, the more it is liable to be assumed to hold all 'responsibility' for quality. Quality is **everyone's** responsibility, and the role of QA in achieving it is more easily understood if people transfer into and out of the QA department, and if the department is small enough that it has sometimes to co-opt help and facilities from outside.

Remember that as a Quality Manager you don't want the QA department to be seen as a dead-end, but as a place where people have a unique opportunity to bring their qualities to top management's notice.

Summary of conclusions

Although quality must be managed, there does not have to be a single quality manager, nor an autonomous quality assurance department – quality is everyone's business! However, since what is seen to be everyone's business can quickly become no-one in particular's business, there must be a management representative, someone with authority exercising a coordinating role in maintaining the quality management system.

If these are in a central QA department the functions it should perform are those which are peculiarly QA specialisms; Stebbing has given a good account of what these are. If the QA department embraces tasks lying outside this specialism it is liable to suffer the fate prophesied by Crosby – of being held responsible for any bad quality the organization creates. In contrast, the QA department should be in a state of constant interaction with other groups, forming and disbanding cross-functional teams as circumstances dictate.

22

Quality in service industries

Introduction

In Parts One and Two of this book we have examined:

1. 'What managing quality means'; the achievement of quality throughout the organization, and at the different stages of the production cycle;
2. The organizational structure necessary to achieve quality, and the techniques and disciplines associated with quality assurance.

In Part Three we have been looking at the effective use of people in furthering quality. However in different industries, different aspects of quality management and quality objectives assume greater importance. In part this depends on the size of a typical organization, in part on who its customers are, and the extent to which its workforce is 'visible' to the customer. The technical basis of the industry also plays a part.

In this chapter we shall stress some distinctive aspects of organizations which set out to offer a service direct to the public. Such enterprises have two special features in common:

1. Their 'product' is less tangible, and less obviously measurable than that of manufacturing organizations on which we concentrated our attention in Part One of this book.
2. A significant proportion of their workforce is uniquely exposed to direct contact with the external customers. The customers' perception of the company will be largely determined by the behaviour of these employees.

Acknowledgement of these facts led to two matters which we shall examine in this chapter:

1. A guidance document for use with ISO 9000, directed at service companies using the standard as a model;

2. The concept of 'customer care'.

The discussion will help to show how concepts and methods pioneered in manufacturing industry can be applied beneficially to a wide range of situations.

Material and personal services

A distinction is often made between 'material' services and 'personal services'. Material services are seen as being to do with 'things' such as retailing, distributing and delivering goods, a telephone or postal service, etc.

Personal services are seen as being to do with the human interface; interaction between suppliers and customers, communication between people. The barber or hairdresser might be a good example, but there is no hard and fast dividing line; all services have both a material and a personal dimension.

The material content of a service

An automobile dealership is heavily product-oriented; its business is the preparation and sale of motor cars to customers, and it frequently accepts the customer's present vehicle in part-exchange.

Consider a criminal law practice, on the other hand. Though it may collect and document numerous statements, generate barrister's briefs and other forms of paper work it has only slight tangible output. The end product, delivered to the customer's satisfaction, may be no more than a jury's verbal 'not guilty' verdict.

In between those two cases, a restaurant delivers meals, but these are immediately consumed. What brings customers back is the ambience and level of attention they receive, as much as the actual food. You might go back repeatedly yet never duplicate the choice of dishes you ate on your first visit.

The personal content: customer care

Where the service is delivered directly by a person to a person, the customer's perception of the supplier, and the supplier's attitude to themselves is the dominant factor in determining customer satisfaction. Whereas a product or material service is expected to be correct, personal service is the area which offers the greatest opportunities for exceeding customer expectations.

Not only that, but it is the supplier's response to a problem or error which often makes the most memorable impact, and can turn a mishap into a public relations achievement. The author remembers a badly delayed international flight which was five hours late leaving Charles de Gaulle Airport, Paris following a technical fault, and delivered him at Heathrow, London well after midnight when all public transport had ceased. His destination was Bristol, and after complaining to British Airways they provided a car and chauffeur for his hundred-mile journey.

Gaining market share is critical, but so is retaining it. This is the reason why personal service organizations try hard to please customers once they have captured them. An article in *The Harvard Business Review* (Sept./Oct. 1990) claims that improving customer retention by 5% can raise profits by 25–85%.

On the other hand, according to a report to the US Office for Consumer Affairs, only 4% of dissatisfied customers actually complain. However, many of them will not do business with the offending supplier again, and on average will tell 13 other people of their experience.

These are the kinds of consideration which make service companies invest in 'customer care'. But customer care is more than simply 'Have a nice day', or 'Hi! I'm Valerie. I'm your waitress for this evening.' Customer care means dedication to the customer. It requires a committed investment in time, money and effort if it is to achieve success. Customer care is, in short, quality.

ISO 9000 Guidelines for Services

Guidelines for Services is the title of a section of ISO 9004, given the number ISO 9004–2:1991. (In the United Kingdom this was originally issued as BS 5750 part 8:1991.) The document sets out to describe a quality system embracing all processes needed to provide an effective service, from marketing to delivery. It also includes advice on the analysis of services provided to customers.

It is interesting to note, as the emphasis on Total Quality management increases, that ISO 9004–2 stipulates explicitly that its provisions can be interpreted in relation to **internal** customers, such as different groups within a large organization.

The service as a product

ISO 9004–2 correctly points out that services vary in the degree to which they are product-related. On the other hand, ISO 9004–2 also recognizes that a 'service' which is not product-oriented does not necessarily mean a person-to-person service. The delivery of the service to the customer may be done through the agency of a machine, such as a vending machine, an automatic banking machine, or a telephone modem, for example.

It is possible for the service itself can be defined and treated as a product. For example, a UK guidance document on the application of ISO 9000 to education and training suggests that the **product** can be treated as being the skills and understanding gained by the student. This implies that the student is 'customer supplied product' (ISO 9001 4.7) presented by the sponsor who pays the fees; subjected to examination (inspection and test, ISO 9001 4.10) and review of poor results (control of non-conforming product, ISO 9001 4.13), etc. but who must be properly housed and fed while attending the course (Handling and storage, ISO 9004 4.15)!

Measurable characteristics of a service

If there is no tangible product, the quality of the service must be gauged by other means. Examples of characteristics that might be specified in customers' requirements, suggested in ISO 9004–2 include:

1. Facilities, capacity, number of personnel, quantity of materials;
2. Waiting, delivery and process times;
3. Hygiene, safety, reliability and security;
4. Responsiveness, accessibility, courtesy, comfort, aesthetics of environment, competence, dependability, accuracy, completeness, state of the art, credibility and effective communication.

Note that while points 1 and 2 are clear objective standards, point 3 and especially point 4 are essentially subjective, and the standards would need careful definition.

Key aspects of a service quality system

The diagram (Figure 22.1), taken from ISO 9004–2 is reminiscent of the representation of TQM given by Oakland, and reproduced in Chapter 20 of Part Three of this book. In the present case, notice the central position given to 'Interface with customers'. Bear in mind also that 'resources' emphasizes personnel resources more so than material; their motivation, training and development, and communication skills are discussed at much greater length than are material resources.

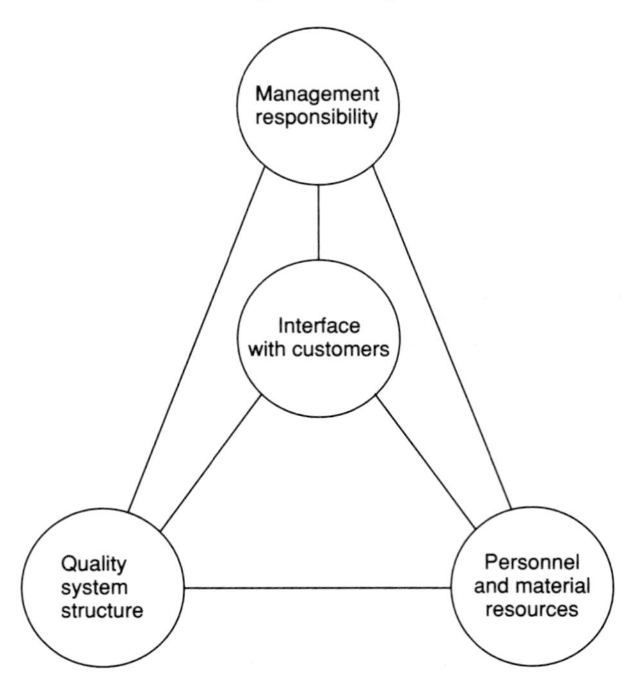

Fig. 22.1 Key aspects of a quality system. (Source, ISO 9004–2 :1991)

The service quality loop

You will recall that in ISO 9004–1 there was represented a 'quality loop' which was discussed in Chapter 1, Part One of this book. ISO 9004–2 offers a more detailed 'service quality loop' which shows the activities within a representative service organization. The diagram (illustrated in Figure 22.2) emphasizes the marketing, design, delivery and use of feedback from the customer in the analysis and improvement of service performance. Two-way communication with customers is a critical part of ensuring service quality.

Quality in marketing a service

Another area of activity which is critical is the role of marketing. Just as we stressed much earlier in the book for consumer sales of products, the customers may not be able to articulate their demands clearly. Even more so when offering a less clearly definable 'service', in order to avoid dissatisfaction and demands for compensation, the supplying organization has to be quite clear what it is contracting to supply.

Quality in market research and analysis

ISO 9004–2 advises that quality-related activities in service market research should include:

1. The establishment of customer needs and demands, such as consumer tastes, grade of service sought, reliability, unexpressed customer biases or expectations;
2. Complementary services;
3. Competitive activities, and standard of performance;
4. A review of pertinent legislation, e.g. relating to health, safety and the environment; standards and codes of practice;
5. Analysis and review of the customer requirements, data and contract information that has been collected, to be passed on to the personnel engaged in design and service delivery;
6. Consultation with all areas of the organization that would be involved, to confirm their ability to meet the service requirements;
7. Ongoing research to monitor changes in market needs, new technology, and competitors' progress;
8. The application of quality control to the marketing process.

Supplier obligations

Supplier obligations to customers may be explicit or implicit, but should be documented by the supplier so far as they are known. This includes:

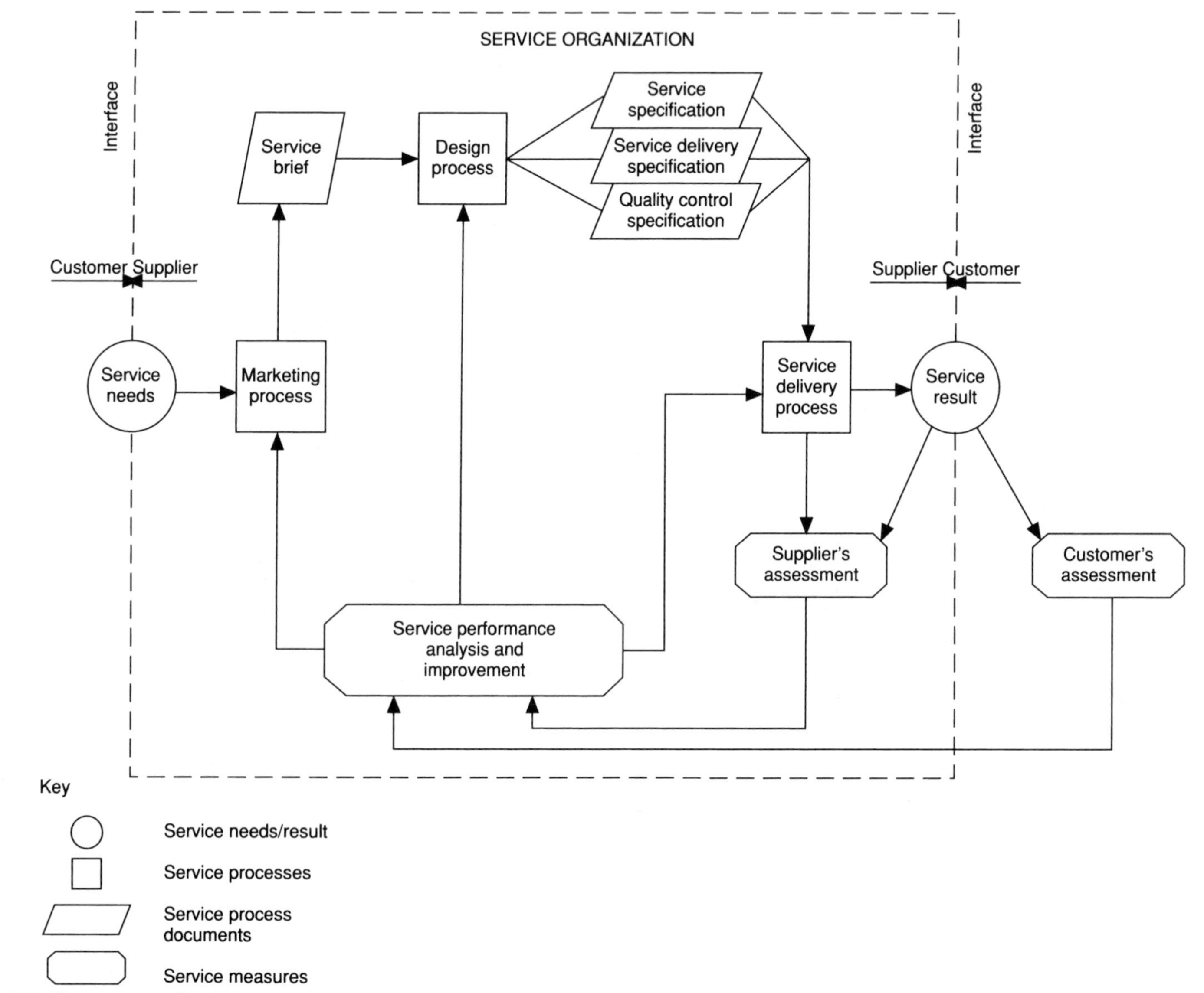

Fig. 22.2 The service quality loop. (Source: ISO 9004–2:1991)

1. Warranty conditions;
2. Related quality documentation;
3. Supplier's capability;
4. Relevant regulatory and legal requirements.

These have to be defined in conjunction with the customer, and when complete incorporated into the service brief.

The service brief

The brief, built up from what has been learned from market research, analysis, and obligations agreed with the customer, will provide the basis for the design of the service. As such, it is analogous to the marketing specification for a product, which is the basis of the design input for the product.

Service management

Before the service is developed, management procedures for planning, organizing and implementing the launch of the service must be established, together with its eventual withdrawal where applicable (e.g. continued customer support for purchasers of a service, even when it is no longer available to new subscribers).

As in the quality system for products (refer to ISO 9001 4.1.2.2 for example), resources, facilities and technical support must be planned for and furnished by management; with particular reference in the quality context, to minimizing safety risks to employees, customers and the environment.

Quality in advertising

Advertising should be consistent with the service specification, and avoid making exaggerated or unsubstantiated claims.

Analysis and improvement of service performance

It follows from the definitions of quality that the customer's assessment of the service is the ultimate measure of its quality. Customer reaction may be immediate (walking out of a restaurant because of poor service), or it may be delayed and retrospective (simply not patronising it again). Frequently customers do not complain and give the supplier the chance to correct their error, so monitoring the level of customer complaints can give a misleadingly optimistic measure of service quality.

Data collection and analysis

In service organizations the collection of data is frequently haphazard, and its interpretation subjective. It may rely on the contents of a 'visitors' book', a document which tends to collect only the favourable comments, casual remarks or enquiries ('was everything all right?') plus the few unmistakable complaints forthrightly expressed (but probably verbal rather than written).

Proper data collection, by contrast, must be purposeful, disciplined and planned. The aim of data collection will be the identification and prevention of systematic errors. Errors ascribed to personnel (or to customers!) often arise from flaws in the way the service system is operated.

Service quality improvement

There is a need for a conscious programme of continuously improving service quality, including the effectiveness and efficiency of the whole service operation. This should embrace:

1. Improving the characteristic(s) of the service which would most benefit customer and also service organization;
2. Changing market needs, and the grade desired;
3. Deviations from specified service quality due to inadequate quality controls;
4. Opportunities for reducing costs whilst maintaining service quality.

Summary of conclusions

This chapter has contrasted, but also drawn parallels between, service and product-creating industries. I hope that the distinction appears less clear-cut now than it may have done initially. Every service has, to a greater or lesser extent, a material content; every manufacturing organization contains many support functions and internal service supplier – customer relationships.

What does tend to be uniquely present in service organizations is a tendency for their outside users to react favourably to strenuous efforts to correct errors or other problems. So how is the unobtrusively efficient organization rewarded? By cost savings. It has been estimated that in some service organizations between one third and two thirds of all employee time is taken up with rectifying or anticipating problems ('Will Mr Jones' suit be ready when he calls for his fitting tomorrow?'). We shall examine the cost savings of good quality organization in Part Four.

In a service organization the discipline of having a tangible, measurable product in front of you is often absent. In these cases personal dedication and a good quality system can be even more critical to business success than in a manufacturing organization. ISO 9004–2 gives guidance on the application of an ISO 9000 quality system to a service organization.

Part Four

Quality – the Key to Prosperity

Introduction

In Parts One, Two and Three of this book we have discussed:

1. The nature and management of quality;
2. The specialist 'quality assurance' functions and their techniques;
3. And the need to involve all employees in the search for continual quality improvement.

Part Four looks at the commercial imperatives; only those companies which can meet or exceed their customers' quality expectations economically, are the ones that will survive and flourish.

To do this demands:

1. Identifying and minimizing your quality-related costs;
2. Understanding the strengths and weaknesses of your international competitors;
3. Identifying and emulating the 'world-class' companies;
4. Anticipating and responding to changes in customer expectations;
5. Complying with consumer and environmental protection legislation;
6. Developing and making the fullest use of the skills of your suppliers, workforce and agents.

We will look at these issues in Part Four.

23

Ways of looking at quality-related costs

Introduction

Senior management evaluate all issues in terms of costs and revenue, so if quality issues are not presented in monetary terms they will not receive their due attention and commitment to action.

In this chapter we shall review various models for analysing quality-related costs, and see how suitable they are for our purpose of expressing factual information, monitoring performance trends and decision-taking on quality issues, based on measured costs.

The importance of measuring quality

We stressed right at the beginning of this book, in Chapter 1, Part One, the necessity of being able to measure quality level. Without an objective standard by which to compare achievement with intention, it would be impossible to measure quality level. Without this ability it would be impossible to manage or control quality.

Fortunately, the establishment of a definition of quality level as 'meeting the customer's needs', when translated into an unambiguous specification does make measurement a possibility. A quality level of 100% means complete conformance to the specification, every time. Shortfalls in quality level can be detected in parametric deviations, or by the number and/or frequency of deviations.

In our discussion of *Total Quality Management* we showed how any organization can be perceived as 'chains of quality' where people are linked with each other as suppliers and customers. Each person in the chain may analyse the demands they have to place on their suppliers and the demands their customers place on them.

Thus each can develop a personal 'quality performance indicator' which measures the quality level they achieve while performing their job.

Why quality measurement must be expressed in monetary terms

We have several alternative ways in which the level of quality being achieved can be expressed numerically, and through which employees can monitor their own or other people's performance. Are these not sufficient to provide a basis for action? We have seen that Crosby declared as one of his 'quality absolutes' that 'The measure of quality is the price if non-conformance'. But why this insistence on quantifying quality in financial terms?

Anyone who is self-employed, is trying to build up a business, or has considered how to make money from a spare-time activity quickly realizes that enterprises are built on borrowed money. Much expenditure has to be invested, e.g. in tools, stock, personnel, advertising before anything can be created and sold. The company's energies and attention, at board level, are directed to surviving while under a burden of debt. Far from paying off its debts, the company is usually trying to expand, and so accumulating additional borrowing commitments.

Hence, everything which comes under scrutiny from top management is evaluated in monetary terms, and decisions made in relation to earning opportunities and cost penalties. For example:

1. Staff levels in terms of payroll costs;
2. Stock in terms of money 'tied up' in inventory;
3. New machinery for productivity improvement in terms of return on investment;
4. Advertising in terms of the value of new business it should bring in.

The more we analyse the search for quality in monetary terms, the more clear it will become that improving quality and preventing errors are sources of cost savings, whereas mistakes, delays and substandard products are wasters of money.

However, unless quality issues; that is to say, opportunities to improve quality, as well as quality-related problems can be accurately expressed as monetary costs or investments, the facts cannot be accurately assessed. They will not assume their true importance, and any decisions made will not be based on sound assumptions.

Use of quality costs

Although most companies still have little idea of their total quality costs, this information would be of great value. If quality cost figures are available you can:

1. Compare the quality cost performance of one plant or production line with another;
2. Monitor quality cost trends;
3. Budget future quality conformance costs, and set improvement goals;
4. Decide which quality costs need to be reduced, and where more investment in prevention could produce savings.

Quality cost estimates

When manufacturing organizations have carefully examined their quality costs, they have typically found them to be as much as 25% of sales value, and this value is quoted by Crosby (1985).

A similar pattern applies to organizations which provide a service rather than manufacturing a product. In the same reference, Crosby quotes a typical quality-related cost figure of 30% of operating costs.

If anything, the figures put forward as 'typical' by experts who have studied them have become higher, rather than lower, in the years since Crosby proposed the figures we have quoted. This appears to be due to the fact that, as people have become more expert at examining company practices and records to discover quality-related costs, more and more sources of wastage have become recognized.

The first attempt to estimate quality-related costs is notoriously prone to underestimation. Crosby suggests 30% of the true value. Particular reasons for this are:

1. Information has not been collected previously, so the figures are difficult to find.
2. There is sometimes a presumption that quality costs are solely those incurred in operating the QA department.
3. Costs indirectly due to rejected and scrapped work-in-progress are overlooked.

The sources of quality costs will be discussed later; you will see that many of them could easily be overlooked at first consideration.

Different ways of looking at quality-related costs

We shall consider four ways of looking at the 'costs' associated with quality. These are:

1. The 'loss to society' model;
2. The 'process cost' model;
3. The 'economic balance' model;
4. The 'prevention/appraisal/failure costs' model.

Our judgements on these alternative approaches will be based on their value to an organization in:

1. Identifying the quality-related costs which are under its control;
2. Establishing priorities for corrective action, on the basis of the cost savings achievable;
3. Monitoring progress in cost-saving.

BS 6143

There is a British Standard, BS 6143, which is entitled *Guide to the Economics of Quality*. This addresses two of the models we have mentioned, since BS 6143:part 2:1990 deals with the prevention, appraisal and failure model, while BS 6143:part 1:1991, deals with the process cost model.

Total loss to society

This is in its origin a Japanese perception of quality-related costs, promoted by Taguchi. An important dimension of the quality of a manufactured product is the total loss to society generated by the shortcomings of that product. Taguchi's definition of quality is 'The loss imparted to society from the time a product is shipped.'

In Taguchi's view loss can be one of two things: either loss caused by variability of function (of the product), or loss caused by harmful side-effects. This is an unusual definition first, in that it defines quality in a negative sense – lack of quality. Second, that it only measures the impact of the finished product: avoidable costs and wastage occurring within the manufacturing plant are not included by Taguchi.

How does Taguchi visualize the loss to society? His measure of quality is essentially its cost. An illustration from one of his books provides a simple illustration. Taguchi draws on the example of an imaginary crease-resistant shirt.

Suppose, says Taguchi, sending a shirt to the laundry costs 250 Yen, and the typical shirt is washed 80 times during its lifetime, then the lifetime laundry costs for the shirt are 20 000 Yen. If a new kind of shirt could be made that soiled and wrinkled only half as fast, the buyer would save 10 000 Yen on laundry bills. If the new shirt cost 1000 Yen more to produce and sold for 2000 Yen more than an ordinary shirt, the manufacturer would gain 1000 Yen and the consumer 8000 Yen, from this saving in quality costs to society of 9000 Yen.

Not only that, but omitting half the laundering would produce an environmental benefit to society in terms of less quantifiable but real savings in energy to heat the water, reduced release of detergents, and less noise.

There are clearly problems in applying Taguchi's approach as a basis for decisions. The environmental benefits would be hard to quantify, and who decides what is the appropriate 'profit- sharing' between shirt manufacturer and society as a whole? Should the manufacturer set the price difference between the standard and improved shirt at 1500 Yen and so offer society a saving of 8500 Yen? Or would the market accept a premium of 4000 Yen; the consumer would still save 5000 Yen over the life of the shirt, and the extra profit would encourage research into additional 'environmentally friendly' products.

It seems, therefore, that within the practical terms of reference we set earlier in this chapter, there is little use for Taguchi's definition. It is nevertheless a very helpful way of looking at an organization's responsibilities towards society, and how it can express its environmental policy. However, Taguchi goes further than this in quantifying his ideas, and the extension we are going to examine now does have applications within the internal supplier–customer chains of manufacturing companies.

The quadratic loss function

As we have recognized while discussing process capability and statistical process control, a process is subject to random variation about its design value. To ensure that assemblies can tolerate the variation in their components, tolerance limits are set, and the assemblies designed to accept variations up to those limits.

In terms of cost, traditional models have assumed that units within their set tolerance were good, and that those outside tolerance were bad, and had a quality cost impact related to the wastage incurred in scrapping, repairing or down-grading them. Taguchi, on the other hand, claims that only the part which reproduces exactly the designed values is free from quality loss, and that the cost of variance is proportional to the square of its deviation from the design value.

The quantitative basis for Taguchi's claim may be questionable, but he does use it as a measure for judging the costs and benefits of process adjustment and improvement within a workshop, with a view to optimizing the overall costs.

The process cost model

We have already seen that it can be very helpful to visualize any activity in terms of a 'process'. This is true not only if a physical manufacturing process is involved but also in the case of a service process or transaction. This is emphasized in the 'Total Quality Management' (TQM) approach to quality in support departments, which involves breaking down all the activities of the company into interlinked 'processes'. Each process has inputs and outputs, both desirable and unwished-for; the desired inputs come from 'suppliers' and the desired outputs are delivered to 'customers'; see Figure 23.1. The suppliers and customers may be outside the organization being addressed, but in the case of many processes they are internal suppliers/customers.

Process cost elements

These derive from, and can be recorded under people, equipment, materials and environment. The process cost model categorizes quality costs as conformance and non-conformance costs, though you must be cautioned that these expressions will be defined differently when we come to consider the prevention/appraisal/failure model.

The process cost model defines the terms as follows:

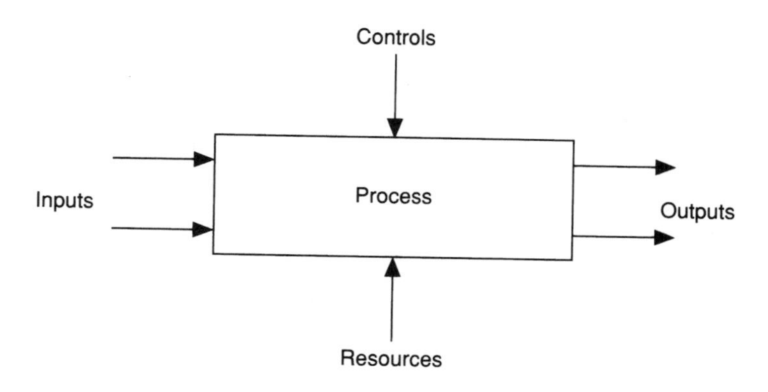

Fig. 23.1 The process model.

Cost of conformance (COC)

The cost of operating the process as specified in a 100% effective manner. This does not imply that it is an efficient, nor even necessary process, but rather that the process, when operated within its specified procedures, cannot be achieved at a lower cast. This is the minimum cost for the process as specified.

Cost of non-conformance (CONC)

The cost of inefficiency within the specified process, i.e. over resourcing or excess costs of people, materials and equipment arising from unsatisfactory inputs, errors made, rejected outputs and various other modes of waste. These are considered non-essential process costs.

Strengths and weaknesses of the process cost model

An advantage claimed is that costs do not have to be identified and categorized as prevention, appraisal or failure, (the approach adopted in BS 6143:part 2:1990, and examined in Chapter 24), a process which BS 6143:part 1:1991 claims can be difficult and unsatisfactory.

The author feels that the labelling of quality costs as relating to prevention, appraisal or failure may be unnecessary, but that their identification is essential. Categorization is helpful even so, because it reminds the analyst that costs classed as failure costs are all potentially avoidable, and as that situation where failure costs are disappearing is approached, so also do appraisal costs become avoidable.

The author also suspects that the complete analysis of a company's activities into interlinked 'processes', accurately and without duplication and consequent double-counting of costs, is likely to be more onerous than the traditional categorization of quality costs. Furthermore, the classification of the running costs of 'inefficient or unnecessary' processes running exactly according to the book into the COC hides the inefficiency.

A repair station with no waiting time, wasted materials or ineffective repairs would incur no PONC, yet the only need for its existence would be the creation of a copious flow of rejected work from the line. To make matters worse, BS 6143: part 1 indicates that 'standard' times, yields, costs, etc. can be taken as the basis of the POC. Thus even the POC can contain hidden yield losses, etc.

In terms of highlighting opportunities and needs for quality improvement, the author considers the process cost model is a retrograde step.

The concept of economic quality

BS 4778 defines economic quality as:

> The economic level of quality at which the cost of securing higher quality would exceed the benefits of the improved quality.

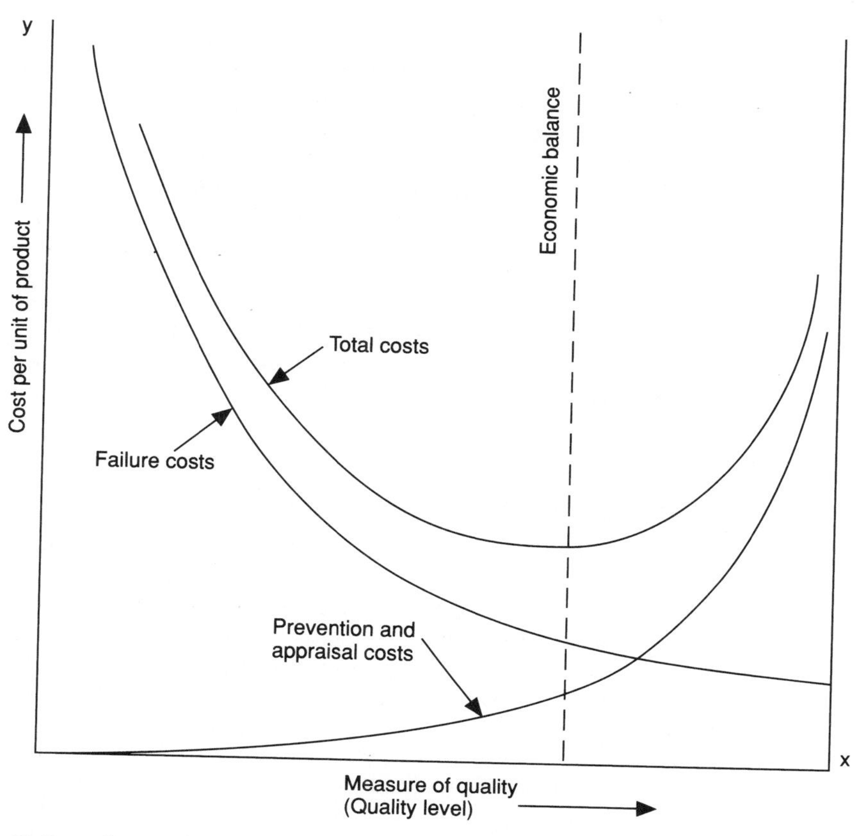

Fig. 23.2 Quality costs model.

This concept is illustrated by the diagram reproduced in Figure 23.2. which appeared in the original version of BS 6143.

Although some well-respected commentators deny that a point of minimum quality cost exists, the concept of an 'economics of quality' warrants discussion since the idea is still prevalent, and is harmful. It will be recognizable after consideration of the following chapter as the result of a perversion or over-simplification of the prevention/appraisal/failure model.

Its justification as presented in Figure 23.2 is ambiguous, misleading, and in practical situations erroneous. I will first present this model, and then analyse its content.

The traditional presentation

The model requires two forms of quality-related costs to be accepted; non-conformance costs due to failures, and conformance costs due to appraisal and preventive activities. No distinction is made between the mode of action and the outcomes of prevention and appraisal.

The orthodox argument proceeds as follows:

1. If there is little or no investment in conformance costs, quality is likely to be poor, i.e. non-conformance costs will be high.
2. If more expense is invested in conformance activities; non-conformance costs will decrease, quality will improve.
3. The more quality is improved, the harder it will be to produce further improvement. One might find for example that it costs £1.20 to eliminate each residual defective component, where the sales value of the component is only £1.05.
4. At such a point, the conformance cost introduced is greater than the non-conformance cost it is intended to eliminate, and the improvement activity becomes uneconomic.
5. Hence there exists a point at which the sum of conformance and non-conformance costs is a minimum. This is the point of 'economic quality'. Producer and supplier must accept the economic necessity for the number of defective items which remain at this point.

Limitations of the model

The first thing to note is that the 'measure of quality' (quality level) is not defined. What is the measure being represented? A variant of this diagram shows 'proportion defective', from 100% at the origin to 0%, and we will assume this to be the intention of Figure 23.2.

Notice also that examples used to illustrate the argument, like the £1.05 component one which we have given above, assume that the loss cannot exceed the cost of creating the rejected item. But suppose the defective unit was installed in an oil-rig, resulting in its shut-down and a service engineer having to be dispatched to Saudi Arabia? Or in some medical instrumentation, the failure of which resulted in the death of a patient? What do you imagine the costs of these failures could amount to in expenses, damages, loss of business and reputation?

Another point to be noted is that the model ignores the time element. The method for preventing a particular type of failure cost may be very simple to install. It may involve changing a written procedure, training a specific individual, or installing a 'poka-yoke' device. Thus preventive actions may be very cheap to introduce. Although having no instant impact on non-conformance, once they are in place they can save non-conformance costs throughout the product life cycle immeasurably greater than their cost of introduction, without further investment.

Finally, the persuasiveness of the argument and diagram is only maintained through the failure to distinguish between appraisal and prevention, and between internal and external failure costs.

It is timely at this point to examine the significant cost differences between internal and external failures, and between the activities of appraisal and prevention. This examination will also be crucial to our presentation of the prevention/appraisal/failure model.

Failure costs

External failure costs are the costs of failures which are first identified in the field, usually by the customer. There is no upper limit to the possible cost of such a failure, for consider the examples I have given earlier.

Internal failure costs are the costs of failures which are identified on the supplier's premises. The identification is normally the result of appraisal; a simulation, inspection or test. Once internal failures have been identified and segregated there is the possibility of replacing them without delay to shipment, without the customer's knowledge of them, and without him/her incurring costs directly as a result of the failures.

You might be tempted to assume that the costs of an internal failure cannot exceed the cost of producing the items up to the point in the process where it is scrapped: this is not the case, as we shall show in the following chapter, but nonetheless the cost of internal failures is typically lower than that of external failures, and importantly all this cost is under the control of the manufacturer.

Appraisal

Appraisal includes all those activities of test and inspection which the manufacturer carries out in order to detect and remove internal failures. It costs money to do appraisal, and the more you do, the more failures you find. So what is the economic justification for appraisal? In fact there are two.

1. The cost of appraisal together with the cost of the internal failures found can still be less than the potential cost of the failures if they were allowed to escape and become external failures.
2. The data from the appraisal activity can be used as the basis for discovering a preventive action to stop the failures recurring.

Prevention

Preventive activities, unlike appraisal, do not attempt simply to minimize the cost-impact of the failures. They attempt to formulate and implement the means to recognize and circumvent potential problems, and to prevent existing problems from being repeated. Monetarily they are an investment, in the expectation of regular future savings.

An alternative approach to 'economic balance'

The above analysis of appraisal and prevention will have shown you that the 'economic balance' model can only be generally valid where conformance costs derive solely from appraisal activities. This is not a realistic assumption.

As I have also pointed out, another serious limitation of the 'economic balance' representation is that it takes no account of changes happening over the course of time. Also, there is really no need to define or plot a quality level as such; we can say that **the external failure cost is itself a measure of quality as perceived by the customer**.

Consequently the form of presentation most favoured in practical situations is simply to plot the total cost of quality, and its four components against time. I give

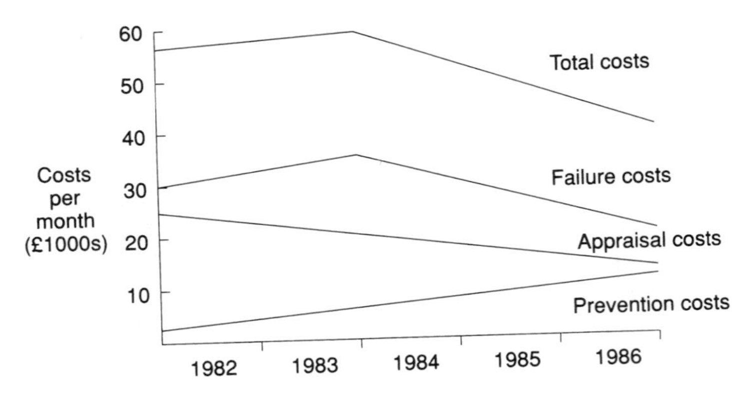

Fig. 23.3 Quality costs trends.

an example in Figure 23.3. Such a plot allows improvements and setbacks to be easily assimilated visually, for this is the information the organization is looking for:

1. Are our **external failure** costs being eliminated, since these represent our quality as seen by the customer?
2. Are our total failure (non-conformance) costs being reduced, since this represents our internal measure of quality?
3. Are we achieving the two previous goals increasingly effectively, i.e. reducing total quality-related costs, by using prevention rather than appraisal?

Summary of conclusions

In this unit we have seen the following.

1. It is important to quantify quality costs in order to be able to:
   (a) Make comparisons;
   (b) Budget and set targets;
   (c) Observe trends.
2. In the broadest view, the cost of quality can be seen as the total loss to society caused by quality failures.
3. At a more restricted level, it can be helpful (especially in a service organization, or one the activities of which have already been analysed along the lines of TQM) to look at quality costs from the 'process cost' model.
4. Even so, the 'process cost' model must be used with caution. So must the idea of 'economic quality'. The rule that 'it is cheaper to do it right first time' still holds, and every opportunity to prevent failures should be accepted.

24

The prevention-appraisal-failure model

Introduction

To know your quality costs, and to act in accordance with that knowledge, is crucial. The costs incurred as a result of errors or failures are the measure of your success in achieving quality – the lower the costs, the better the quality.

Having examined various ways of looking at quality-related costs in the last chapter, we are going to concentrate on the most popular approach, the prevention-appraisal-failure model. We will show how to use it and how to present the results for maximum impact.

The cost of quality

Quality costs fall into one or the other of two major categories. On the one hand is the cost deliberately incurred in efforts to maintain or improve quality. This is the cost of conformance. On the other hand is the cost suffered as a result of bad quality. This is the cost of non-conformance.

I want to stress that the costs of bad quality normally heavily outweigh the costs invested in quality maintenance and improvement. I shall continue to emphasize this throughout the discussion, and show the opportunities it offers for investing in preventing errors.

Origins of the prevention-appraisal-failure model

The prevention-appraisal-failure model was, in effect, introduced during the 1950s in the books and lectures of Juran and Feigenbaum as well as others. The list of elements under each of the categories was formalized in a document *Quality Costs: What and how?* which was issued by the American Society for Quality Control (ASQC) in 1967. Much of this text was adopted some time later for the original British Standard on the subject, which was BS 6143:1981.

Categories of quality cost

Quality-related costs comprise both voluntary costs of achieving a desired level of quality, and also the involuntary costs of failing to achieve it. These are respectively called the costs of conformance, and of non-conformance; sometimes also 'cost of quality' and 'cost of un-quality'. Both pairs of names are worth remembering, because they remind us to distinguish between the 'good' costs which are incurred in actively pursuing quality, and the 'bad' costs which are the penalty of failure to achieve the required quality.

Each category of quality cost can be broken down into two further divisions as shown in Figure 24.1.

The definitions of these four kinds of cost are taken from British Standard 4778, *Quality Vocabulary*, part 2. Although these terms are defined in respect of a manufacturing organization, they are nevertheless equally applicable to a service organization. The notes included in the definitions indicate some, though not all, of the less obvious elements of each cost category.

Costs of prevention

The cost of any action taken to investigate, prevent or reduce defects and failures.

Note: Prevention costs can include the cost of planning, setting up and maintaining the quality system.

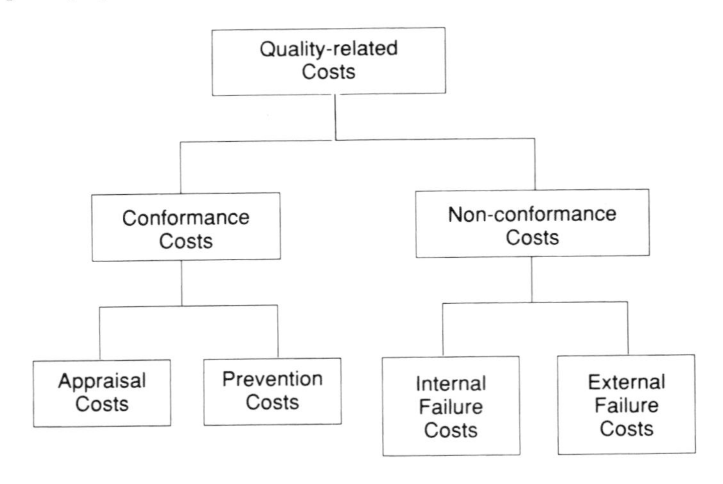

Fig. 24.1 Categories of quality cost and their subdivisions.

Costs of appraisal

The cost of assessing the quality achieved.

Note: Appraisal costs can include the costs of inspecting, testing, etc. carried out during and on completion of manufacture.

Failure costs – internal

The costs arising within the manufacturing organization of the failure to achieve the quality specified.

Note: The term can include the cost of scrap, rework and reinspection, and also consequential losses within the organization.

Failure cost – external

The costs arising outside the manufacturing organization of the failure to achieve the quality specified.

Note: The term can include the costs of claims against warranty, replacement and consequential losses of custom and goodwill.

Somewhat differently worded definitions for the four cases are given in BS 6143 part 2:1990, but without any significant change to the meaning in each case.

Elements within quality cost categories

Several contributing elements can be identified within each of the cost categories, some of them alluded to in the footnotes to the definition. Lists of suggested elements are put forward in various works, e.g. the original ASQC document, Feigenbaum and Oakland. However the version we shall quote here is from BS 6143 part 2:1990.

Prevention costs

These are costs incurred in reducing failure and appraisal costs to a minimum. They can include the following.

Quality planning

This is the activity of planning quality systems and translating product design and customer quality requirements into measures that will ensure the attainment of the requisite product quality. It includes the broad array of activities that collectively create the overall quality plan, the inspection plan, the reliability plan and other appropriate specialized plans. It also includes the preparation and vetting of manuals and procedures necessary to communicate these plans to all concerned. Such quality planning may involve departments other than the quality organization.

Design and development of quality measurement and test equipment

These include the costs of designing, developing and documenting any necessary inspection, testing or proving equipment (but not the capital cost of the equipment in question).

Quality review and verification of design

This includes the quality organization's monitoring activity during the product's design and development phase to assure the required inherent design quality. The quality organization's involvement with design review activities and in verification activity during the various phases of the product development test programme including design approval tests and other tests to demonstrate reliability and maintainability are included.

This includes quality organization effort associated with that part of process control which is conducted to achieve defined quality goals.

Calibration and maintenance of quality measurement and test equipment

This should include the cost of calibration and maintenance of jigs, templates, fixtures and similar items.

Calibration and maintenance of production equipment used to evaluate quality

This should include the costs of calibration and maintenance of jigs, fixtures, and similar measuring and evaluating devices, but not the cost of equipment used to manufacture the product.

Supplier assurance

This embraces the initial assessment, subsequent audit and surveillance of suppliers to ensure that they are able to meet and maintain the requisite product quality. This also includes the quality organization's review and control of technical data in relation to purchasing orders.

Quality training

This includes attending, developing, implementing, operating and maintaining formal quality training programmes.

Quality auditing

This activity involves the appraisal of the entire system of quality control, or specific elements of the system used by an organization.

Acquisition, analysis, and reporting of quality data

This is the analysis and processing of data for the purpose of preventing future failure.

Quality improvement programmes

This includes the activity of structuring and carrying out programmes aimed at new levels of performance, e.g. defect prevention programmes and quality motivation programmes.

Appraisal costs

These costs are incurred in ascertaining initially the conformance of the product to quality requirements; they do not include costs from rework of reinspection following failure. Appraisal costs normally include the following.

Pre-production verification

This is the cost associated with the testing and measuring of pre-production for the purpose of verifying the conformance of the design to the quality requirements.

Receiving inspection

This is the inspection and testing of incoming parts, components and materials. Also included is inspection at the supplier's premises by the purchaser's staff.

Laboratory acceptance testing

These costs relate to tests for evaluating the quality of purchased materials (raw, semi-finished or finished) which become part of the final product, or that are consumed during production operations.

Inspection and testing

This represents the activity of inspecting and testing; first during the process of manufacture, and then as a final check to establish the quality of the finished product and its packaging. Included are product quality audits, checking by production operators and supervision and clerical support for the function. It does not include inspection and testing made necessary by initial rejection because of inadequate quality.

Inspection and test equipment

These include the depreciation costs of equipment and associated facilities; and the cost of setting up and providing for maintenance and calibration.

Materials consumed during inspection and testing

This represents the costs of materials consumed, or destroyed during the course of destructive tests.

Analysis and reporting of test and inspection results

This is the activity conducted prior to release of the product for transfer of ownership in order to establish whether quality requirements have been met.

Field performance testing

This is the cost of testing performed in the expected user environment, which may be the purchaser's site, prior to releasing the product for customer acceptance.

Approvals and endorsements

These are the costs of mandatory approvals and endorsements by other authorities.

Stock evaluation

This is the cost of testing and inspecting stocks of products and spares which may have limited shelf life.

Record storage

These costs relate to the expense of storage of quality control results, approval and reference standards.

Failure costs

These are subdivided into internal and external failure costs: internal costs arising from inadequate quality discovered **before** the transfer of ownership from supplier to purchaser and external costs arising from inadequate quality discovered **after** transfer of ownership from the supplier to the purchaser.

The internal costs include the following.

Scrap

Materials, parts, components, assemblies and product end items which fail to conform to quality requirements and which cannot be economically reworked. Included is the labour and labour overhead content of the scrapped items.

Replacement, rework and repair

The activity of replacing or reworking defectives to make them fit for use, including requisite planning and the cost of the associated activities by material procurement personnel.

Troubleshooting or defect/failure analysis

The costs incurred in analysing non-conforming materials, components or products to determine causes and remedial action, whether non-conforming products are usable and to decide their final disposition.

Reinspection and retesting

This is applied to previously failing material that has subsequently been reworked.

Fault of subcontractor

The losses incurred due to failure of purchased material to meet quality requirements and payroll costs incurred. Credits received from the subcontractor should be deducted, but costs of idle facilities and labour resulting from product defects should not be overlooked.

Modification permits and concessions

The costs of the time spent in reviewing products, designs and specifications.

Downgrading

Losses resulting from a price differential between normal selling price and reduced price due to non-conformance for quality reasons.

Downtime

The cost of personnel and idle facilities resulting from product defects and disrupted production schedules.

The external failure costs include the following:

Complaints

The investigation of complaints and provision of compensation where the latter is attributable to defective products or installation.

Warranty claims

Work to repair or replace items found to be defective by the purchaser and accepted as the supplier's liability under the terms of the warranty.

Products rejected and returned

The cost of dealing with returned defective components. This may involve action to either repair, replace or otherwise account for the items in question. Handling charges should be included.

Concessions

Cost of concessions, e.g. discounts made to purchasers due to non-conforming products being accepted by the purchaser.

Loss of sales

Loss of profit due to cessation of existing markets as a consequence of poor quality.

Recall costs

Cost associated with recall of defective or suspect product from the field, including the cost of preparing plans for product recall.

Product liability

Cost incurred as a result of a liability claim and the cost of premiums paid for insurance to minimize liability litigation damage.

Collecting quality cost information

The collection and reporting of the costs incurred in running an operation are recognized as part of the duties of the finance department, usually of a specialist section known as 'cost accounting' or 'management accounting'. However, since historically, quality-related costs have not been treated as a key issue for management attention, costs have not been collected in the form I have just outlined. Indeed, you may well find it difficult to extract them and apportion them under the kind of heading we have been discussing.

Because of this is it often the QA department which takes the first steps towards producing an estimate of quality-related costs. I say 'estimate', because the initial attempt is bound to be approximate, and you should treat it as no more than a pilot study. The objectives should be to:

1. **Involve the finance department** in order to:
   (a) demonstrate what the exercise is trying to achieve;
   (b) discover where the difficulties in presenting data in the desired form lie;
   (c) ensure the figures finally arrived at, will be supported by the finance department;
   (d) prepare them for future responsibility for collecting the figures on a routine basis.
2. **Produce figures** which are valid within accepted limits of uncertainty:
   (a) but without wasting effort establishing an unnecessary level of accuracy;

   (b) with the objective of establishing the major areas of cost (which can then be re-examined if greater accuracy is needed).
3. **Present the findings** to senior management, in order to:
   (a) give some measure of the cost of quality, and the savings potentially achievable.
   (b) give a means of comparison between product and product, unit and unit, and possibly between the company and its competitors.
   (c) give a base line against which future goals can be set, and improvements measured.
   (d) encourage the acceptance of quality cost data gathering as a routine finance activity.
   (e) propose actions to control and limit quality-related expense.

Sources of quality-related cost data

BS 6143 suggests that the following are valuable source documents:

1. Payroll analyses;
2. Manufacturing expense reports;
3. Scrap reports;
4. Rework or rectification authorizations or reports;
5. Travel expense claims;
6. Product cost information;
7. Field repair, replacement and warranty cost reports;
8. Inspection and test records;
9. Material review records.

Not all of these will exist under the same names in your organization. You will have to ascertain what is available that will be useful, and the name each source goes by. Other useful sources **not** mentioned by BS 6143 include:

1. Organization charts;
2. Job descriptions;
3. Departmental budgets;
4. Standard costs at all stages of manufacture;
5. Standard of historic yields at all stages of manufacture.

Using the data to calculate costs

It is not possible for me to give you detailed rules for calculating quality costs. Activities, organization and accounting practices will vary too much from company to company. That is one of the reasons you should involve your own accounting department as much as possible.

You can identify the personnel involved in quality-related activities using departmental organization charts and by job titles and job descriptions. These will also indicate whether their activities are preventive, appraising, or dealing with failures. In cases where the role is mixed or unclear interviewing the people concerned will clarify this, e.g. a quality engineer could tell you how much time was taken in writing quality procedures (prevention), setting up test equipment for inspectors (appraisal) and verifying rejections from inspectors/customers (internal/external failure respectively).

There are some areas where differences of opinion can exist, e.g. is calibration of measurement equipment a prevention or an appraisal cost? You may decide the answer depends on the exact use to which the equipment is put. Measured costs of external failure will come from sources such as:

(a) Customer service department records of returns and replacements;
(b) Legal department records of warranty claims and liability costs;
(c) Failure analysis laboratory records of investigations and sources of fault.

Together, these will establish the number of returns, suppliers' liability (if any), source of error and value. To this must be added (for example) the costs of running and manning the sections involved, but the value of unjustified returns should be ignored.

Stages in collecting quality cost data

BS 6143 recommends collecting quality cost data in five stages, starting with the QA department. This is sound advice inasmuch as if the collector is the quality manager or a delegate, all the raw data should be readily to hand. It may well be that the majority of the final total will originate outside the QA department, but the collector will have come to appreciate the problems of collecting the data and be more skilled before he/she has to seek cooperation from other departments.

So the five recommended stages are:

1. Calculate the costs directly associated with the quality function;
2. Calculate the quality-related costs of functions performed by personnel outside the QA department;
3. Internal cost of 'budgeted failures' or yield allowance;
4. Internal cost of 'unplanned failures' i.e. those not allowed for in the yield allowance;
5. Cost of failures after change of ownership, i.e. after delivery to, or acceptance by the customer.

Problems in establishing quality-related costs

Oversights

The full cost impact of some aspects of quality may only become evident after careful thought and enquiry or observation, even though the BS 6143 list of quality elements is a valuable check list. Factors which may be easily overlooked include the costs of secondary factors and activities relating to internal failures, e.g. scheduling delays, the need to keep reserve stocks, time spent confirming and analysing failures, etc. These are examined in more detail in the course of the next chapter.

Over-elaboration

A major problem you need to guard against is over-elaboration, especially in the early stages of developing a quality-cost gathering system. It may quickly become clear, to take an example, that the cost of consumable materials used up in performing quality appraisal tests is far less than the cost of external rejects; perhaps even less than the uncertainty in the value of the latter. It would be sensible to let the estimate of the former stand, and so permit more time to be spent quantifying the latter more exactly.

Double-counting and related overestimates

It is easy to introduce this kind of error, especially when evaluating the cost of internal failures. Take the imaginary case we looked at under 'oversights'.

It would not be valid to record as quality costs both the value of the scrapped items and the value of the replacements. If the rejected items can be salvaged, repaired or sold as 'seconds', then the nett loss is the difference between the full value and the reduced value. On the other hand, all the associated extra labour and administration costs of the recovery operation must be included.

Allocation of overheads

This is a related area where it is easy to introduce a double count. If manufacturing costs already carry an overhead allowance which makes provision for the cost of maintaining the QA department, then it may not be valid to identify a separate quality cost for extra inspection incurred. In fact, it gives a clearer picture of the impact of the cost of poor quality if the overhead is ignored and the inspection recognized as a charge precisely where and when it actually occurs.

This is one of the reasons why it is important to seek the advice of, and agree the rules with, the finance department before finalizing and presenting your results.

Built-in yield allowances

This is yet another area which must be treated with care. Standard manufacturing costs may have an allowance built-in for 'standard yield', i.e. a certain level of losses may be assumed and built into the price.

Equally, batch starts may be automatically made a certain number or percentage of units oversize, to compensate for expected losses. This tends to hide and legitimize the cost of scrap. Again, the way to deal with and present these issues should be agreed with the finance department.

Monitoring quality costs

As I have stated earlier, the purpose of collecting quality-related costs is:

1. To make comparisons between group and group, to indicate how well an operation is controlling costs, and where attention is most needed;
2. To provide a baseline against which to set goals and to measure improvements.

To make it easier to identify these patterns, there are convenient ratios and figures of merit which can be calculated.

Figures of merit

If comparing quality costs in different plants, or on different occasions, a figure of merit helps you to make valid comparisons. The particular ratio you use will depend on the factor you want to highlight, e.g. quality costs in relation to labour utilization, sales value, production costs, etc. Some of these ratios are:

1. Labour-based: $\dfrac{\text{internal failure costs}}{\text{direct labour costs}}$
2. Cost-based: $\dfrac{\text{total failure costs}}{\text{manufacturing costs}}$
3. Sales-based: $\dfrac{\text{total quality costs}}{\text{nett sales}}$
4. Unit-based: $\dfrac{\text{total quality costs}}{\text{units of production}}$
5. Added value-based: $\dfrac{\text{total quality costs}}{\text{value added}}$

Failure-appraisal-prevention ratio

This is very often presented in the form of a pie chart. Typically, failure costs are the largest proportion, then appraisal, with prevention costs the least. The major part of the failure costs may be internal or external, depending on the industry and its market.

In some cases appraisal costs may be higher than failure costs. This should be treated as a serious danger signal, and the reason investigated. It may be that:

1. Some of the failure costs have been overlooked.
2. Appraisal results in internal failures which cost less (at that stage) than the appraisal activity. However, if undetected and allowed to become external failures, they would cost much more than the appraisal does currently. Therefore appraisal is deliberately held at a high level.
3. Too much reliance is placed on appraisal rather than prevention – the organization is trying to 'inspect in' quality.

4. The organization is performing expensive trials and acceptance tests which are not necessary for the organization's confidence in its product, but which are demanded by the customer.

You should examine the circumstances to establish which situation applies, and whether some action is necessary. In almost all cases, one conclusion will be to strengthen the preventive actions.

Presenting and using quality cost information

Quality-related costs are worth collecting only if there is the intention of taking some action as a result. Such actions may be aimed directly at reducing the costs, or at making people aware of the costs and their importance, as a step towards the same end.

The first quality cost report

When presenting a first report it is important:

1. To involve all senior managers who have a direct interest in the financial well-being of the company;
2. To ensure that your figures have the blessing of the finance department, and acknowledge their advice and assistance;
3. To point out that your presentation is based on the format recommended in BS 6143;
4. To stress that it is only a pilot study, and that regular updates will require the involvement of the finance department;
5. To relate your figures to an accepted norm for your industry, if this is known;
6. To highlight the main points, and the areas which seem most amenable to quality improvement and cost reduction;
7. To invite comment and suggestions.

It is unlikely that your pilot study will disclose the full extent of quality-related costs. Crosby suggests that a typical first estimate rarely arrives at more than 30% of the true value. Even so, it is possible to create first disbelief, and then alarm, and hopefully a breakthrough in quality awareness as people face the facts that.

1. Quality is not vague, but is measurable in monetary terms.
2. It impacts the whole company, not just the QA department.
3. Bad quality hits where it hurts most – on the bottom line.
4. Everyone's budget suffers to some degree.
5. Substantial savings are waiting to be made.
6. Insufficient resources have been invested in prevention in the past.

Looking for cost savings

The savings may well be the result of corrective actions, but the easiest way to identify these actions is by investigating the causes of failure costs. Setting up a quality improvement team or task force is often advocated. This should be multidepartmental and multidisciplinary. Some skill and insight is necessary to select the priorities for action, particularly since the first task tackled should be an unambiguous success if the momentum is to be maintained. Points you should bear in mind are:

1. As stated above, the first task undertaken should be one where complete success is confidently expected.
2. Don't direct much effort to obsolescent processes or products near the end of their lifespan.
3. More benefit may be gained by smaller savings achieved on recently launched products for which a long and successful future is predicted; not least because their acceptance and future sales will be influenced by customer satisfaction in the early stages.
4. The likely ease or difficulty of solving a problem must be taken into account as well as its absolute magnitude.
5. Even if a cure is found to be impracticable due to expense or disruption of production the objective should not be abandoned – an opportunity to implement a solution may arise later, e.g. when a product is redesigned or a production line re-equipped.
6. A small improvement in yield on a high-volume line will offer more savings than a larger percentage improvement on a lower-volume line.
7. Taskforce membership does not have to be fixed – its personnel can be changed, its scope broadened, or it can be disbanded when it has served its purpose.

Keeping the momentum going

In a small company it may be possible only to calculate and review quality costs annually. This will show the progress being made during the past year, and where effort should be concentrated during the forthcoming 12 months.

In a company with larger resources, where typically every manager having budgetary responsibilities receives a detailed monthly account of his/her expenses and manufacturing costs are reviewed on a similar basis, then it should be possible to report and monitor quality cost data on a similar basis.

Quality cost reporting does not in itself solve any problems. It supports and intensifies other company-wide quality improvement activities by converting 'quality problems', liable to be seen as the province of the QA department, into 'profitability problems' recognized as everyone's priority concern.

Summary of conclusions

In this chapter we have seen the following:

1. Quality costs can be distinguished as conformance costs and non-conformance costs.
2. Conformance costs fall into appraisal and prevention categories.

3. Non-conformance costs fall into internal and external failure cost categories.
4. These categories are equally applicable to manufacturing and service organizations. The four categories are further subdivided into elements, and a scheme for these is included in BS 6143 part 2.
5. The initiative in collating quality costs is often taken by the QA department, but the exercise should always have the blessing, and preferably the active cooperation of finance.
6. The prime purpose of collecting quality costs is to achieve management action to prevent failures.

25

Just-in-time and variability reduction

Introduction

The previous chapter has analysed the sources of quality-related costs, and has indicated areas in which savings are likely to be possible. In most cases these opportunities are instances where deliberate investment in preventing errors occurring bring far greater savings in failure and appraisal costs.

This chapter will focus on some of the more spectacular applications of this principle; notably the concept of 'just-in-time' production management pioneered by the Japanese. As you will appreciate as you read about it, this technique can only operate in an environment of total quality management, where all quality problems are addressed as they appear, with the aim of their permanent eradication.

Other issues I want to introduce to you are:

1. The economic importance of correcting design faults at the earliest possible stage;
2. The paramount importance of ensuring the compatibility of a design and the capability for manufacturing it;
3. The quality costs absorbed in finding and dealing with defects found during manufacturing inspection, and hence the savings available if they can be reduced or eliminated.

The above are all interrelated topics.

Cost savings of prevention

What is the cost of errors? Not just the cost of the errors themselves, but also all the costs arising from detection and correction. Once an organization loses confidence in its workforce, and expects errors to arise, it will invest a great deal of effort in double-checking and 'second-guessing' activities. It has been estimated that in some service organizations checking, correcting and follow-up activities account for two-thirds of total staff time!

In the preceding paragraph I referred to lack of confidence. To comprehend all the costs that are incurred by **not** 'getting it right first time' you have to ask yourself: 'If I was **sure** I was not generating any errors or defects what activities would become unnecessary?'

Eliminating redundant processes

Because of the need to detect and correct defective products, without disrupting process flow, the typical production line includes many steps which do not add any value to the product.

We will take an example from a manufacturing production line. Imagine the following simple process flow (Figure 25.1). Because of the proportion of defective items produced by process A it is necessary for the output to be tested before it passes on to process B. The reason for this is that after process B, defects produced by process A can no longer be repaired.

Each defective item from test has to be analysed to verify the failure and determine what form of repair is required. After repair each repaired item has to be retested to verify that it now functions correctly. Because of the disruption to product flow due to fluctuating losses incurred at process A, and waiting for units to be repaired, a buffer stock or reserve inventory has to be held between processes A and B.

Let me elaborate further on the point I have just demonstrated. This time suppose we are looking at the end of a process sequence (Figure 25.2). Before the items are accepted into finished goods stock they are tested. We will suppose that the customer demands an incoming quality of 100 p.p.m. defective maximum, but the process average is 0.25% defective, i.e. 2500 p.p.m.

You might suppose that 100% final inspection would be sufficient to guarantee whatever incoming p.p.m. the customer might demand; but remember that '100% inspection is never 100% effective'.

In this instance, suppose that the test is 90% effective; then after 100% inspection the outgoing quality will still be 250 p.p.m. (a tenth of the value before test, since only 90% of the defective items will have been found). To achieve the

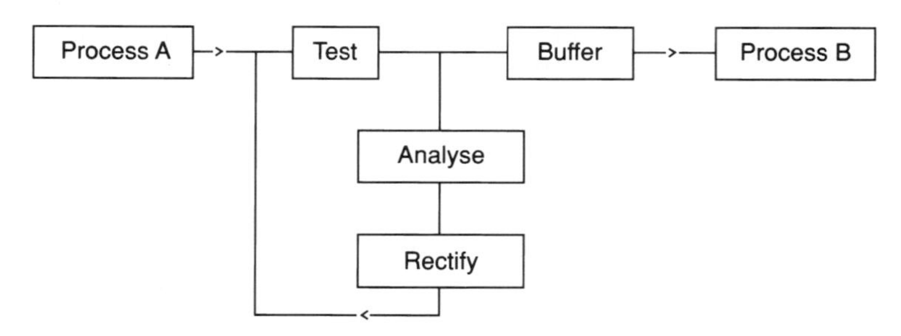

Fig. 25.1 Flowchart for a simple process sequence.

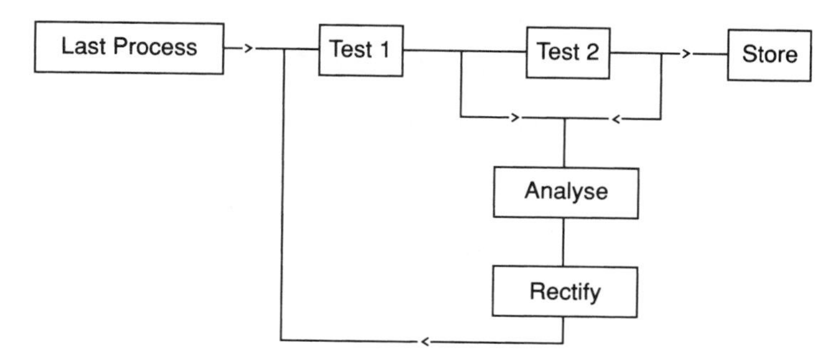

Fig. 25.2 Flowchart for an end process sequence.

required p.p.m. a second test will be required, and this can be expected to bring the outgoing reject level down to 25 p.p.m. (once again, a tenth of the previous level), so meeting the customer's requirement.

Savings from eliminating defects

If defects or errors can be reduced, certain costs can be eliminated. As we have seen above, whole inspection, rectification and storage stages in a process can be eliminated. This releases equipment and space. It obviates maintaining the equipment, and servicing the space which is no longer needed; alternatively the space can be used for another purpose which could avoid the need for building expansion. Bottlenecks will be removed and cycle times will be accelerated, so improving the return on capital invested.

No less impressive is the more efficient use which can be made in that most valuable of resources, people. People's time can be used more productively, concentrating on value-adding and preventive actions instead of dealing with the impact of failures. We can illustrate this by the following example.

Consider the imaginary but perfectly feasible chain of events resulting from the rejection of a batch of product at an inspection station such as the one assumed in Figure 25.1. We will suppose that it is a 'gate' based on a lot sampling inspection. (The sources of monetary wastage are shown in parentheses.)

1. The acceptance number of rejects for the sample is exceeded, so a second larger sample is drawn (extra QA inspector time).
2. The acceptance number is exceeded again, so the lot is quarantined and the rejects referred, with a reject advice note, for engineering confirmation of failure (form-filling, waiting time).
3. The engineer measures the rejects independently and confirms the inspector's decision (setting up test equipment, form-filling, scrap).
4. The batch is rejected and returned to production for re-screening (extra production labour).
5. About 15% are thrown out, the reduced batch is resubmitted to QA, sampled and passed (scrap, extra QA inspector time).
6. The lot is now too small to fulfil the customer order; it is held in inventory while a small new batch is started to complete the required quantity (production

control administration, extra materials, extra production labour, storage space, uneconomic batch size).
7. Customer service contact the customer to advise rescheduled delivery time (customer service time, loss of goodwill).
8. The extra lot is submitted to QA and passes (extra setting-up and QA inspector time).
9. The two lots are separately packed, separately recorded on a release note, invoiced and shipped to the customer (extra clerical time).
10. Customer pays late due to delayed receipt of goods (lost interest on cash received).

Were all these sources of wastage already obvious to you before you read this?

You will recognize that all of the activities I have just described are examples of 'non-value adding processes' as introduced in Chapter 20, Part Three. It is helpful whenever you are considering quality-related costs, to pause and consider each process, asking yourself 'If we were sure we were performing correctly, would this process still be necessary?'. If the answer is 'no' then, regardless of whether we are following the 'PAF' or the 'process cost' model, the cost of running that process is a quality-related cost. It applies to all manner of review, rescheduling and repair activities.

Just-in-time

In Part Three (Chapters 15 and 16) I explained so-called 'Taylorism' and the development of 'assembly-line' organization. I discussed aspects of this work philosophy which were contrary to good employee-relations practice as found by research, and also to the maintenance of good quality. In particular you will remember that line workers were divested of authority or responsibility to initiate quality improvement or corrective actions. If defective work was produced it had to be passed on to be screened out by inspectors.

Moreover, because of the investment in machinery there were seen to be heavy expenses involved in the machinery remaining idle. This is why the goal had to be to keep the line running at all times, even while problems of faulty work-in-progress were being investigated. Inventory stocks were built up at key stages (including bought-in starting materials, and finished goods) in order to provide a buffer of good material to keep the rest of the line running if there was a breakdown somewhere.

Production flow: pushed or pulled?

Thus you can say that production control was driven by the 'push' of new material waiting to be started or to progress along the line, even when the line itself suffered from problems or bottlenecks at some point. The Japanese concept of JIT (just-in-time) production turns this principle on its head – it insists that production control must be driven by the 'pull' of material being required at the next stage; nothing is to be started or elaborated further until there is a need for it at the next stage in the process.

In JIT, work is built for the 'customer' (in the TQM sense of the next in line), not for stock. This is immediately conducive to 'ownership' and eradication of quality problems, since both 'supplier' and 'customer' are acutely aware that if what is handed on is faulty, and cannot be used, the line will stop. Working on this principle demands sophistication, cooperation and mutual trust from end to end of the 'chain of quality', but the payoff in terms of cost savings, flexibility and speed of response (both to production problems and to changes in market demand) is enormous.

A definition of just-in-time

In his book on Total Quality Management, Oakland describes JIT as:

> A programme directed towards ensuring that the right quantities are purchased or produced at the right time, and that there is no waste.

But he also stresses the quality-improvement implications by representing it as:

1. A series of operating concepts that allow systematic identification of operational problems;
2. A series of technology-based tools for correcting problems following their identification.

This dual aspect of JIT (inventory elimination; visibility of, and action on, operational problems) is natural enough if you recognize that a series of operations carrying inventory can continue to limp along in the presence of chronic problems, whereas in a JIT environment the production process will cease until good material can be generated. In the JIT case the problems come to light immediately, and can be investigated in 'real time'; in the conventional buffered system the problem will take time to surface, defective material will be more difficult to trace, and the circumstances of its creation more difficult to establish.

Aims of JIT

The aims and advantages of JIT, which I have implied in the introduction to JIT, are summarized as:

1. To meet the customer's requirements exactly; without waste, and immediately on demand.

JIT methods identify problems by tracking:

1. Material movements; and also any delays, stoppages or diversions which under JIT can only be the result of some operational problem.
2. Material accumulations; these are also symptomatic of problems, covering up excessive process variability and other difficulties.

JIT methods encourage and enable process flexibility through reduction of batch time and acceleration of turn-round time.

They also improve productivity and costs by reducing unproductive activity (working on material which will subsequently have to be reworked or scrapped).

Perhaps even more obviously, there are both direct and indirect cost savings in inventory reduction (not just the value of goods held in inventory, but elimation of inventory storage space, racking, clerical records, etc.).

Tools for JIT

Here is a list of standard tools which are valuable in implementing JIT. Some will be familiar to you from Part Two of this book.

The techniques mentioned by Oakland are:

1. Value analysis/engineering;
2. Flow charting;
3. Statistical process control (SPC);
4. Method study and analysis;
5. Preventive maintenance;
6. Plant layout methods;
7. Standardized design.

Oakland gives another list of techniques which he says are more directly associated with JIT. These are:

1. Standardized containers;
2. Kanban or cards with material visibility;
3. Foolproofing;
4. Batch or lot size reduction;
5. Pull scheduling;
6. Set-up time reduction;
7. Flexible workforce.

I contrasted 'push' and 'pull' scheduling at the beginning of this chapter. You will also remember that I mentioned reduction of set-up time, and 'foolproofing' operations when we discussed the ideas of Shigeo Shingo in Chapter 18 of Part Three. We have commented on the importance of a trained and flexible workforce at several points in the book.

What we must do now is consider how JIT is made to work – this involves the other factors: of standardized containers, 'kanban' cards, and lot size reduction.

'Kanban'

The 'kanban' is so central to the operation of just-in-time than the 'kanban' system is sometimes used in the West as a synonym for JIT.

However, Oakland points out that the use of kanbans, as we shall describe the practice in the next two sections, will only work in the context of JIT, i.e. using all the disciplines listed in the preceding section. He says that 'A JIT programme can succeed without a kanban-based operation, but kanbans will not function without JIT'.

Kanban is a Japanese word meaning' 'visible record', normally a card which signals the need to produce more units. How many units to build at a time is a carefully considered decision; an aim of JIT is to make the lot size as small as possible; in the case of large items, the ideal batch size to be sought is one.

In the case of small items there will be a finite batch size, but to ensure that it is not exceeded all material movements are made in a specified container which can only hold the desired number. There are two kanban cards for each container, the P-kanban (Production) and the C-kanban (Conveyance). When work is transferred between two stations the kanbans are exchanged. No fresh work may be started until the P-kanban has been returned by the using station.

The kanban square

The system is easiest to visualize in the case where a 'kanban square' is used. The square is a storage or parking area reserved for the work being passed on from one work station to the next. It may simply be an area of the floor marked out with tape or paint. When a unit (e.g. an assembly) has been completed, or a container (e.g. of components) filled by the 'supplier' station it is moved to the kanban square. When the 'customer' station needs the material, it will remove the unit or container from the kanban square. The 'supplier' station **must not start** any more work until its operator sees that the kanban square has been emptied. Once it **is** empty, that is the signal to recommence work.

Waiting time

To production engineers or supervisors trained under the tyranny of the production line, the most startling aspect of JIT is the amount of operator waiting time that can potentially be involved. This concern arises from two sources:

1. The 'Taylorist' conviction that money is being wasted if the line is not running.
2. The 'Theory X' belief that the workforce must be kept busy or they will create mischief.

However, these deeply-engrained ideas are both fallacies. Under JIT if the line has stopped it must be because somewhere downstream there is a stage which cannot accept more work. This in turn must be because either:

1. No more output is required;
2. Or else a problem has been identified.

If no more output is required yet, there is no point in investing extra value ahead of time, when it will simply stand in inventory at some stage before it can be sold.

If a problem has been detected, its origin is either downstream or upstream of the stage we are focusing on. If downstream, we don't want work done on the good work-in-progress at our station until the problem is corrected. If, as may be the case, the source of the problem is upstream, the work at our station is likely to be defective already. Passing it on wastes the value added at succeeding stations, and may also make the fault more difficult to identify or repair.

Theory 'X' implies that it is important to keep the workforce busy at all costs, even if it is only digging holes and then filling them up again. JIT assumes a trained, flexible and motivated workforce. The sorts of tasks that Japanese operators would be expected to do if the line was on hold could include:

1. Housekeeping their workplace (cleaning, tidying, routine maintenance);
2. Updating their SPC charts;
3. Helping find the source of the problem;
4. Studying;
5. Working on a quality circle project;
6. Switching temporarily to another job, as directed by their supervisor.

Vendor relationships, under JIT

Companies which take JIT seriously extend it to the receipt of materials and subassemblies from their suppliers and subcontractors, for this is an area in which the greatest savings can be made, through eliminating large incoming goods stores. As an illustration Toyota in Japan receive scheduled deliveries from local suppliers every twelve hours. In other cases deliveries of small components might be scheduled to arrive as frequently as every 2 hours.

It will be evident that this is a situation in which the greatest degree of mutual trust and confidence between 'customer' and 'supplier' is essential. The external supplier is, in effect, simply an extension of the internal JIT flow. Regular small shipments will be received, and accepted straight onto the production line, even though 'parts-per-million' reject levels are demanded. The onus is totally on the supplier to provide the right quantity at the right time, at the right quality level.

Thus, features of supplier quality assurance especially characteristics of JIT are:

1. Careful selection of vendors on the basis of in-depth evaluation of their systems, process control and quality levels;
2. A preference for a single supplier for each part, and the establishment of a close long-term relationship. (In Japan, major customers often take a minority share-holding in their key suppliers);
3. Selection of suppliers on the above criteria in preference to lowest price;
4. A preference for geographically local suppliers;
5. Mutual openness about quality problems, with problem-solving by joint teams from supplier and customer.

Contrast this approach with the more common 'adversarial' one where a company with strong purchasing power likes to keep several vendors competing for a share it its orders, playing off one against the other in order to bargain the lowest possible price.

Further reading

This chapter contains no more than a simplified introduction to JIT. For more detailed reading, including more information on the kinds of problems that arise in trying to introduce and operate JIT, I recommend you to read a more specialized text. One such is *Just-in-time Management of Manufacturing* by Ian Graham.

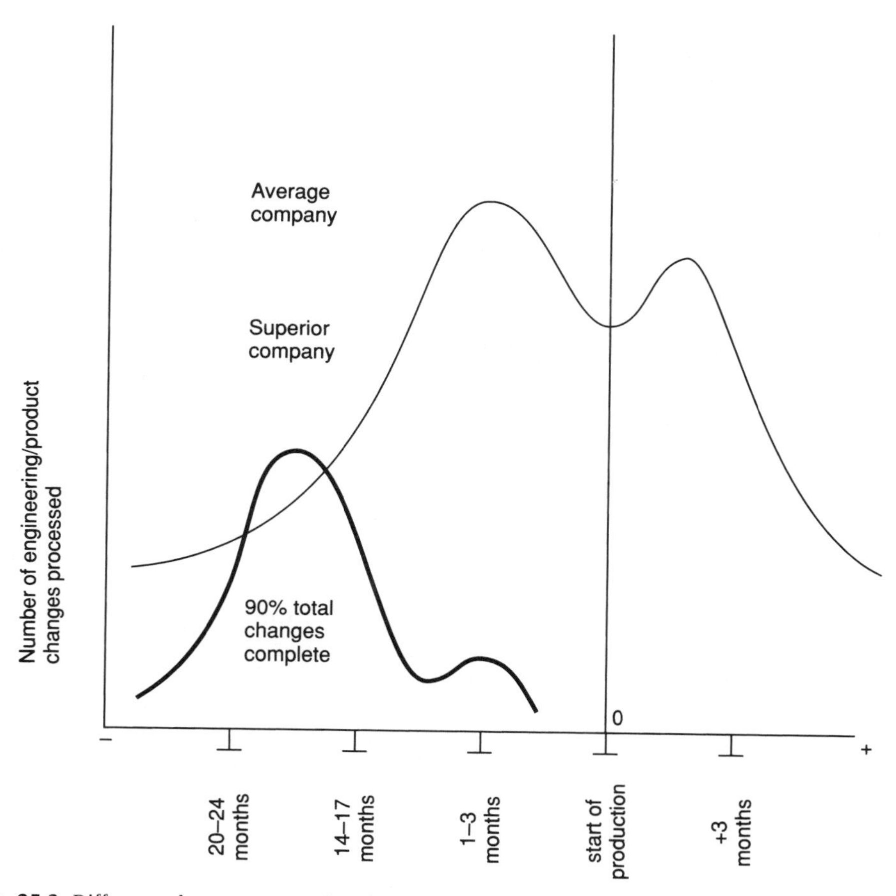

Fig. 25.3 Difference between a good and a typical new design introduction.

Compatibility of design with manufacturing capability

This is another area which is critical both for quality and for cost; or as we would say, for quality-related costs. One manifestation of problems in harmonizing the design (what you want to make) with manufacturing capability (what you can actually make) is the number of design modifications which have to be made.

Figure 25.3 illustrates the difference between a good and a typical new design introduction.

The 'good' situation, where the design is finalized before production commences and no further modifications are needed is superior in two respects.

1. No customer receives an imperfectly designed product which does not include all the design modifications later found to be necessary.
2. It is cheaper for the manufacturer because the further into the design cycle, the more expensive it is to modify the design (Figure 25.4).

Many, though not all design and process changes are generated through the discovery that the specified design is beyond the manufacturing process capability.

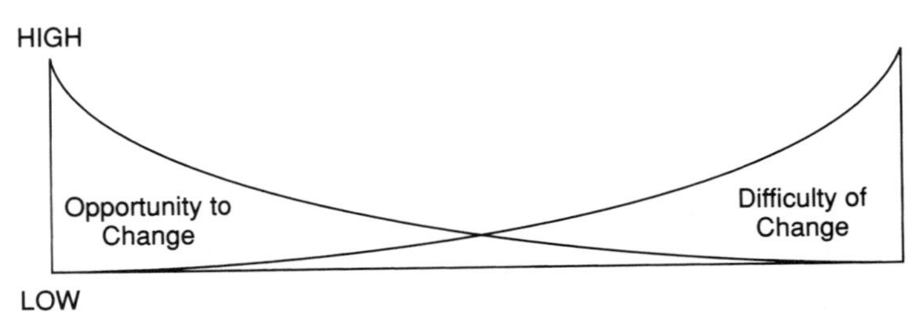

Fig. 25.4 Design modifications at different stages in the design cycle.

Process capability

You will recall that the idea of 'process capability' was explained to you in Chapter 12 of Part Two, as part of the discussion of statistical process control. When using the traditional approach of SPC in conjunction with Shewhart control charts action is taken on the parameters controlling the process if a measurement falls outside the $\pm 3\sigma$ limits.

If the $\pm 3\sigma$ limits fall within the specification tolerance limits (as shown in Figures 25.3 and 25.4 as the upper and lower specification limits, USL and LSL), then the process is said to be under statistical control, or 'capable'.

It will be clear from Figure 25.5 that $\pm 3\sigma$ only represents a 'capable' process if the mean or target value is accurately held. If the mean is skewed a tighter process

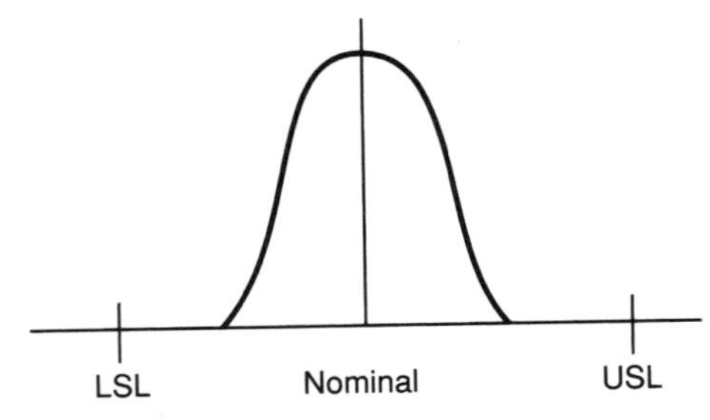

Fig. 25.5 Process capability – mean or target value accurately held.

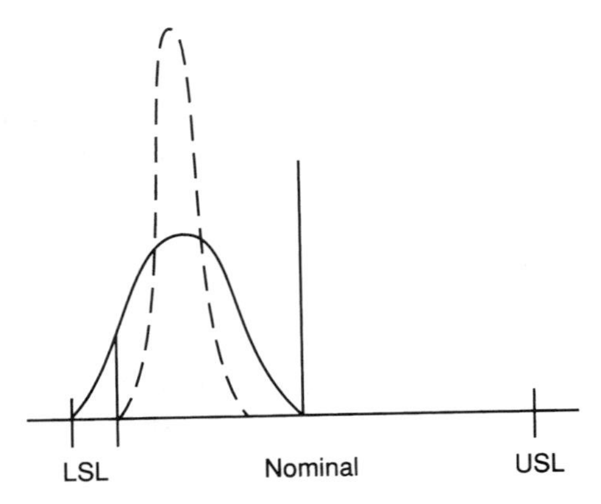

Fig. 25.6 Process capability – skewed mean.

distribution is needed, as in the example Figure 25.6. Thus it is found in studies of Western industrial plants that a process capability of ± 4 to $\pm 4.5\sigma$ is typical.

Significance of Taguchi's loss function

Once again, the practices in most Japanese companies (and the best Western ones) is different. Being highly conscious of SPC techniques, the Japanese do not treat the specification limits as being a hard-and-fast border between good and bad, as in Figure 25.7. Taguchi's loss function says that the customer suffers some inconvenience or loss whenever a parameter differs from its design value in any degree, even if within specification, and that the degree of loss is often proportional to the square of the deviation (see Figure 25.8).

As a result, Japanese factories try harder to aim at the design value rather than accept anything within specification limits and, applying the Kaizen (continuous improvement) philosophy, they attempt to continually reduce the process variability. The same study which established the USA norm as ± 4 to 4.5σ found that the Japanese were achieving $\pm 6\sigma$. Especially in the case of complex products involving a long sequence of operations the difference in yield and cost starting from these two baselines is enormous, as we shall show in a moment. But first, a well-known and often quoted illustration.

Ford and Toyota collaborate jointly in some markets, and Ford were having some transmission units manufactured for them by Toyota. Although both companies were working to Ford drawings, over the course of time Ford found they were receiving significantly less complaints or warranty claims relating to the Toyota-manufactured transmissions, which were also often perceived by customers as quieter.

Investigation showed that Toyota controlled their machining processes well inside the Ford specification as a matter of course. Thus any two gears assembled

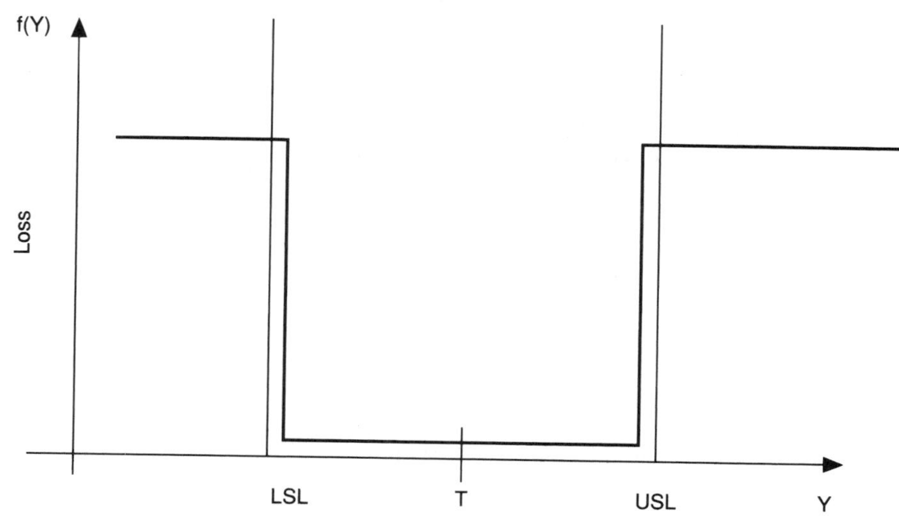

Fig. 25.7 There is no loss to the customer provided the performance characteristic is within the specification limits.

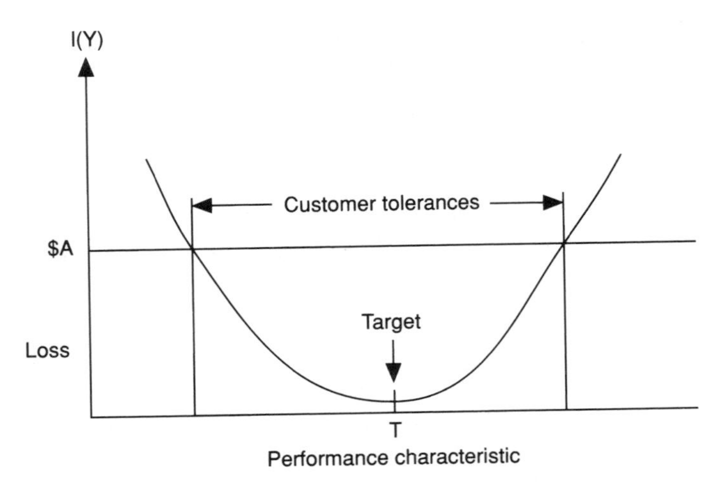

Fig. 25.8 Loss to the customer increases as the performance characteristic deviates from the target value.

at random would fit better, with less risk of poor performance due to the unfortunate combination of a pair of gears both at the extreme limits of tolerance. Moreover, Toyota experienced less scrap transmission components and fewer reject assemblies.

Cumulative effect of errors and tolerances

An operation that produces a defect or out-of-tolerance incidence of one part in 1000 (0.1% or 1000 p.p.m., may sound very satisfactory. It represents a yield of 99.9%; but if it is one of a sequence of 1000 similar operations or one of a combination of 1000 similar components the overall yield of good assemblies is $(99.9)^{1000}$% or only 36.77%.

This may surprise you, and illustrates the importance of a high yield for each individual process or component. The yield improvements that Japanese companies have achieved through 'kaizen' are very relevant to their price competitiveness, and their ability to recoup investment costs of a new model very quickly.

Robust design

This is not to say that improving process yield is easy, hence much attention is paid to what Taguchi calls 'robust design', or which we might describe as 'tolerant' design; this involves detailed examination of the behaviour of the process or design to determine which variables it is sensitive or insensitive to. It is the sensitive variables which have to be optimized and the most carefully controlled.

That is the reason why Taguchi pays so much attention to design and interpretation of the results of statistical experiments where process parameters are varied.

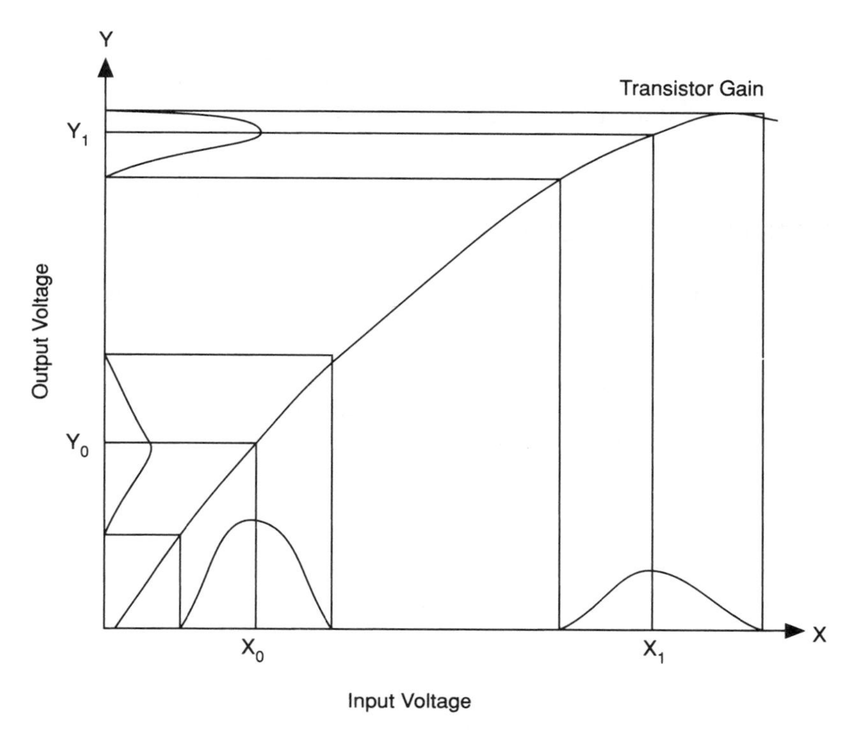

Fig. 25.9 Effect of transistor gain on output voltage. Use of operating point X_1 Y_1 on the characteristic in preference to the operating point X_0 Y_0 is a simple example of a robust design.

It is also the basis of Shainin's precept of 'talking to' the part or process in order to understand, as intimately as possible, how it can be controlled.

Figure 25.9 is an illustration of a design made more robust through understanding the behaviour of the device being used. Since the gain of the transistor is non-linear, designing a circuit so that the transistor is operating with input voltage X_1 is much more 'robust' than operating with input voltage X_0; an equivalent variation in input voltage will produce much less variation in output voltage Y_1 than in Y_0.

Matching design with the process capability

Compatibility between product design and process capability can be achieved by the following steps:

1. Identify which product characteristics are critical to satisfying customers' performance and environmental needs;
2. Determine which product elements help to achieve the critical characteristics;
3. Determine which process steps or options control the critical characteristics, through the product elements;
4. Determine a nominal design value and maximum allowable tolerance for each critical characteristic;

5. Determine the statistical 'capability' of the components and process elements that control the critical characteristics;
6. If the process capability is insufficient for the specification limits, improve the design, or process, or process control.

Summary of conclusions

This chapter dealt with JIT and other related matters from the point of view of variability reduction, and quality cost saving. It will have helped to illuminate Taguchi's concept of quality costs as 'loss to society', and shown how prevention of errors can enable whole sequences of processes to be re-engineered, with great saving in non-value adding costs. It also re-emphasizes (refer back to Chapter 3, Part One) the importance of getting the right balance between robust design and capable manufacturing – 'problems are either designed out or designed in'.

26

Quality, safety and product liability

Introduction

An implied need of any customer is that the product or service purchased should not cause any harm or damage to the purchaser, his/her property or any third party. Just as in recent years the concept of providing 'quality' has focused more and more on the customer's needs, so have the legal constraints on a supplier, and the interpretations or judgments made by courts of law.

This chapter will discuss some of these developments, and the ways in which an effective quality system, and a 'total quality management' approach to achieving and refining quality, can help a company protect the safety of its customers and society at large.

Quality, safety and consumerism

You may be a little surprised if it has occurred to you that I have put the unit on quality and safety in the middle of this particular section of the book, whose objective is to emphasize the link between quality and profitability.

In fact, it is as appropriate a place for it as any. Profitability always depends on providing what the customer wants. During the last decade or so customer expectations have been evolving and becoming more exacting, throughout the developed world. Not only is there increasing unwillingness to accept inferior goods or poor service but also there is increasing demand for products which are safe to use and products which do not harm the environment. Thought should convince you that compliance with society's expectations, and with the laws on personal and environmental safety are of key importance for your company's prosperity both in:

1. Maintaining or improving sales in the face of shifting consumer priorities;
2. Avoiding heavy damages, and loss of reputation (and hence sales) due to liability hearings in court (or cases settled out of court).

As with other aspects of quality, safety and environmental expectations are evolving and becoming more exacting.

Quality and the law

We have talked of quality as something which the customer expects to receive from a supplier; briefly a product or service which is fit for its purpose, appropriate to the price paid for it, and delivered on time. The law of the land is involved when either:

1. An appeal to a court is made on the grounds that the contract between supplier and customer has been breached, or
2. A prosecution is made, claiming that a statute relating to the supply of goods has been breached;
3. A prosecution is made, claiming that a statute relating to the protection of the safety of people at large, or of maintenance of the environment, has been breached.

Statutory legislation directed towards ensuring that customers receive products or services of adequate quality (in the board sense in which we interpret quality in this book) can be categorized as dealing with either:

1. Correct quantity;
2. Correct quality;
3. Fair trading;
4. Safety.

Whereas nearly all that this book has to say about quality is universally applicable, when it comes to the legal aspects this is no longer the case. The legal system, and the laws enacted, differ from country to country. If I took my examples from English law, they would not apply exactly even in Scotland.

For the reason just explained, I shall limit discussion to the general issue of **consumer safety**. I shall ignore specialized legislation such as that which exists in the food and pharmaceutical drugs industries. Nor shall I detail current environmental pollution regulations, or regulations regarding employee safety.

Despite these omissions I believe I can make a case which would apply equally in those other categories. The recipe for successful achievement of safety is found within the same framework as for ensuring other aspects of quality; namely by having an effective **quality system** within which total quality management can safeguard the customers' needs.

The European Community Directive on Product Liability

Because of the size of the Community, the European Product Liability Directive of 1985 is the legislation which, of its kind, affects the largest number of people in many different countries. Its provisions were made law in the United Kingdom by Act of Parliament (Consumer Protection Act of 1987). Therefore we will use the requirements of this directive in order to examine how companies can safeguard the consumers of their products or services.

The Directive deals with compensation for death, personal injury and damage to property. It acknowledges divergences between member States' existing legislation, and claims that the directive is necessary because:

> The existing divergences may distort competition and affect the movement of goods within the common market and entail a differing degree of protection of the consumer against damage caused by a defective product to his health or property.

Features of the Product Liability Directive

There are several features of the European Community Product Liability Directive which warrant emphasis and examination. Some of its implications, as embodied within the UK in the Consumer Protection Act 1987, are highlighted here:

Scope

The Act is concerned with **product safety**. It allows anyone, whether the direct purchaser or not, who has suffered damage (death, injury, or damage to personal property) as a result of a defective product to sue the producer or importer directly, i.e. it establishes 'strict liability'.

The injured party is not required to prove negligence; he/she only has to show causal connection, i.e. that the damage did happen as a result of the defect.

The Directive allowed member states discretion on certain points, e.g. certain categories of product supplies such as agricultural products could be exempted, a maximum limit to damages awarded could be set. In the UK agricultural produce is excluded, no upper limit to damages is set.

Defences under the directive

There exist only six defences under the Act; the one demanding most attention from the quality assurance point of view is 'development risk' known in the USA as 'state-of-the-art'.

Development risk

The producer will not be liable if the state of scientific and technical knowledge at the time when the product was put into circulation was not such as to enable the existence of the defect to be discovered.

Other defences

The producer will not be liable if:

1. He did not put the product into circulation (e.g. if it was a pirated copy mistaken by the plaintiff for the original brand);
2. It is probable that the defect which caused the damage did not exist when the product was put into circulation;
3. The product was not manufactured for sale;
4. The defect was due to compliance with mandatory regulations;
5. In the case of the manufacturer of a component, the defect was due to the design of the product into which it was incorporated.

Reduced liability

The producer's liability may be reduced or disallowed when, having regard to all the circumstances, the damage is caused both by a defect and by the fault of the injured person or anyone for whom the injured person is responsible.

Risk minimization

The EC Directive on Product Liability has focused business attention on exposure to litigation. A trading company, manufacturer or supplier needs to be acutely aware of the issues involved, and must operate in such a way as to:

1. Minimize the risk of injury caused by one of its products;
2. Mitigate its liability should such an incident occur.

This is known as a risk minimization policy.

How can a business minimize the risk of finding itself defending a law suit over breach of contract, negligence or consumer protection, arising out of faulty goods?

The law concerns itself with matters of:

1. Merchantable quality;
2. Fitness for intended use;
3. Avoidance of negligence in design and manufacture;
4. Safety of the end user, and all others coming into contact with the goods.

You will have realized, I am sure, that these are all aspects of 'quality' as we have defined it and discussed at length at several places earlier in the book. Thus a risk minimization policy will employ many of the techniques of quality assurance explained in Parts One and Two of this book.

We can say that a risk minimization policy is simply one aspect of a general quality policy, and to be successful it has to employ the same disciplines. A quality management system is essential if a risk minimization policy is to be effective.

Elements of a risk minimization policy

These are associated with:

1. Designing for safety;
2. Prototype testing;
3. Volume production;
4. Presentation;
5. Feedback from customers;
6. Records;
7. Product recall.

Let us look at each of these in turn.

Designing for safety

Disasters resulting from the malfunction of products, especially those which make the newspaper headlines, tend to be associated in people's minds with manufacturing faults. In reality, most safety failures (as with quality problems in general) arise at the design stage. It is here that problems are either designed **in** or designed **out.**

At the specification stage it is necessary to define the following parameters, amongst others, with or on behalf of the customer:

1. Intended use;
2. Environmental conditions to be experienced;
3. Maintenance requirements;
4. Performance requirements;
5. Critical failure modes;
6. Required lifetime and reliability.

You can use techniques such as reliability prediction, FMECA analysis, and other techniques mentioned in Chapter 14 of Part Three of this book, in order to assess whether the product will meet the design criteria on safety.

At each **design review** when design progress is appraised, it is essential to verify that the relevant information is available in which to judge whether the defined safety standards are being achieved.

Prototype testing

At the prototype or pilot-production stage one, a few, or a pilot production run (depending on the complexity of the item) of the product will be built. These are not for sale, but for assessment of the product, and the processes that will be used to produce it. Before the design is finalized and approved for production, data gained from the prototypes must be used to verify that the product will conform to its safety standards.

Volume production

When the prototype has shown itself fit for production, the design is frozen and placed under 'configuration control'. Production for sale can then commence. Thereafter, the product will typically be tested before release, and its overall performance reappraised periodically by more detailed tests on a sample.

Any further changes in the process or product specification which seem desirable in the light of production experience may only be introduced under the discipline of a formal engineering change notice. These disciplines will continue to safeguard the safety provisions as well as all other aspects of product quality.

'Presentation'

I have put this term in quotes as it may be less familiar to you. As used in the EEC Directive, this expression covers the packing, transport, installation, etc. of the product (compare ISO 9004–1, *Handling and Post-production Functions*).

However, and this is most important, it also covers the guidance, written or otherwise, given to the customer on the proper use and maintenance of the product. This information must adequately define the correct installation, operation and maintenance procedures to be followed, and it must also warn of any hazards or restrictions on the use of the product.

Presentation covers an even wider field than that. It covers pre-sales advertising, and education on potential risks (e.g. a UK Gas industry TV commercial on safe and unsafe ways of dealing with suspected gas leaks). An article entitled *'Product liability: who holds the baby now?'* Mackmurdo (1988) includes the statements:

> It is becoming more and more obvious that, since the new legislation, marketing and sales departments now have as significant a role as technical and quality departments in ensuring that a product is without defects. The function of market research in discovering safety expectations is every bit as important as technical R&D . . .
>
> The market oriented rather than technically oriented safety standard set in the law for assessing the defectiveness of a product sets a moving target . . .
>
> The important message for managers is that a serious and imaginative approach to avoidance of product defects, as defined, is probably the most effective form of product liability prevention. Such initiative, far from being seen as a cost and a burden can, if carried out properly, be regarded as an investment in all the opportunities available from the customer awareness, quality and safety programmes comprised in the initiative.

Feedback from customers

The product's maker must be kept informed of the customers' experiences with the product. All agents of the company manufacturing the product – salesmen, distributors, retailers – have a part to play in collecting and feeding back this information so that, where necessary, action can be taken as a result.

In particular, a customer may report a product as being the cause (or a contributory cause) of an accident. Especially if the customer intends litigation on the issue, it is obviously essential to seek and gather as much information concerning the circumstances as is possible.

It will be obvious that such information is essential if a case is to be successfully defended, but the most basic purpose in establishing the facts is to ensure that the hazard is eliminated by redesign, modified production/test methods, amended installation instructions, etc.

Records

As with all elements of quality assurance, it is essential to keep records to establish the level of quality being provided, where problems exist, and the nature and magnitude of the problems. This is a pre-requisite of correcting the problems. One of the essentials of recording is to be able to apply the highest practicable degree of **traceability,** so that if a problem is reported and analysed:

1. You can identify all other products which might be affected;
2. You can identify and warn the recipients, and impound any which you or your agents still hold in stock;
3. If necessary you can recall suspect product from purchasers.

Records which are relevant to safety should be kept for a long time, since under the EC Directive an injured party has 10 years in which they may start proceedings.

Product recall

If a product is suspected of being unsafe, it may be necessary or advisable for the manufacturer to initiate a product recall programme. This enables suspect product to be examined and, if necessary, corrected or taken out of circulation. This requirement implies the need for traceability, records, and effective recall procedures.

Role of the quality manager

What role is appropriate for the QA department, and in particular the quality manager, in developing and implementing a risk minimization strategy?

He/she can use the quality system as the framework within which to develop the strategy. In fact, a good exercise in assessing the effectiveness of an existing quality system, is for the quality manager to see how well it anticipates and fulfils the needs of risk minimization. He/she could have this exercise incorporated into the next management review of the company's quality system, if this issue has never been examined before. If deficiencies are apparent, the quality system must be refined.

The decisions on which the safety of the end product depends are under the control of numerous different individuals, but the QA department plays a key role; not least in monitoring other people, and in particular by:

1. Devising and auditing the quality system;
2. Taking part in design reviews;
3. Approving procedures and specifications;
4. Approving new products and processes for production use;
5. Monitoring in-process quality;
6. Releasing completed product;
7. Maintaining records;
8. Customer quality liaison;
9. Investigating customer complaints and returned product;
10. Initiating product recalls;
11. Administering corrective actions;
12. Auditing distributors and warehouses.

You will appreciate how the activities just listed tie in with the items discussed under the heading 'risk minimization'. Perhaps the areas needing greatest attention (i.e. those most likely to have been neglected in the past) are:

1. Defining, safeguarding and validating safety considerations at the specification, and design stages.
2. The documentation aspects of 'product presentation'; giving the customer all the necessary guidance on safe installation, use and maintenance of the product, plus the limitations placed on its use by safety considerations.

Value of a recognized quality system

The primary purposes of the company quality system in relation to product safety are:

1. To prevent accidents happening;
2. To minimize the seriousness of any accident that might still occur;
3. If an accident does occur, to prevent recurrence of similar mishaps.

There are two other possible advantages of having a recognized quality system, though their relative importance should not be overstated:

1. A better defence if taken to court;
2. Lower indemnity insurance premiums.

For the purposes of preventing and reducing the seriousness of accidents, and limiting their recurrence, the only criterion of the quality system is that it should be **effective**. To gain the secondary advantages, it also needs to be perceived by outsiders such as lawyers and insurance brokers as being acceptable. For this reason it is valuable to have gained an independent approval of the quality system, e.g. under ISO 9000.

The existence of the quality system would not of itself be a defence in a no-fault liability case, neither would a demonstrable low incidence of defects of the particular type which caused the damage.

> The law, however, is designed to provide remedy in any circumstances of a defect causing personal damage, an intention which includes all cases where a rogue product causes injury, no matter how high or low the relevant statistical rate may be. (Mackmurdo, 1988)

Reiterating, the purpose of the quality system is to prevent accidents involving your products arising. The fact that it is registered to ISO 9001 is insufficient defence if injury does occur.

Using your quality system in defending a liability case

If despite all your vigilance a case does arise, having maintained an effective quality system can aid your defence in many situations, nonetheless. Examples are all the defences we listed earlier in this chapter as points 2–5 under 'other defences'.

Suppose, for example, that you suspected that the defect was a result of an unauthorized modification introduced by an intermediary who sold it on to the plaintiff.

For your defence to succeed you might need to show that:

1. The same identified unit was tested before dispatch, and passed without exhibiting the fault described.
2. The shipping records trace its destination to the suspect intermediary.
3. Its configuration when recovered after the accident differed from that given in your controlled drawings.

Thus the success of your defence would depend on the effectiveness of the traceability, drawing control and quality record-keeping aspects of your quality system.

Equally, a successful defence based on 'development risk' would depend on your having records from design review meeting minutes etc. showing:

1. When the product was designed and released;
2. That specification and design reviews had been undertaken, taking account of customer requirements and legislation existing at the time;
3. That a conscious attempt had been made to discover and take account of any known or suspected hazards.

Product safety and liability in ISO 9000

Regrettably, the ISO 9000 documents are inconsistent. ISO 9004–1 rightly recognizes product safety and liability as an element of the quality system. However, there is no mention of this topic as a system element in the contractual documents ISO 9001–9003, except a mention under Design Review in ISO 9001.

ISO 9001 on design safety

The only reference comes under clause 4.4.4 Design Output, where 4.4.6 (c) says that the design output shall:

> Identify those characteristics of the design that are crucial to the safe and proper functioning of the product.

Clause 4.4.4 also implies reference to safety and environmental friendliness since it states that:

> designed input requirements relating to the product including applicable statutory and regulatory requirements shall be identified, documented and their selection reviewed by the supplier for adequacy.

I think you will agree with me that these statements alone, with no reference to safety matters elsewhere in ISO 9001, or anywhere in ISO 9002 and 9003, are inadequate.

ISO 9004–1 on safety

This states that consideration should be given to identification of the safety aspects of the product, with the aim of enhancing safety. Steps can include:

(a) Identifying relevant safety standards in order to make the formulation of product or service specifications more effective;
(b) Carrying out design evaluation tests and prototype (or model) testing for safety, and documenting the test results;
(c) Analysing instructions and warnings to the user, maintenance manuals and labelling and promotional material, in order to minimize misinterpretation, particularly regarding intended use and known hazards;
(d) Developing a means of traceability to facilitate product recall (see 11.2, 14.2 and 14.6);
(e) Considering development of an emergency plan in case recall of a product becomes necessary.

QUENSH management

For the present I don't know how general this term will become, but it has been coined at the University of Paisley Quality Centre to emphasize the convergence of Quality, Environmental and Health and Safety management principles. (Renfrew D., Muir G. and Cumming R. 'QUENSH Management – the future of quality?', *Quality World*, June 1994, and 'Quensh management, a strategically healthy environment', *Quality World*, August 1994.)

The papers note the obvious similarities between the standards being formulated for managing these various aspects of customer satisfaction, in three British Standards where the similarities in content are reflected in related standard numbers; BS 5750 (now BS EN ISO 9000, etc.) on Quality management systems, BS 7750 on Environmental management systems, and BS 8750 on Health and Safety management systems.

The elements common to all of these documents include:

- Management responsibility;
- A management system;

- A management plan;
- A management manual and supporting documentation;
- Maintenance of records;
- Internal audits;
- Verification, measurement and testing;
- Corrective action;
- Training.

Already in this book we have also stressed that management systems for reliability assurance, and for the administration of calibration laboratories are simply subsets of a quality management system. The authors of the 'QUENSH' papers predict that ISO 9000 will develop into an international standard embracing the management of all these aspects of quality.

Summary of conclusions

As the scope assigned to Total Quality Management is expounded, and broadened in successive expositions, it becomes increasingly difficult to distinguish from the general issue of 'enlightened management practice'. Every aspect of good management depends on making the best use of subordinates' experience and intuition, within the framework of a system which empowers rather than inhibits these people.

This is particularly noticeable if one compares the models proposed in British or ISO standards for Quality, Reliability, Calibration, Health and Safety, and Environmental Systems management. Indeed, the expression 'QUENSH Management' has been coined for the set of disciplines which can be applied equally to safeguarding quality, environment, safety and health.

Discussion of environmental protection, health and safety in this chapter has focused attention on the legal aspects of quality conformance. Once again, this has enabled emphasis to be placed on the preventive aspects of quality improvement; the value of quality management is not in defending a legal case, so much as in preventing it from arising in the first place.

27

The international market-place

Introduction

Chapter 27 of Part Four of this book looks at various aspects of international competition. The terms of the Single European Market Act of the European Community are cited as an example of how transnational legislation can affect domestic businesses. Under such legislation, all companies wishing to trade within its area of jurisdiction will be required to fulfil the same safety and quality standards which, for the most part, are likely to be modelled on the highest standards presently obtaining in member countries.

The fittest companies will prosper, the less able are likely to perish. How do you know what standards you have to aim for? The technique of finding out is called 'best practice benchmarking', measuring your performance against the best in your class.

I shall also say something about another way of comparing your company with the best in its field. That is by entering competitions such as the Baldrige Award scheme.

Winner takes all!

Survival is not compulsory. (Dr W. Edwards Deming)

This chapter will stress some of the international aspects of industrial quality. Increasingly, more and more companies are competing with international rivals, for what is essentially an international rather than a local market. Winner takes all!

As we have already observed, the past 40 years has proved a testing time for industries throughout the world. Since 1945 the UK has lost traditional markets for manufactured products with the break-up of Commonwealth trading patterns which used to be enshrined in the policy of 'Imperial Preference'; the British Empire depended on Great Britain for its manufactured goods, and in turn its produce and raw materials had tariff preference in the UK. This changed with UK entry into the European Common Market; now, for example, much New Zealand lamb is exported to the Arab countries, and Australian beef and iron ore goes to Japan. Thus Britain's main trading partners currently (the European Community) are in many respects its direct competitors, and its former clients have found new partners.

Continental Europe had to reconstruct its industry after the end of the war, and slowly re-establish its former position. Japan was in a similar, or rather even worse position, since it did not have the pre-war economic base and reputation of, for example, Germany.

The position in the USA was, as we have noted before, that from 1945 it was the only nation whose industry had survived intact and had not, to the same extent as in Britain, been turned over totally to war production. The prosperity brought about by being the only nation able to respond immediately to a post-war seller's market, plus the prestige of US nuclear and aerospace achievements brought overconfidence. Thus it also brought a failure to respond energetically to overseas competition when it did arrive; inevitably arrive, since the US market was the richest prize available.

Many economies, and more especially many individual companies, have found themselves vulnerable to competitors from overseas offering more attractive features or better-priced products.

Not only is competition more fierce, but there is less ability to hold on to your local market because of your special knowledge of its unique characteristics; local currency and financial institutions, local language, local tastes and customs, distribution systems. The world is becoming a smaller place!

Communications are improving, people become more familiar with, and emulate the lifestyle of more fortunate neighbours. From Ecuador to Pakistan to Zimbabwe, the TV aerials sprout up and the Coca Cola sign is ubiquitous.

Other than actually improving quality two kinds of defence against foreign competitors can be employed:

- **Tariff barriers** (taxes and duties on imported products, especially those competing with locally made ones).
- **Non-tariff barriers** (such as technical or safety standards different from those set in other parts of the world).

Over the course of time different nations group together to establish larger free markets, as in the case of the European Common Market. In 1992 the European Community largely established its goal of a Single European Market.

The way to compete with foreign and local competition, in your own market and in challenging competitors in their own local market, is to offer equivalent or superior quality, in its broadest sense and all its aspects.

The single European market

This part of the chapter will deal with some aspects of the impact of the single market on businesses both within and outside the EC. The intention is to illustrate the impact that all existing and future developments of a similar kind are likely to have.

The desire for a single market

The opening lines of the Treaty of Rome, signed by the six original member countries, and later by other countries on joining the Community, identify the goal of a single European market:

> The Community shall have as its task, by establishing a common market and progressively approximating the economic policies of member states, . . .

The concept of a European single market without internal barriers to trade may go back to the Treaty of Rome, but it received its impetus from the Single European Act agreed in June 1985. The Act committed the Community to adopting:

> Measures with the aim of progressively establishing the Internal Market over a period expiring on 31 December 1992.

The Act defines the Single Market as:

> An area without internal frontiers in which the free movement of goods, persons, services, and capital is ensured in accordance with the provisions of the Treaty of Rome.

The need for a single market

Without a common economic framework the Community's members suffered due to their national fragmentation, dissimilar regulations, and barriers to mutual trade. The net result was that this 'uncommon market' was burdened with heavy costs, one of the major factors making European business less competitive than its Japanese and American counterparts, despite a population of 320 million. The major ongoing costs resulting from a disunited European were claimed by Levy (1989) to be:

1. High administrative costs of dealing with different national bureaucratic requirements;
2. Higher transport costs because of border formalities;
3. Increased costs resulting from different national standards hence smaller product runs;
4. High costs of non-competitive and heavily regulated state activities, as exemplified by national public procurement policies;
5. Ultimately, high costs and reduced choice for consumers confined to their national market.

Thus the single market is concerned with sweeping away obstacles which had hitherto prevented the development of a genuinely free market in goods and services throughout the 12 member states. A home market of 320 million consumers does not mean a homogeneous market. But it does mean that many goods and services hitherto restricted to particular countries are able to move freely across national borders within the Community, provided they meet the essential requirements for safety, etc. established by the Community on behalf of the member states.

Content of the Single Market Act

The main elements in the programme for completing the single market are listed here as examples of what such a programme can embrace:

1. Removal of fiscal barriers to trade, such as differences in the rates of indirect taxation such as value added tax (VAT).
2. Removal of technical barriers to trade, e.g. those created by different product standards, and testing and certification requirements;
3. Removal of physical barriers to trade, including customs posts and other frontier controls on the free movement of goods;
4. Free trade in financial services, including banking, securities, insurance, etc;
5. Free trade in telecommunications and information technology services, e.g. through the setting of common standards;
6. Freedom of capital movement, such as stocks, shares, bonds and credits;
7. Free movement of professionals for employment in any of the member countries, with mutual recognition of equivalent qualifications;
8. Deregulation of transport services; including road haulage, air and water transport services;
9. Opening of public procurement to competitive bidding by contractors throughout the market;
10. Uniform intellectual and industrial property rights, e.g. in the areas of protection of micro-circuit designs and patenting of biotechnological inventions.

Public procurement policy

Community policy has been to ensure that all companies operating within the Community have the opportunity to bid for public sector contracts, and that such contracts should not be let on the basis of single tenders, and also that all public contracts should be properly advertised.

Nonetheless there is a marked tendency to 'buy national', even of refusing to consider bids from non-national companies. Information about public contracts is still inadequate, specifications discriminatory, and tendering procedures complex and hard to comply with.

Levy (1989) quoted government procurement within the EC as £200 billion annually. Of this, he claimed only 2% was in practice accessible to firms from outside the member state, and 75% went to 'national champions' around whose capabilities the tenders had been written. Professor Levy cited a study which had put the cost of these practices as £57 billion. We could interpret this as a 'quality-related cost' of over 25% from the one cause alone!

The arrangements for fairness in public purchasing are of great importance for any company now seeking to supply a public body or major utility in the transport, energy, water and telecommunications fields. One example is the insistence in many cases that any bidder must hold an ISO 9000 certification of its quality management system; this has acted as a spur to many companies and plants outside the community to seek registration. Compliance with the new regulations may need extra work in some areas but the effort should be worthwhile; purchases by governments and other public bodies amount to no less that 15% of the Community's total gross domestic product.

Issue of the 'Single Market' directives

For many years the Community sought to remove technical barriers to trade by adopting Directives setting out detailed requirements which products had to satisfy before they could be sold freely throughout the community. Moreover the Treaty of Rome required Directives to be adopted unanimously.

However in 1985 a 'new approach' to technical standards and harmonization was adopted by the Council of Ministers, of the European Community. Under the new approach, directives would be simpler and merely set out 'essential requirements' relating to safety, etc. European standards would then be drafted to fill in the detail and for business to demonstrate compliance with. All products sold anywhere within the European Community must satisfy specified essential requirements even if they are not being sold across national boundaries.

The first three 'new approach' Directives to be adopted were for:

1. Simple pressure vessels (June 1987);
2. Toy safety (May 1988);
3. Construction products (December 1988).

European standards

European standards giving effect to the New Approach will usually be prepared by CEN (Comité Européen de Normalisation (European Standards Committee) or CENELEC (European Committee for Electrotechnical Standardization). Many European standards will be based on any already agreed at a world level within ISO (International Standards Organization) or IEC (International Electrotechnical Commission). For example, we have seen that the ISO standard for quality management systems (ISO 9000) has been adopted by CEN as EN 29000, for use in all its member countries.

In addition to the establishment of European Standards it is necessary to establish harmonization of the way in which test laboratories and certification bodies operate. This is being done within the framework of ISO/IEC guide 25 *General Requirements for the technical competence of testing laboratories* which provides for:

1. Recognition of the competence of Certification Bodies and Test Laboratories;
2. Agreement between Certification Bodies and Test Laboratories
3. Agreements between National Bodies responsible for recognizing Certification Bodies and Test Laboratories;
4. The nomination of Certification Bodies and Test Laboratories for regulatory purposes by member Governments.

The challenge of the Single European Act

This will open up new markets for many businesses, but it will also result in more intense competition within those markets. Because goods and services can move freely within the single market, there will be more players in the game, offering the consumer wider choice and the supplier new competitors.

All firms need to ask themselves which 'new approach' Directives are likely to affect their businesses. Firms with important customers in central or local government or in the transport, energy, water or telecommunications sectors in particular, should be thinking about the implications of new European standards for future business.

For a very few companies, for example those offering essentially local services, completion of the single market may indeed make very little practical difference. But for most firms, whether they trade in the local domestic market or export; whether they are in manufacturing or offer services; whether they are large or small, public or private; the advent of the single market is their biggest challenge since it will inevitably bring new standards to meet, and new competitors to meet on more equal terms than heretofore.

Best practice benchmarking

As many businesses have found to their cost, it is dangerous to become complacent and assume that one's current market is secure. Every company is vulnerable to a competitor who can provide better quality, provide the same level of quality as yourself but cheaper, or offer a product of superior grade. There is no room for complacency.

One way in which progressive companies attempt to be alert to such challenges, and respond to them, is called 'best practice benchmarking'.

The expression 'benchmarking' derives from the benchmark, literally a mark on a bench, against which a craftsman could measure an item he was making, or a retailer could measure bulk materials such as lengths of cloth that he was selling.

Hence 'best practice benchmarking' is measuring the performance of your company against the yardstick (a similar metaphor) of the best operation of a similar kind known to you. You may compare yourself with different companies for different aspects of your business, but in each case the benchmark will be the associate, rival or other relevant company which, in your opinion, is best at that particular aspect of your business.

Thus, Rank-Xerox in the United Kingdom, which makes photocopiers, benchmarks its distribution performance against:

1. 3M at Dusseldorf, Germany;
2. Ford Motors, Cologne, Germany;
3. A UK depot of Sainsbury's supermarket chain;
4. Volvo parts distribution depot, Gothenburg, Sweden;
5. IBM's French warehouse and international warehouse.

Note in particular the food and motor industry warehouses in this list. They are in no way competitors of Rank-Xerox, and for that reason it may have been easier to exchange information with them. They were chosen simply because they were considered to be coping with essentially similar problems, and had proven themselves leaders in the distribution field. Notice also that Rank did not limit themselves to comparison with local companies. They see their market-place as Europe, and try to measure themselves against the best in Europe.

Similarly Motorola Semiconductor Products Group's factory in Scotland benchmarks itself against:

1. Toshiba (with whom it has a technology transfer agreement) on memory chip yield and performance characteristics;
2. All other Motorola SPG plants world-wide on cycle time, scrap, yield, etc.;
3. Automated assembly performance against Motorola Japan, and it in turn against a Japanese subcontractor;
4. Warehouse performance (cycle time, quality, productivity and space usage) against other Motorola warehouses;
5. Purchasing performance against other Motorola plants and 'friendly' companies outside Motorola;
6. Salary and staff benefit packages against other manufacturing companies who have operations within central Scotland's industrial belt.

The above examples are also quoted in a booklet on *Best Practice Benchmarking*, which is published by Britain's Department of Trade and Industry under their 'Enterprise Initiative'.

This DTI booklet singles out the following features as characteristic of all best practice benchmarking (BPB) exercises:

1. Establishing how the customer distinguishes between an excellent and an ordinary supplier;
2. Setting standards for each of the factors on which customers make their judgement of excellence;
3. Finding out how the best companies achieve the standards;
4. Applying their own and other people's ideas towards achieving the standards.

Learning through other companies' experience

> Learn from other people's mistakes, you don't have time to make them all yourself (Anon)

Notice how the first step in BPB echoes our very first theme; of finding out, and giving the customer what he wants. Notice also that you need to approach your customers and discuss their needs before you can start BPB. You may wonder who you are going to be able to benchmark yourself against. If your customers are sympathetic they may well volunteer some part of their own organization for exchanging information with, or may put you in touch with other suppliers.

Your own suppliers may be another source of benchmarking information. If you are a good customer they will be predisposed to cooperate, and if they feel they are particularly good at some aspect of their business they will be glad of the chance to show it off to you.

The third source of information is non-competing local companies. They may well be interested in sharing information. Note my emphasis on sharing; if you are to gain information you must be prepared to give out an equivalent amount of information as well.

Finally . . . the competition! You can well imagine that semiconductor manufacturers buy rival silicon chips through distributors, characterize and analyse them in great detail, and that car manufacturers purchase, test-drive and strip down rival models. At a more local level, if you own a restaurant, why not occasionally pay some of your staff to dine out on their day off at a rival establishment, and give you their impressions?

Even though it's unlikely that you will be able to get data direct from a competitor, you may be able to get an averaged data from an industry sector from the appropriate trade association.

Benefits of BPB

The DTI booklet lists the benefits to be gained from BPB as:

1. Better understanding of customers and competition;
2. Fewer complaints, more satisfied customers;
3. Reduction in waste, quality problems and reworking;
4. Faster awareness of important innovations, and how they can be profitably applied;
5. A better reputation in the market-place;
6. As a result of all the above, increased profits and sales turnover.

In other words, you will recognize that BPB is both a marketing tool, and a contribution to 'Total Quality Management'.

Competing for quality awards

Introduction

Earlier in the book, and particularly in Chapter 8, Part Two, we examined schemes such as ISO 9000 for assessing and certifying quality management systems. There also exist various competitive award schemes, and some of these have also been used by certain companies as the basis for evaluating and developing their quality management systems. A prime example is the Malcolm Baldrige award scheme which was, for example, adopted by IBM as the basis of its quality improvement initiatives throughout the corporation.

Nonetheless, there are fundamental differences between the objectives of these award schemes on the one hand and models for assessment on the other. This section will emphasize some of the distinctions so that, despite their considerable overlap, the strengths and complementary nature of the two approaches can be clarified. ISO 9000 will be the reference 'Quality System Model' used, and it will be compared with key features of the Malcolm Baldrige Award. The Deming Prize, and the EOQC Award will also be mentioned in passing.

The Malcolm Baldrige National Quality Award

The Baldrige Award was established by the US Department of Commerce through the Malcolm Baldrige National Quality Improvement Act of 1987, as a US national award scheme. This is its first difference from ISO 9000; it is designed for use on a purely national basis, as a government initiative. Its organizing committee was established on that basis, namely to create a vehicle to encourage improved quality performance in American industry. It has been competed for annually since 1988, and up to six prizewinners may be announced, at the judges' discretion. These are for the following categories (maximum two winners in each):

- manufacturing
- service
- small firms.

Organizations are awarded points on a scoring scheme, which can be amended for each successive year's competition. The scheme used in 1994 was as follows:

1.0 Leadership (95 points)
  Senior executive leadership
  Management for quality
  Public responsibility and corporate citizenship

2.0 Information and analysis (75 points)
  Scope and management of quality and performance data and information
  Competitive comparisons and benchmarking
  Analysis and usage of company-level data

3.0 Strategic quality planning (60 points)
  Strategic quality and company performance planning process
  Quality and performance plans

4.0 Human resources development and management (150 points)
  Human resources planning and management
  Employee involvement
  Employee education and training
  Employee performance and recognition
  Employee well-being and satisfaction

5.0 Management of process quality (140 points)
  Design and introduction of quality products and services
  Process management product and service production and delivery processes
  Process management business process and support services
  Supplier quality
  Quality assessment

6.0 Quality and operational results (180 points)
  Product and service quality results
  Company operational results
  Business process and support service results
  Supplier quality results

7.0 Customer focus and satisfaction (300 points)
  Customer expectations, current and future

Customer relationship management
Commitment to customers
Customer satisfaction determination
Customer satisfaction results
Customer satisfaction comparison

1000 points are available in total.

Participants in the MBNQA scheme are required to make their own initial presentation based on self-assessment, and their own estimate of the score they have achieved. Different ranges of score are assigned 'bronze', 'silver' and 'gold' levels. Only organizations claiming high scores, which appear justified on the basis of the details cited in their submissions, are assessed independently by the MBNQA examiners. The award emphasizes quality improvement, evaluated over a five-year period. Following on from this, winners are debarred from re-entering for the award for a period of five years.

This brief overview identifies several of the ways in which competitions such as the MBNQA differ from the philosophy of ISO 9000 even though there is a considerable degree of overlap in what they are seeking to evaluate or, if you prefer, in the content of the model they are putting forward as an ideal.

Comparing elements of MBNQA and ISO 9001.

Virtually all commentators have identified the agreement between MBNQA section 5 and the ISO 9000 elements, though many who are more familiar with MBNQA assessment than with ISO 9000 assessment assume that there is little relevance in ISO 9000 to other areas of MBNQA. I hope that even the brief introduction to ISO 9000 given in this book will have made it evident to you that it has a considerable bearing on the issues listed in MBNQA topics listed under other headings, in particular what 1.0 implies in regard to senior management responsibilities.

At the same time, ISO 9000 undoubtedly does not focus so strongly on such topics as:

- Public responsibility and corporate citizenship;
- Competitive comparisons and benchmarking;
- Employee involvement; their education, recognition, well-being and satisfaction;
- Customer's future expectations, good customer relations, monitoring customer satisfaction.

Contrasting aims of MBNQA and ISO 9000

The differences between MBNQA and ISO 9000 are to a large extent the reflection of their different objectives, and I will try to highlight some of these by means of the following table.

	MBNQA	*ISO*
range:	US national	International
model:	Process based	Life-cycle based
focused on:	continuous improvement	current capability
seeking:	leaders in excellence	basic capability
evaluation:	once-off	continuing assessment
scope:	all aspects of quality management	quality assurance
measure:	degree of excellence, points score	auditable compliance or non-compliance
goal:	excellence, delight, best-of-breed	100% compliance with contractual conditions
method:	continuous improvement and refinement	continuous correction and updating

Deming Prize (Japan)

The Deming Prize was established in Japan in 1951 by the Japanese Union of Scientists and Engineers (JUSE) in honour of Dr W. Edwards Deming. It is awarded in three categories:

- For an individual person;
- For an application;
- For quality control within a factory.

The subject of the award is the successful implementation of a set of principles or techniques related to process analysis, SPC, Quality Circles or the like. In the case of the company (factory) award, the following factors are taken into account:

- Company policy and planning;
- Organization and its management;
- Quality control education and dissemination;
- Collection, transmission and utilization of quality information;
- Analysis;
- Standardization;
- Control;
- Quality assurance;
- Effects;
- Future plans.

The prizes are awarded on an annual basis.

The European Quality Award

The European Foundation for Quality Management's Award focuses on the issues of:

- Leadership;
- Policy and strategy;
- People management;

- Resources;
- Processes;
- Customer satisfaction;
- People satisfaction;
- Impact on society;
- Business results.

From this it can be sensed that the concerns of the EQA are similar to those of the MBNQA but with stronger focus on issues such as energy conservation, working conditions and employment stability, preservation of the environment, and shareholder satisfaction.

The schemes in action

Schemes like the MBNQA and EQA schemes take a broader view of what quality represents than do management system models such as ISO 9000; it could be argued that they are more 'TQM' oriented. Competitive schemes are becoming more popular, as witness, for example, the number of schemes sponsored by individual States within the USA.

Nevertheless, the more modest and pragmatic scope of ISO 9000 makes it, I believe, more objective and its conclusions more reliable. In its wide register of approved firms it is therefore a more valuable guide on potential vendors to customers who, after all, may find no prizewinners who are in the business of supplying the particular materials or service they need.

ISO 9000 has been criticized as 'bureaucratic' but Baldrige also has its critics. I put forward typical comments that I have heard:

- 'Baldrige is a beauty competition';
- 'ISO 9000 put teeth into our TQM programme';
- 'We spend too much management time preparing our Baldrige submission';
- 'ISO 9000 is good because it leaves management no place to hide'.

Summary of conclusions

- International regulations begin to affect even businesses which only have a local market.
- Therefore, for future prosperity, a company has to be capable of competing in quality with the best in its field.
- Ways of testing its competitiveness include:
  best practice benchmarking
  competing in quality award competitions.
- The first step is still the establishment of an effective quality management system.

28

Quality trends, challenges and opportunities

Introduction

The purpose of this final chapter is to tie up a few loose ends and to reiterate, and in some cases expand on, what I feel are some important messages of the book. In so doing, I shall make some closing remarks on trends and challenges quality assurance is likely to offer in coming years. Hence the 'summary of conclusions' at the end of this chapter represents a conclusion to the whole book, not just to this chapter alone.

Before this summing up I wish to put down a few words about some techniques which have received recent emphasis, especially those related to improving quality in design. They relate in particular to making design and development work more responsive to customer needs and wishes. These include the techniques which go by the names of 'the seven new tools of quality assurance', 'quality function deployment' and 'the house of quality'. These terms are becoming widely used and each is a potentially puzzling or misleading title. This is regrettable – quality management too easily becomes loaded with jargon or emotive phraseology, so I am taking the opportunity to explain very broadly what each of these phrases means.

The 'seven new tools of quality'

I mentioned confusing and inhibiting phraseology and the 'seven new tools' represent a case in point, in that I imagine few readers will consider them all to be 'new'. In isolation, if not in combination, all have been used for some time in particular industries and companies. You will certainly recognize that some of them have already been introduced to you in this book. They are often contrasted with seven so-called 'old' tools, which are presented as being more concerned with problem solving within manufacture, whereas the 'new tools' are seen as being deployed when defining and implementing improved designs. Logically, I should introduce the old took-kit first.

The old tools

The traditional tools of a methodical approach to quality improvement are frequently represented as being seven in number, each with its own part to play:

1. **Process flow charting** to display the sequence of what is done.
2. **Check sheets (or tally charts)** to record how often things are done incorrectly.
3. **Histograms** to display the overall variation.
4. **Pareto analysis** to identify the most significant problems.
5. **Brainstorming** with **cause and effect analysis** to determine the causes.
6. **Scatter diagrams** to show up relationships between factors causing variation.
7. **Control charts** to show when to intervene in order to control variation.

You will appreciate that all of these methods have been touched on, at appropriate points in the book. Process flow charting was a technique actually used in Chapters 3 to 6, Part One, in order to elucidate the manufacturing life-cycle (Figure 3.1, etc.). A further example is given in Figure 28.1, with the key to its symbology in Figure 28.2. The other techniques were dealt with at various points in Part Two.

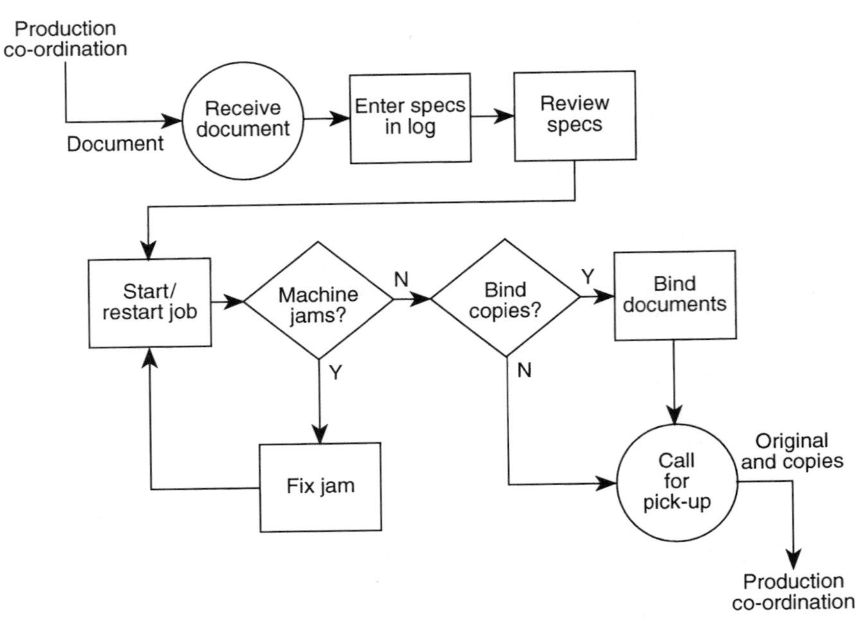

Fig. 28.1 Flowchart (for document reproduction). Source: BS 7850:part 2:1992.

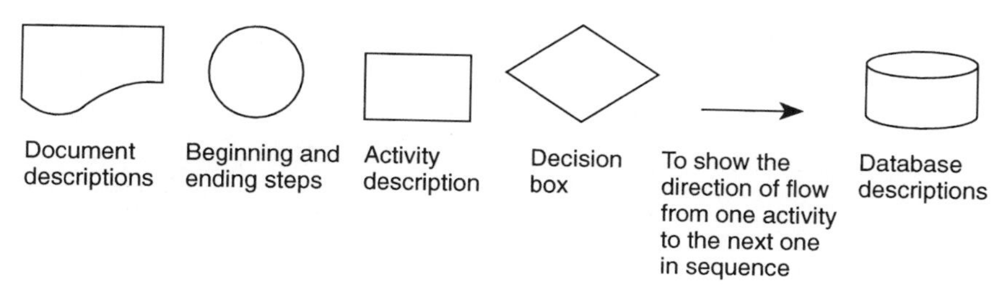

Fig. 28.2 Flow diagram symbols. Source: BS 7850:part 2:1992.

The tools of quality improvement in design

It is more than likely that different authorities will present the list of 'seven new tools' somewhat differently, so I will explain that the treatment I am following is that given in Oakland, *Total Quality Management*, 2nd edition. He considers them to be:

1. Affinity diagram;
2. Interrelationship graph;
3. Tree diagram;
4. Matrix diagram or charts;
5. Matrix data analysis;
6. Process decision programme chart;
7. Arrow diagram.

Oakland sees the affinity diagram and interrelationship graph as representing the creative and logical starting points respectively, of gathering data or ideas. By combining the information gathered by these methods, a tree diagram can be constructed. This leads to the development of a matrix diagram which is developed by means of matrix data analysis; and process decision programme (PDP) charts and arrow diagrams can be constructed for the implementation of a programme of work (Figure 28.3).

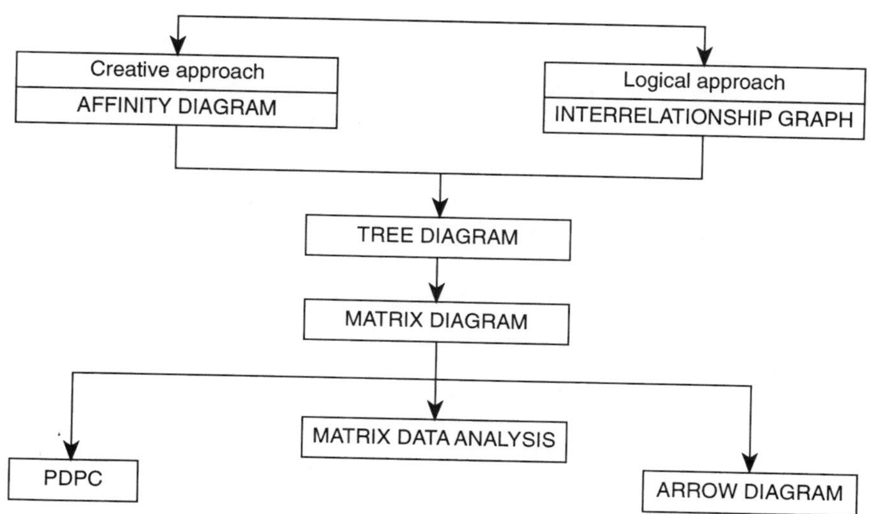

Fig. 28.3 The 'seven new tools' of quality design. Based on diagram in Oakland, *Total Quality Management*.

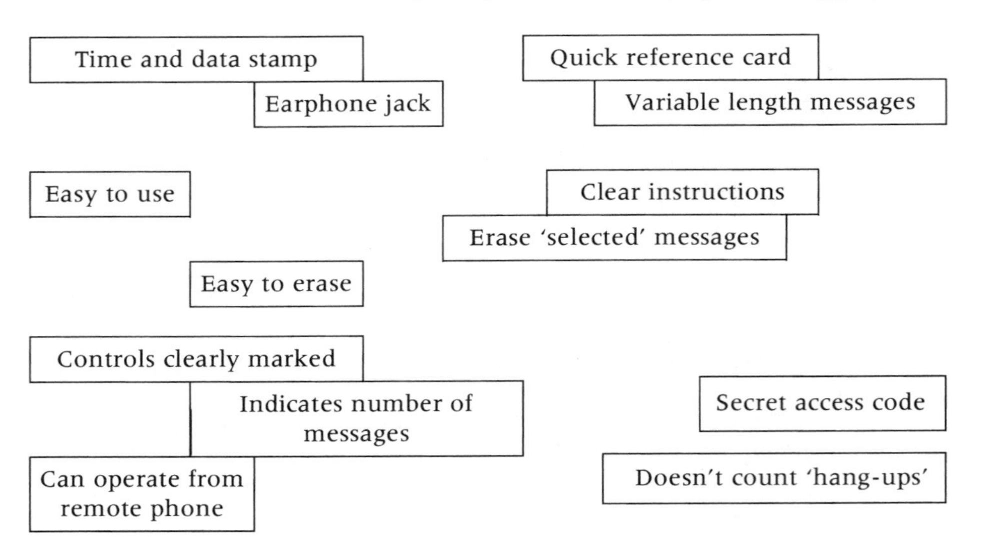

Fig. 28.4 Affinity diagram for a telephone answering machine. Source: BS 7850:part 2:1992.

Affinity diagram

An affinity diagram is valuable when a large number of ideas or alternative opinions are being collected. Figure 28.4 shows how the individual ideas put forward for desirable features of a telephone answering machine are grouped.

Interrelationship graph

The interrelationship graph takes a somewhat similar approach. A central idea or problem is taken and the links between various factors contributing to it can be

Variable length messages Time and date stamp Doesn't count 'hang-ups' Indicates number of messages	Incoming messages
Secret access code Earphone jack	Privacy
Clear instructions Quick reference card	Instructions
Controls clearly marked Easy to use Can operate from remote phone	Controls
Easy to erase Erase 'selected' messages	Erasing

Fig. 28.5 Interrelationships in features of a telephone answering machine. Source: BS 7850: part 2:1992.

sketched out. Figure 28.5 is essentially the interrelationship graph, but presented in tabular form, equivalent to the affinity diagram in Figure 28.4. The interrelationship graph can then be developed into a tree diagram (for a concept) or indeed into a fishbone diagram (for a problem).

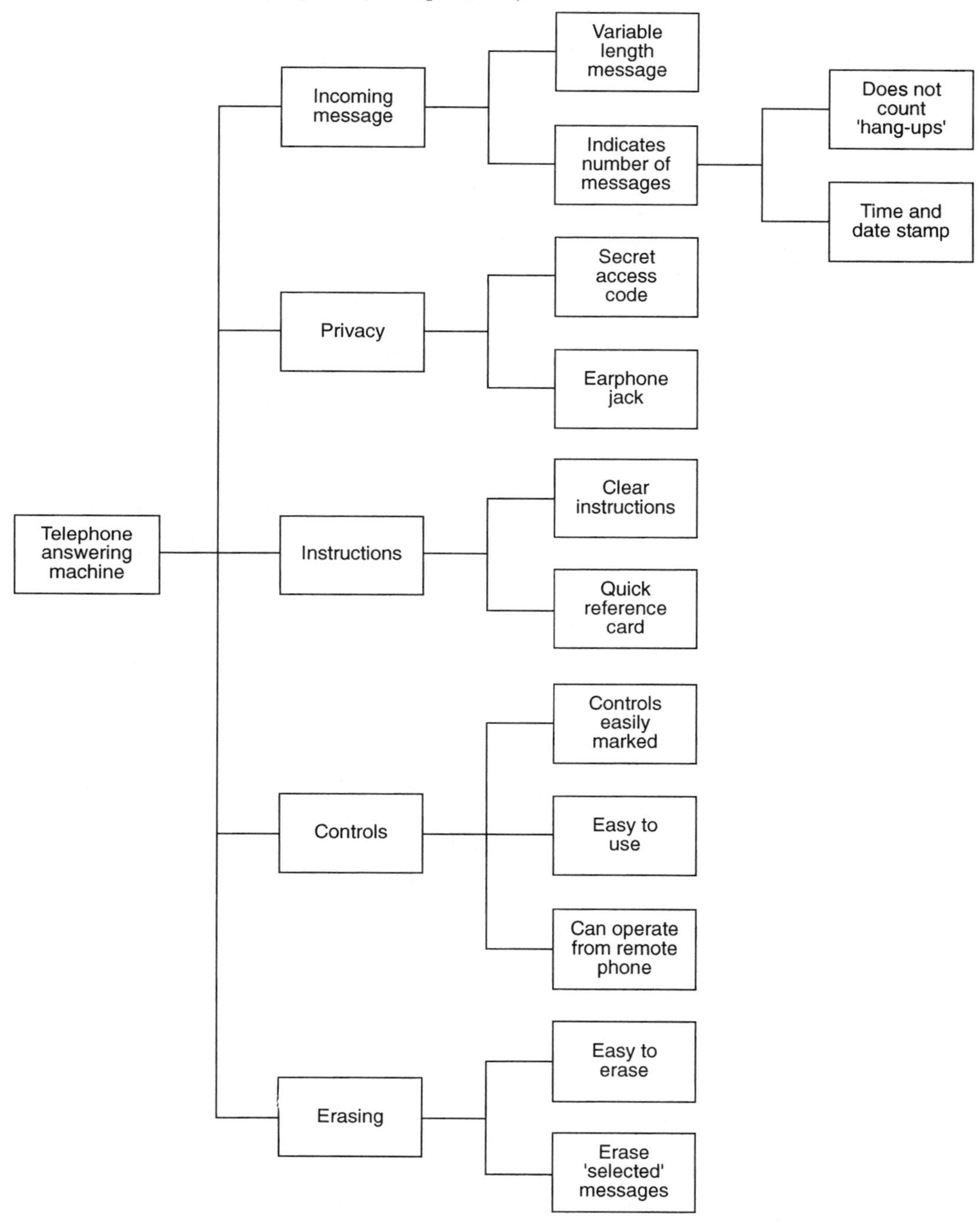

Fig. 28.6 Tree diagram for a telephone answering machine. Source: BS 7850:part 2:1992.

Tree diagram

A tree diagram can be seen as an extension of the interrelationship graph exercise. Figure 28.6 shows a tree diagram for the desired features of the same telephone answering machine.

Matrix diagram

Oakland sees this as the central 'new tool'. Its function is to delineate the relationship between different characteristics or tasks, so that interactions and conflicts can be recognized, and an optimized solution found. In the basic 'L-shaped' matrix shown in Figure 28.7, customer requirements 'B' ($b_1 \ldots b_n$) are related to the product or service features which could influence them, 'A' ($a_1 \ldots a_n$). A weighting can be given to the importance of each customer feature, and the degree of correlation between each factor a and customer need b can be recorded as strong, moderate, weak or non-existent. This will become clearer with an example, and a simple one is given in Figure 28.8. This relates to an educational project and shows (a) WHAT assistance students want, and (b) HOW lecturers might meet these needs better. The ORDER in which $b_1 \ldots . b_n$ were listed could have represented the students' order of priority, e.g. exam practice as a 'must' and documented objectives 'nice to have, but not essential'.

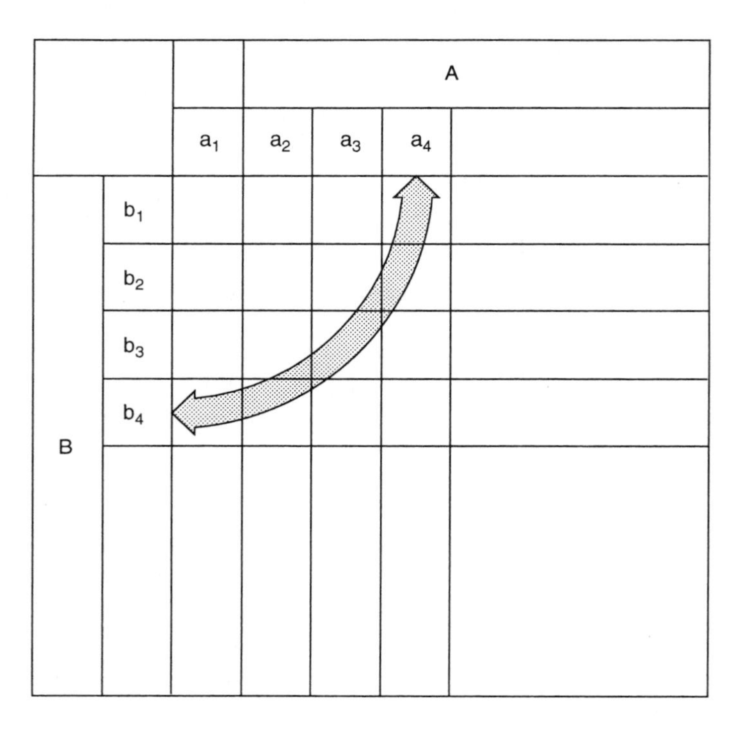

Fig. 28.7 L-shaped matrix. Source: Oakland, *Total Quality Management*.

RELATIONSHIPS:
Strong ☻
Medium ☺
Weak Δ

WHAT course delegates want: / HOW course can be improved:	Check room and facilities on the evening before start	Draw up course materials check list	Commission a new video	Devise better role-plays/case studies	Match instructors to delegates' background	Train instructors in teaching techniques	Draw up a course timetable	Compile guidelines on how to answer exam questions	Analyse delegate feedback sheets after each course run
Knowledgable instructors		Δ			☻	☻			☺
'Hands on' practice			☻	☻		Δ			
Comprehensive course notes		☺							
Varied and interesting exercises	Δ	☺	☻	☻		Δ			
Examples of exam questions								☻	
Finish on time each day	☺	Δ					☻		
Informal environment					☻	☺			Δ
Effective use of time	☻	☻				☺	☺		

Fig. 28.8 Matrix diagram; 'Improving auditor training course'.

Matrix data analysis

This takes the data from a matrix diagram and rearranges them to graph the magnitude of the correlation between variables, as illustrated in Figure 28.9.

Process decision programme chart (PDPC)

This is used to show events and contingencies which can arise between stating a problem and arriving at a solution. It has obvious kinship with tree diagrams, fishbone diagrams and FME(C)A analysis (Figure 28.10).

Arrow diagram

Under this heading Oakland includes two types of project planning graphs which may already be familiar to you as Gantt Charts and PERT (or Critical Path Analysis) charts (Figures 28.11 and 28.12)

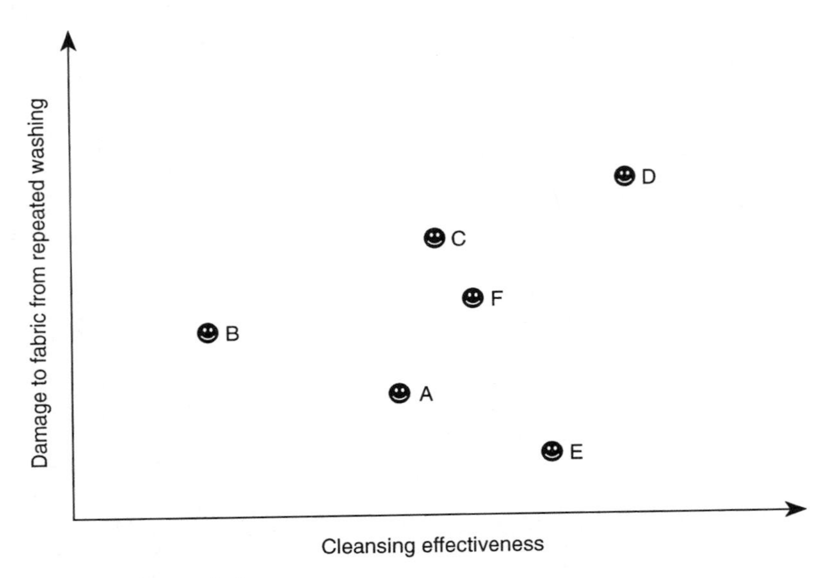

Fig. 28.9 Matrix data analysis.

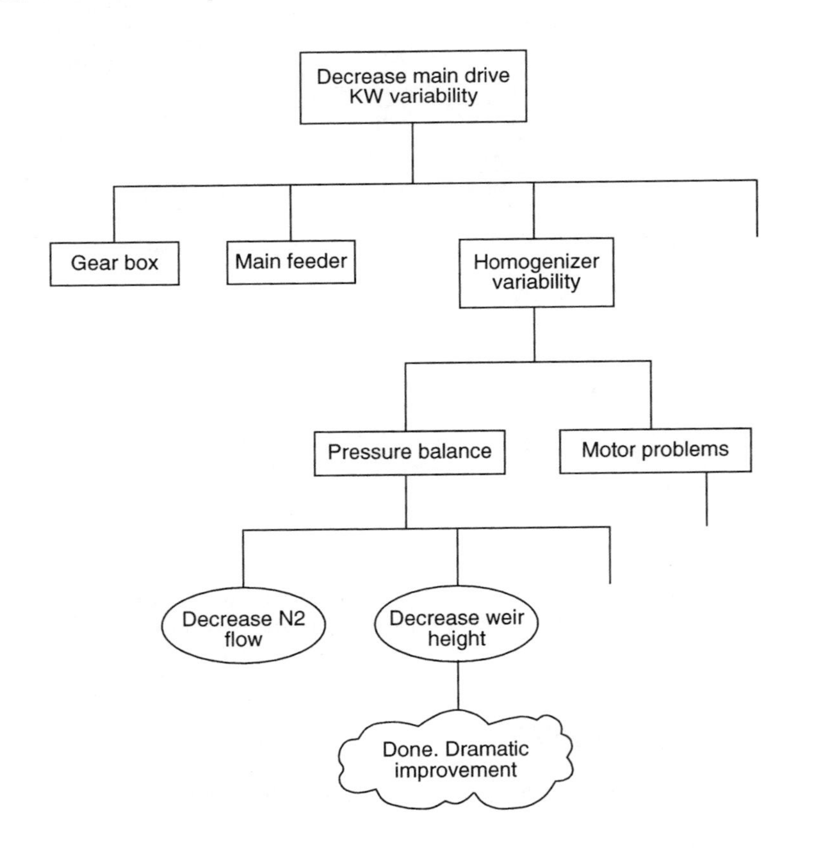

Fig. 28.10 PDP chart. Source: Oakland, *Total Quality Management*.

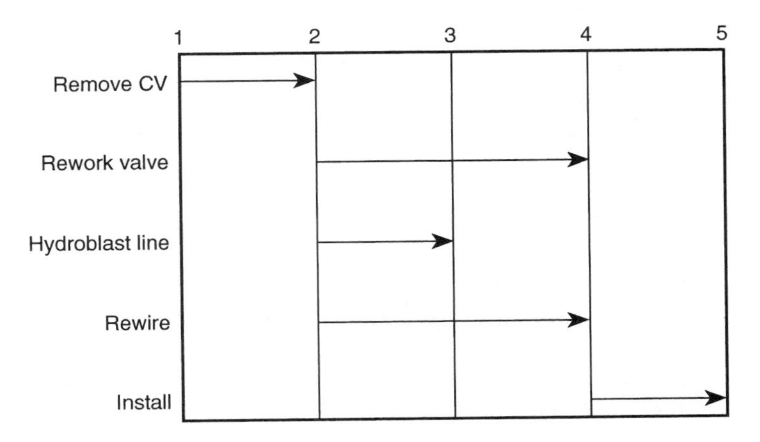

Fig. 28.11 Gantt chart. Source: Oakland, *Total Quality Management*.

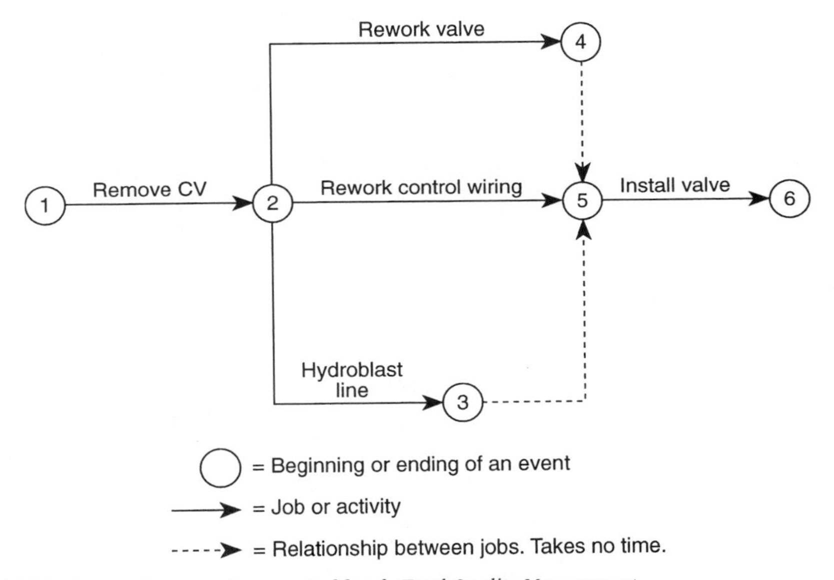

Fig. 28.12 Arrow diagram. Source: Oakland, *Total Quality Management*.

Quality function deployment

The basic matrix analysis can be elaborated in a number of ways. For example two independent sets of customer needs ('WHAT's) could be plotted, transforming the matrix from 'L-type' to 'T-type' in shape. Additionally, the correlation between the ('HOW's) can be plotted. All of these embellishments can be incorporated into a single diagram which may become very complex.

This activity was pioneered in Japan in ship and automobile design where it was given a name normally translated as 'Quality function deployment' or QFD, though I am told that the original Japanese phrase does not contain the literal word 'quality' at all.

Activities required for a fully developed QFD exercise include:

Market research
Technical research
Invention
Design
Prototype testing
Production-version testing
After-sales problem investigation.

The groups made responsible for these exercises are referred to QFD teams, but one of the reasons I have discussed the term QFD is to stress that it does NOT mean 'how to deploy the QA department'.

A schematic sketch of what might be included in a developed QFD diagram is shown in Figure 28.13.

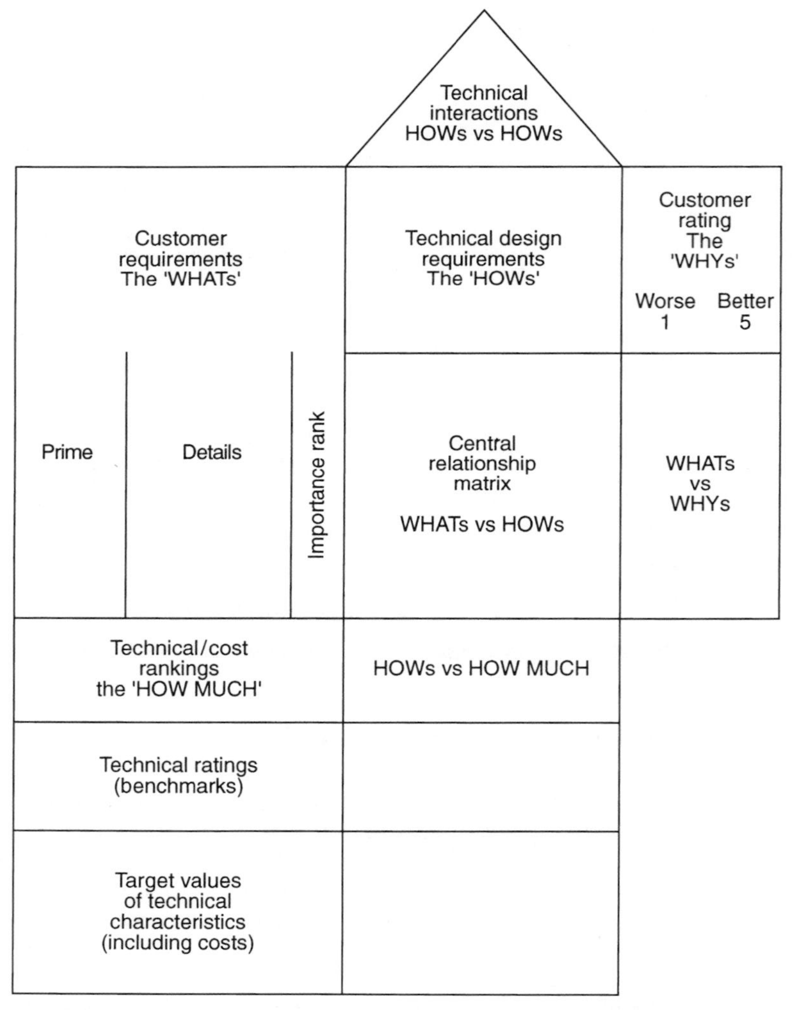

Fig. 28.13 QFD diagram schematic or 'house of quality'. Source: Oakland, *Total Quality Management*.

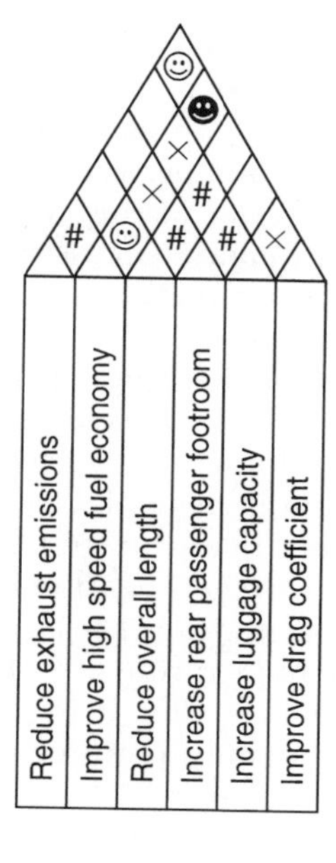

Fig. 28.14 How the 'house of quality' arises.

The house of quality

A QFD diagram is often referred to as a 'house of quality'. The reason for this is its general box-like shape, but with a gable 'roof' resulting from the 'correlation matrix' part of the diagram. This is seen more clearly in the example presented as Figure 28.14, which examines the correlations, both positive and negative, existing between some proposed changes to the design of a motor car.

Summary of conclusions

It is now some time since you read the opening chapter of the book or, indeed, since I wrote it. The book has been prepared for you by an industrial scientist with practical quality management experience, not primarily an academic. I hope it has been interesting and will be of lasting value to you as you pursue your careers. The following notes will be just a brief summary of some of the points which linger in my own mind. They are, therefore, not so much a summary of this chapter, as of the whole book.

You may remember Frank Price's epitome of quality assurance as a despised disciple, practised by despised people. How did that come about?

I think you can see now that the activity of quality control (or quality assurance) and the way it was managed until recently arose from the division of responsibility arising from scientific management. Along with this, valuable techniques were developed; sampling procedures, statistical control, quality audit, FMECA and others but the division between the doers and the experts remained.

The damaging result of this was that certain people, 'QA', were seen as being 'responsible' for quality. Yet apart from exhortation, after-the-fact inspection and making decisions on the releasability of product, these people could do nothing directly to enhance quality. As Crosby, himself a former 'hands-on' quality manager has observed, there was a high attrition rate amongst managers, who first eagerly accepted their management's perception that 'they were responsible' for quality and then found that its creation was outside their hands.

Deming and Juran began by lecturing, demonstrating and consulting on industrial statistics; but over the course of 40 years they too came more and more to emphasize the 'win their hearts and minds' aspects of successful quality assurance.

Statistical process control is a powerful technique. It has tremendous applications within manufacturing industry and many more companies in the West now appreciate its power. This is in large part due to customer pressure from Ford, whose supplier quality requirements, formulated in their procedure Q101, demand that vendors maintain and supply Ford with process control records. Thus SPC has a high profile, but ideas of total quality control have spread further, into support functions and into service industry where the SPC methods are less relevant.

Another idea which attracts much interest, and which many companies in the West have sought to apply, is quality circles. Some introductions have been successful, many others have failed. The failures are not the fault of the concept of quality circles but through them having been introduced with the wrong motives – e.g. for 'motivation', or as a 'first step' in total quality control. Japanese companies view quality circles as the culmination of employee involvement in quality improvement, not as its beginning.

I have frequently cited successful Japanese practices as a model of what is possible. This is not defeatist or unpatriotic, but is done in the spirit of 'best practice benchmarking'. There are currently some things which, over the past generation, the Japanese and other oriental races have achieved with more success than westerners. As a European who has spent much of his life working, not necessarily in the USA but largely for American-owned companies I say:

> Let us look at these achievements, see what we have to equal and if we can, learn how we can match them. In those areas of invention and innovation where we have our strengths, let us do all we can to maintain our lead.

Here in the United Kingdom we have, in BS 5750 (now BS EN ISO 9000), the currently best developed and most widely used national quality system standard. Commercial pressures to gain BS 5750 have made all company managements more aware of quality issues as a result. Multinational companies with a manufacturing base in the UK have found the performance of their UK plants to be the equal of those anywhere else in the world including the USA and Japan.

The major barriers to quality leadership in the UK seem to be:

- Ignorance of QA principles on the part of the traditional management of small- and medium-sized businesses, especially the fear that quality means 'gold-plating' and will cost more. **You** now know why that is not the case; indeed quite the opposite. **You** can start educating your boss if necessary.

- Industrial relations – 'us' and 'them'. Yet the successful companies don't suffer from bad industrial relations, and studies suggest that UK workforces respond to the strong demands made by Japanese managers even without the security afforded to employees in Japan. Why? Apparently because:
    - Japanese managers are visible;
    - They understand the shopfloor jobs;
    - They communicate the company's ideals;
    - They provide good training;
    - They listen to ideas;
    - They appreciate and acknowledge conscientious work;
    - They practise what they preach.

You could do those things too!

As the world becomes a 'global village' your company will more and more, be competing with the smartest companies throughout the world. It matters not whether you are competing globally or only on your own doorstep, the shock waves will still reach you. What can you do to help your company become 'World-class'? I've given you some clues in this book. **You** can do it!

It's up to **You**!

Appendix: British Standards mentioned in the text

British Standards Institution, Milton Keynes.

BS 4778 *Quality Vocabulary.*
Part 1:1987 International terms (identical with ISO 8402:1986)
Part 2:1991 Quality Concepts and related definitions
Part 3 section 3.1:1991 Guide to concepts and related definitions
Part 3 section 3.2:1991 Glossary of international terms

BS 4891:1972 *A Guide to Quality Assurance.*

BS 5309 *Methods for Sampling Chemical Products.*
Part 1:1976 Introduction and general principles
Part 2:1976 Sampling of gases
Part 3:1976 Sampling of liquids
Part 4:1976 Sampling of solids

BS 5700:1984 *Guide to Process Control Using Quality Control Chart Methods and CUSUM Techniques.*

BS 5701:1980 *Guide to Number-defective Charts for Quality.*

BS 5703 *Guide to Data Analysis and Quality Control Using CUSUM. Techniques*
Part 1:1980 Introduction to Cusum charting
Part 2:1980 Decision rules and statistical tests for CUSUM charts and tabulations
Part 3:1981 CUSUM methods for process/quality control by measurement
Part 4:1982 CUSUMS for counted/attributes data

BS 5750 *Quality Systems.*
NOTE: During 1994 the numbering method for this standard was changed and

when ISO 9000–1, 9001, 9002, 9003, 9004–1 were revised BS 5750 parts 0.1, 1, 2, 3 and 0.2 were also reissued as BS EN ISO 9000–1, 9001, 9002, 9003, 9004–1 respectively.

BS EN ISO 9000–1:1994 *Quality management and quality assurance standards.*
Part 1: Guidelines for selection and use. (Identical with ISO 9000–1:1994)

BS EN ISO 9001:1994 *Quality Systems – Model for quality assurance in design, development, production, installation and servicing.* (Identical with ISO 9001:1994)

BS EN ISO 9002:1994 *Quality Systems – Model for quality assurance in production, installation and servicing.* (Indentical with ISO 9002:1994)

BS EN ISO 9003:1994 *Quality Systems – Model for quality assurance in final inspection and test.* (Identical with ISO 9003:1994)

BS EN ISO 9004–1:1994 *Quality management and quality system elements.*
Part 1 Guidelines. (Identical with ISO 9004–1:1994)

BS 5750:Part 4:1994 *Quality systems.*
Part 4: Guide to the use of BS EN ISO 9001 'Model for quality assurance in design, development, production, installation and servicing' (formerly BS 5750 part 1), BS EN ISO 9002 'Model for quality assurance in production, installation and servicing' (formerly BS 5750 part 2) and BS EN ISO 9003 'Model for quality assurance in final inspection and test' (formerly BS 5750 part 3)
Part 8:1991 Guide to quality management and quality system elements for services (identical with ISO 9004–2:1991)
Part 13:1991 Guide to the application of BS 5750 part 1 to the development, supply and maintenance of software (identical with ISO 9000–3:1991)

BS 5760 *Reliability of Constructed or Manufactured Products, Systems, Equipments, or Components.*
Part 0:1986 Introductory guide to reliability
Part 1:1985 Guide to reliability and maintainability programme management
Part 2:1981 Guide to the assessment of reliability
Part 3:1982 Guide to reliability practices – examples
Part 4:1986 Guide to specification clauses relating to the achievement and development of reliability in new and existing items
Part 5:1991 Guide to failure modes, effects and criticality analysis (FMEA and FMECA)
Part 6:1991 Guide to programmes for reliability growth
Part 7:1991 Guide to fault tree analysis

BS 5781 *Measurement and Calibration Systems.*
Part 1:1979(1988) Specifications for systems requirements
Part 2:1979(1988) Guide to the use of part 1

BS 6001 *Sampling Procedures for Inspection by Attributes.*
Part 1:1991 Specification for sampling plans indexed by acceptable quality level (AQL) for lot-by-lot inspection (ISO 2859–1:1989)
Part 2:1993 Specification for sampling plans indexed by limiting quality (LQ) for isolated lot inspection (ISO 2859–2:1985)
Part 3:1993 Specification for skip-lot procedures (ISO 2859–3:1991)

BS 6002: *Sampling Procedures for Inspection by Variables*

Part 1:1993 Specification for single sampling plans indexed by acceptable quality level (AQL) for lot-by-lot inspection (identical with ISO 3951:1989)

BS 6143 *Guide to the Economics of Quality.*
Part 1:1991 Process Cost Model
Part 2:1990 Prevention, appraisal and failure model

BS 7229 *Guide to Quality System Auditing.*

BS 7229 Parts 1, 2, 3 now renumbered as BS EN 30011–1, –2, –3 respectively.
Part 1:1991 Auditing (identical with ISO 10011–1:1991)
Part 2:1991 Qualification criteria for auditors (identical with ISO 10011–2)
Part 3:1991 Managing an audit programme (identical with ISO 10011–3:1991)

BS 7850:part 1:1992 *Total quality management.*
Part 1: Guide to management principles

BS 7850:part 2:1994 *Total quality management.*
Part 2: Guidelines for quality improvement (identical with ISO 9004–:1993)

References

American Society for Quality Control (1970) *Quality Costs: What and how?* ASQC, Milwaukee.

American Society for Quality Control (1977) *Guide for Reducing Quality Costs*. ASQC, Milwaukee.

Council of the European Community (1985) *The European Product Liability Directive*. Brussels EEC, reprinted in *Quality Assurance, Product Liability* Special Issue Dec. 1986, pp. 115–18, IQA, London.

Crosby, P.B. (1979) *Quality is Free*. McGraw-Hill, New York.

Crosby, P.B. (1984) *Quality Without Tears*. McGraw-Hill, New York.

Crosby, P.B. (1989) *Let's Talk Quality*. McGraw-Hill, New York.

Dehnad, K. *Quality Control: Robust design and the Taguchi method*. Wadsworth and Brooks Cole, Pacific Grove, USA.

Deming, W.E. (1982) *Quality, Productivity and Competitive Position*. MIT, Cambridge, Mass.

Department of Trade and Industry (1986) *Total Quality Management: A practical approach*. HMSO, London.

Department of Trade and Industry *Best Practice Benchmarking*. HMSO, London.

Department of Trade and Industry (1988) *The Single Market – An Introduction*. HMSO, London.

Department of Trade and Industry *Leadership and Quality Management: A guide for chief executives*. HMSO, London.

Feigenbaum, A.V. (1986) *Total Quality Control*, 3rd edn. McGraw-Hill, New York.

Graham, I. (1988) *Just-in-time Management of Manufacturing*. Elsevier Technical Communications, Oxford.

Groocock, J.M. (1974) *The Cost of Quality*. John Wiley, New York.

Groocock, J.M. (1986) *The Chain of Quality*. John Wiley, New York.

Hutchins, D. (1985) *Quality Circles Handbook*. Pitman, London.

Juran, J.M. and Gryna, F.M. (1988) *Quality Control Handbook*, 4th edn. McGraw-Hill, New York.

Juran, J.M. (1964) *Managerial Breakthrough*. McGraw-Hill, New York.

Juran, J.M. *Juran on Leadership for Quality*. McGraw-Hill, New York.

Levy, J.C. (1989) *Europe, 1992: Engineers and engineering. Quality News*, Feb. 1989, IQA, London.

Mackmurdo, R. (1988) Who holds the baby now? in *Quality News*, March 1988, 94–7, IQA, London.

Manns, T. and Coleman, M. (1988) *Software Quality Assurance*. Macmillan, Basingstoke.

Oakland, J.S. (1993) *Total Quality Management*, 2nd edn. Butterworth-Heinemann, London.

Oakland, J.S. (1986) *Statistical Process Control*. Heinemann, London.

O'Connor, P.D.T. *Practical Reliability Engineering*, 2nd edn.

Pascale, R.T. and Athos, A.G. (1986) *The Art of Japanese Management*. Sidgwick and Jackson, London.

Pugh, D.S. (1971) *Organisation Theory*. Penguin, London.

Price, F. (1984) *Right First Time*. Gower, Aldershot.

Price, F. (1990) *Right Every Time. Using the Deming approach*. Gower, Aldershot.

Sayle, A.J. (1988) *Management Audits*, 2nd edn.

Scherkenbach, W.W. (1986) *The Deming Route to Quality and Productivity*. ASQC, Milwaukee.

Shingo, S. *Zero Quality Control*. Productivity Press, USA.

Stebbing, L. (1986) *Quality Assurance: The route to efficiency and competitiveness*. John Wiley, Chichester, New York.

Taguchi, G. (1986) *Introduction to Quality Engineering*. Tokyo Asian Productivity Organization.

Taylor, F. (1912) *Scientific Management*. Testimony to the US House of Representatives Committee, reproduced in *Organisation Theory* (ed. D.S. Pugh), Penguin, London.

Titchenor, A. (1989) *Quality – the Countdown to 1992*. *Quality News*, April 1989, 146–52, IQA, London.

Trevor, M. (1983) *Japan's Reluctant Multinationals*. Frances Pinter, London.

Vorley, G. (1991) *Quality Assurance Management (Principles and Practice)*. Whitehall Communications, Maidstone.

Ware, J.E. (1989) Standardisation and certification. *Quality News*, Nov. 1989, 508–13, IQA, London.

White, M. and Trevor, M. (1983) *Under Japanese Management*. Heinemann, London.

Wickens, P. (1987) *The Road to Nissan*. Macmillan, London.

Wilson, R. (1986) *The Sun at Noon*. Hamish Hamilton, London.

Index